YUVAL NOAH HARARI

'I encourage all of us, whatever our beliefs, to question the basic narratives of our world, to connect past developments with present concerns, and not to be afraid of controversial issues'

Dr Yuval Noah Harari has a PhD in History from the University of Oxford and now lectures at the Hebrew University of Jerusalem, specialising in World History. His books *Sapiens*, *Homo Deus*, and *21 Lessons for the 21st Century* have become an international phenomenon.

ALSO BY YUVAL NOAH HARARI

Homo Deus: A Brief History of Tomorrow

21 Lessons for the 21st Century

YUVAL NOAH HARARI

Sapiens

A Brief History of Humankind

VINTAGE

Vintage
20 Vauxhall Bridge Road,
London SW1V 2SA

Vintage is part of the Penguin Random House group of companies
whose addresses can be found at global.penguinrandomhouse.com.

Translated by the author, with the help of
John Purcell and Haim Watzman

First published in Vintage in 2015
First published in hardback by Harvill Secker in 2014
First published in Hebrew in Israel in 2011 by Kinneret, Zmora-Bitan, Dvir

penguin.co.uk/vintage

A CIP catalogue record for this book is available from the British Library

ISBN 9780099590088

Typeset in Sabon by Palimpsest Book Production Limited,
Falkirk, Stirlingshire

Picture Research by Caroline Wood
Maps by Neil Gower

Printed and bound in India by Thomson Press India Ltd.

The authorised representative in the EEA is Penguin Random House Ireland,
Morrison Chambers, 32 Nassau Street, Dublin D02 YH68

YUVAL NOAH HARARI

Sapiens

A Brief History of Humankind

VINTAGE

40

Vintage
20 Vauxhall Bridge Road,
London SW1V 2SA

Vintage is part of the Penguin Random House group of companies
whose addresses can be found at global.penguinrandomhouse.com.

Translated by the author, with the help of
John Purcell and Haim Watzman

First published in Vintage in 2015
First published in hardback by Harvill Secker in 2014
First published in Hebrew in Israel in 2011 by Kinneret, Zmora-Bitan, Dvir

penguin.co.uk/vintage

A CIP catalogue record for this book is available from the British Library

ISBN 9780099590088

Typeset in Sabon by Palimpsest Book Production Limited,
Falkirk, Stirlingshire

Picture Research by Caroline Wood
Maps by Neil Gower

Printed and bound in India by Thomson Press India Ltd.

The authorised representative in the EEA is Penguin Random House Ireland,
Morrison Chambers, 32 Nassau Street, Dublin D02 YH68

Penguin Random House is committed to a sustainable future for
our business, our readers and our planet. This book is made from
Forest Stewardship Council® certified paper.

In loving memory of my father,
Shlomo Harari

Contents

The Timeline of History ix

Part One The Cognitive Revolution

1 An Animal of No Significance 3
2 The Tree of Knowledge 22
3 A Day in the Life of Adam and Eve 45
4 The Flood 70

Part Two The Agricultural Revolution

5 History's Biggest Fraud 87
6 Building Pyramids 110
7 Memory Overload 134
8 There Is No Justice in History 149

Part Three The Unification of Humankind

9 The Arrow of History 181
10 The Scent of Money 193
11 Imperial Visions 210
12 The Law of Religion 233
13 The Secret of Success 264

Part Four The Scientific Revolution

14 The Discovery of Ignorance 275
15 The Marriage of Science and Empire 307
16 The Capitalist Creed 341
17 The Wheels of Industry 374

18 A Permanent Revolution 392
19 And They Lived Happily Ever After 421
20 The End of *Homo Sapiens* 445
Afterword: The Animal that Became a God 465
Notes 467
Acknowledgements 481
Image Credits 483
Index 485

The Timeline of History

Number of years ago	
13.8 billion	Matter and energy appear. Beginning of physics. Formation of atoms and molecules. Beginning of chemistry.
4.5 billion	Formation of planet Earth.
3.8 billion	Emergence of organisms. Beginning of biology.
6 million	Last common grandmother of humans and chimpanzees.
2.5 million	Humans evolve in Africa. Use of stone tools.
2 million	Humans spread from Africa to Eurasia. Evolution of different human species.
400,000	Neanderthals evolve in Europe and the Middle East. Regular use of fire.
300,000	*Homo sapiens* evolves in Africa.
70,000	The Cognitive Revolution. Emergence of storytelling. Beginning of history. Sapiens spread out of Africa.
50,000	Sapiens settle across Australia. Extinction of Australian megafauna.
30,000	Extinction of Neanderthals. *Homo sapiens* the only surviving human species.

15,000	Sapiens spreads across Americas. Extinction of American megafauna.
12,000	The Agricultural Revolution. Domestication of plants and animals. Permanent settlements.
5000	First kingdoms, script and money. Polytheistic religions.
4250	First empire – the Akkadian Empire of Sargon.
2500	Invention of coins – a universal money. The Persian Empire – a universal political order. Buddhism in India – a universal teaching.
2000	Han Empire in China. Roman Empire in the Mediterranean. Christianity.
1400	Islam.
500	The Scientific Revolution. Humankind admits its ignorance and begins to acquire unprecedented power. Europeans begin to conquer America and the oceans. The entire planet becomes a single historical arena. The rise of capitalism.
200	The Industrial Revolution. Family and community are replaced by state and market. Massive extinction of plants and animals.
The Present	Humans transcend the boundaries of planet Earth. Nuclear weapons threaten the survival of humankind. Organisms are increasingly shaped by intelligent design rather than natural selection.
The Future	Intelligent design becomes the basic principle of life? First non-organic life-forms? Humans become gods?

Sapiens

A Brief History of Humankind

Part One

The Cognitive Revolution

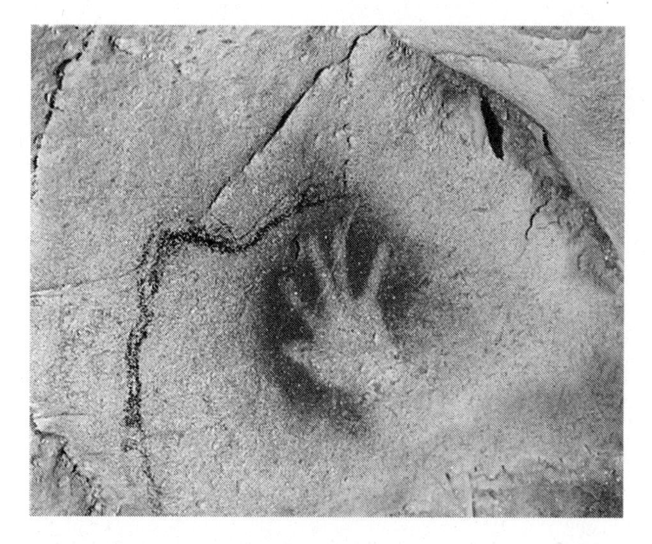

1. A human handprint made about 30,000 years ago, on the wall of the Chauvet-Pont-d'Arc Cave in southern France. Somebody tried to say, 'I was here!'

1

An Animal of No Significance

ABOUT 14 BILLION YEARS AGO, MATTER, energy, time and space came into being in what is known as the Big Bang. The story of these fundamental features of our universe is called physics.

About 300,000 years after their appearance, matter and energy started to coalesce into complex structures, called atoms, which then combined into molecules. The story of atoms, molecules and their interactions is called chemistry.

About 4 billion years ago, on a planet called Earth, certain molecules combined to form particularly large and intricate structures called organisms. The story of organisms is called biology.

About 70,000 years ago, organisms belonging to the species *Homo sapiens* started to form even more elaborate structures called cultures. The subsequent development of these human cultures is called history.

Three important revolutions shaped the course of history: the Cognitive Revolution kick-started history about 70,000 years ago. The Agricultural Revolution sped it up about 12,000 years ago. The Scientific Revolution, which got under way only 500 years ago, may well end history and start something completely different. This book tells the story of how these three revolutions have affected humans and their fellow organisms.

*

There were humans long before there was history. Animals much like modern humans first appeared about 2.5 million years ago. But for countless generations they did not stand out from the myriad other organisms that populated the planet.

On a hike in East Africa 2 million years ago, you might well have encountered a familiar cast of human characters: anxious mothers cuddling their babies and clutches of carefree children playing in the mud; temperamental youths chafing against the dictates of society and weary elders who just wanted to be left in peace; chest-thumping machos trying to impress the local beauty and wise old matriarchs who had already seen it all. These archaic humans loved, played, formed close friendships and competed for status and power – but so did chimpanzees, baboons and elephants. There was nothing special about humans. Nobody, least of all humans themselves, had any inkling that their descendants would one day walk on the moon, split the atom, fathom the genetic code and write history books. The most important thing to know about prehistoric humans is that they were insignificant animals with no more impact on their environment than gorillas, fireflies or jellyfish.

Biologists classify organisms into species. Animals are said to belong to the same species if they tend to mate with each other, giving birth to fertile offspring. Horses and donkeys have a recent common ancestor and share many physical traits. But they show little sexual interest in one another. They will mate if induced to do so – but their offspring, called mules, are sterile. Mutations in donkey DNA can therefore never cross over to horses, or vice versa. The two types of animals are consequently considered two distinct species, moving along separate evolutionary paths. By contrast, a bulldog and a spaniel may look very different, but they are members of the same species, sharing the same DNA pool. They will happily mate and their puppies will grow up to pair off with other dogs and produce more puppies.

Species that evolved from a common ancestor are bunched together under the heading 'genus' (plural genera). Lions, tigers, leopards and jaguars are different species within the genus *Panthera*. Biologists label organisms with a two-part Latin name, genus followed by species. Lions, for example, are called *Panthera leo*, the species *leo* of the genus *Panthera*. Presumably, everyone reading this book is a *Homo sapiens* – the species *sapiens* (wise) of the genus *Homo* (man).

Genera in their turn are grouped into families, such as the cats (lions, cheetahs, house cats), the dogs (wolves, foxes, jackals) and the elephants (elephants, mammoths, mastodons). All members of a family trace their lineage back to a founding matriarch or patriarch. All cats, for example, from the smallest house kitten to the most ferocious lion, share a common feline ancestor who lived about 25 million years ago.

Homo sapiens, too, belongs to a family. This banal fact used to be one of history's most closely guarded secrets. *Homo sapiens* long preferred to view itself as set apart from animals, an orphan who has no family, no cousins, and – most importantly – no parents. But that's just not the case. Like it or not, we are members of a large and particularly noisy family called the great apes. Our nearest living relatives include chimpanzees, gorillas and orang-utans. The chimpanzees are the closest. Just 6 million years ago, a single female ape had two daughters. One became the ancestor of all chimpanzees, the other is our own grandmother.

Skeletons in the Closet

Homo sapiens has kept hidden an even more disturbing secret. Not only do we possess an abundance of uncivilised cousins, once upon a time we had quite a few brothers and sisters as well. We are used to thinking about ourselves as the only humans, because for the last 10,000 years, our species has indeed been the only human species around. Yet the real meaning of the word human is 'an animal belonging to the genus *Homo*', and there used to be many other

2. Our siblings, according to speculative reconstructions (left to right): *Homo rudolfensis* (East Africa); *Homo erectus* (East Asia); and *Homo neanderthalensis* (Europe and western Asia). All are humans.

species of this genus besides *Homo sapiens*. Moreover, as we shall see in the last chapter of the book, in the not-so-distant future we might again have to contend with non-*sapiens* humans. To clarify this point, I will often use the term 'Sapiens' to denote members of the species *Homo sapiens*, while reserving the term 'human' to refer to all members of the genus *Homo*.

Humans first evolved in East Africa about 2.5 million years ago from an earlier genus of apes called *Australopithecus*, which means 'Southern Ape'. About 2 million years ago, some of these archaic men and women left their homeland to journey through and settle vast areas of North Africa, Europe and Asia. Since survival in the snowy forests of northern Europe required different traits than those needed to stay alive in Indonesia's steaming jungles, human populations evolved in different directions. The result was several distinct species, to each of which scientists have assigned a pompous Latin name.

Humans in Europe and western Asia evolved into *Homo neander-*

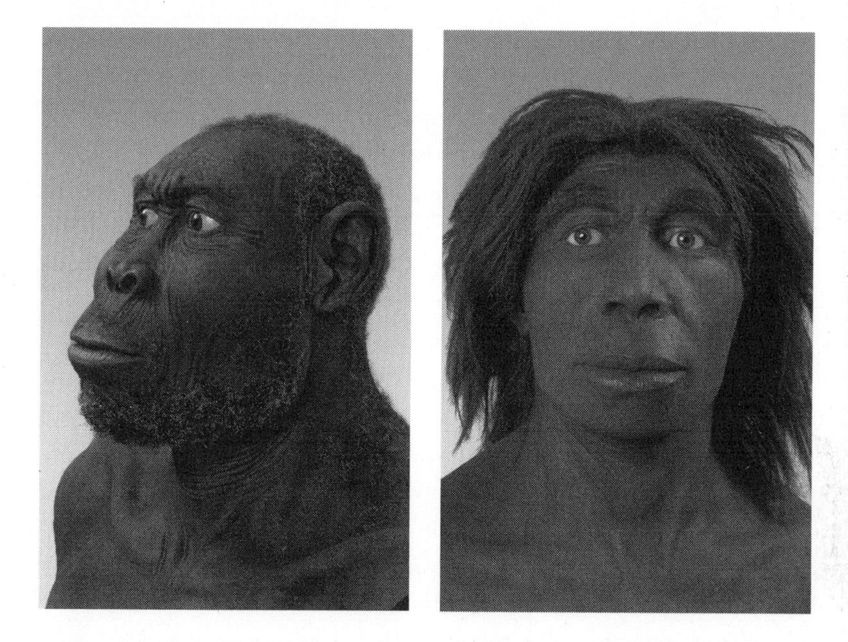

thalensis ('Man from the Neander Valley'), popularly referred to simply as 'Neanderthals'. Neanderthals, bulkier and more muscular than us Sapiens, were well adapted to the cold climate of Ice Age western Eurasia. The more eastern regions of Asia were populated by *Homo erectus*, 'Upright Man', who survived there for close to 2 million years, making it the most durable human species ever. This record is unlikely to be broken even by our own species. It is doubtful whether *Homo sapiens* will still be around a thousand years from now, so 2 million years is really out of our league.

On the island of Java, in Indonesia, lived *Homo soloensis*, 'Man from the Solo Valley', who was suited to life in the tropics. On another Indonesian island – the small island of Flores – archaic humans underwent a process of dwarfing. Humans first reached Flores when the sea level was exceptionally low, and the island was easily accessible from the mainland. When the seas rose again, some people were trapped on the island, which was poor in resources.

Big people, who need a lot of food, died first. Smaller fellows survived much better. Over the generations, the people of Flores became dwarves. This unique species, known by scientists as *Homo floresiensis*, reached a maximum height of only one metre and weighed no more than twenty-five kilograms. They were nevertheless able to produce stone tools, and even managed occasionally to hunt down some of the island's elephants – though, to be fair, the elephants were a dwarf species as well.

In 2010 another lost sibling was rescued from oblivion, when scientists excavating the Denisova Cave in Siberia discovered a fossilised finger bone. Genetic analysis proved that the finger belonged to a previously unknown human species, which was named *Homo denisova*. Who knows how many lost relatives of ours are waiting to be discovered in other caves, on other islands, and in other climes?

While these humans were evolving in Europe and Asia, evolution in East Africa did not stop. The cradle of humanity continued to nurture numerous new species, such as *Homo rudolfensis*, 'Man from Lake Rudolf', *Homo ergaster*, 'Working Man', and eventually our own species, which we've immodestly named *Homo sapiens*, 'Wise Man'.

The members of some of these species were massive and others were dwarves. Some were fearsome hunters and others meek plant-gatherers. Some lived only on a single island, while many roamed over continents. But all of them belonged to the genus *Homo*. They were all human beings.

It's a common fallacy to envision these species as arranged in a straight line of descent, with Ergaster begetting Erectus, Erectus begetting the Neanderthals, and the Neanderthals evolving into us. This linear model gives the mistaken impression that at any particular moment only one type of human inhabited the earth, and that all earlier species were merely older models of ourselves. The truth is that from about 2 million years ago until around 10,000 years ago, the world was home, at one and the same time, to several human species. And why not? Today there are many species of bears:

brown bears, black bears, grizzly bears, polar bears. The earth was once walked by at least six species of man. It's our current exclusivity, not that multi-species past, that is peculiar – and perhaps incriminating. As we will shortly see, we Sapiens have good reasons to repress the memory of our siblings.

The Cost of Thinking

Despite their many differences, all human species share several defining characteristics. Most notably, humans have extraordinarily large brains compared to other animals. Mammals weighing sixty kilograms have an average brain size of 200 cubic centimetres. The earliest men and women, 2.5 million years ago, had brains of about 600 cubic centimetres. Modern Sapiens sport a brain averaging 1,200–1,400 cubic centimetres. Neanderthal brains were even bigger.

That evolution should select for larger brains may seem to us like, well, a no-brainer. We are so enamoured of our high intelligence that we assume that when it comes to cerebral power, more must be better. But if that were the case, the feline family would also have produced cats who could do calculus and frogs would by now have launched their own space programme. Why are giant brains so rare in the animal kingdom?

The fact is that a jumbo brain is a jumbo drain on the body. It's not easy to carry around, especially when encased inside a massive skull. It's even harder to fuel. In *Homo sapiens*, the brain accounts for about 2–3 per cent of total body weight, but it consumes 25 per cent of the body's energy when the body is at rest. By comparison, the brains of other apes require only 8 per cent of rest-time energy. Archaic humans paid for their large brains in two ways. Firstly, they spent more time in search of food. Secondly, their muscles atrophied. Like a government diverting money from defence to education, humans diverted energy from biceps to neurons. It's hardly a foregone conclusion that this is a good strategy for survival on the savannah. A chimpanzee can't win an argument with a *Homo sapiens*, but the ape can rip the man apart like a rag doll.

Today our big brains pay off nicely, because we can produce cars and guns that enable us to move much faster than chimps, and shoot them from a safe distance instead of wrestling. But cars and guns are a recent phenomenon. For more than 2 million years, human neural networks kept growing and growing, but apart from some flint knives and pointed sticks, humans had precious little to show for it. What then drove forward the evolution of the massive human brain during those 2 million years? Frankly, we don't know.

Another singular human trait is that we walk upright on two legs. Standing up, it's easier to scan the savannah for game or enemies, and arms that are unnecessary for locomotion are freed for other purposes, like throwing stones or signalling. The more things these hands could do, the more successful their owners were, so evolutionary pressure brought about an increasing concentration of nerves and finely tuned muscles in the palms and fingers. As a result, humans can perform very intricate tasks with their hands. In particular, they can produce and use sophisticated tools. The first evidence for tool production dates from about 2.5 million years ago, and the manufacture and use of tools are the criteria by which archaeologists recognise ancient humans.

Yet walking upright has its downside. The skeleton of our primate ancestors developed for millions of years to support a creature that walked on all fours and had a relatively small head. Adjusting to an upright position was quite a challenge, especially when the scaffolding had to support an extra-large cranium. Humankind paid for its lofty vision and industrious hands with backaches and stiff necks.

Women paid extra. An upright gait required narrower hips, constricting the birth canal – and this just when babies' heads were getting bigger and bigger. Death in childbirth became a major hazard for human females. Women who gave birth earlier, when the infant's brain and head were still relatively small and supple, fared better and lived to have more children. Natural selection consequently favoured earlier births. And, indeed, compared to other animals, humans are born prematurely, when many of their vital systems are still underdeveloped. A colt can trot shortly after birth; a kitten leaves its mother to forage on its own when it is just a few weeks old. Human babies are helpless, dependent for many years on their elders for sustenance, protection and education.

This fact has contributed greatly both to humankind's extraordinary social abilities and to its unique social problems. Lone mothers could hardly forage enough food for their offspring and themselves with needy children in tow. Raising children required constant help from other family members and neighbours. It takes a tribe to raise a human. Evolution thus favoured those capable of forming strong social ties. In addition, since humans are born underdeveloped, they can be educated and socialised to a far greater extent than any other animal. Most mammals emerge from the womb like glazed earthenware emerging from a kiln – any attempt at remoulding will only scratch or break them. Humans emerge from the womb like molten glass from a furnace. They can be spun, stretched and shaped with a surprising degree of freedom. This is why today we can educate our children to become Christian or Buddhist, capitalist or socialist, warlike or peace-loving.

We assume that a large brain, the use of tools, superior learning abilities and complex social structures are huge advantages. It seems self-evident that these have made humankind the most powerful animal on earth. But humans enjoyed all of these advantages for a full 2 million years during which they remained weak and marginal creatures. Thus humans who lived a million years ago, despite their big brains and sharp stone tools, dwelt in constant fear of predators, rarely hunted large game, and subsisted mainly by gathering plants, scooping up insects, stalking small animals, and eating the carrion left behind by other more powerful carnivores.

One of the most common uses of early stone tools was to crack open bones in order to get to the marrow. Some researchers believe this was our original niche. Just as woodpeckers specialise in extracting insects from the trunks of trees, the first humans specialised in extracting marrow from bones. Why marrow? Well, suppose you observe a pride of lions take down and devour a giraffe. You wait patiently until they're done. But it's still not your turn because first the hyenas and jackals – and you don't dare interfere with them – scavenge the leftovers. Only then would you and your band dare approach the carcass, look cautiously left and right – and dig into the edible tissue that remained.

This is a key to understanding our history and psychology. Genus *Homo*'s position in the food chain was, until quite recently, solidly in the middle. For millions of years, humans hunted smaller creatures and gathered what they could, all the while being hunted by larger predators. It was only 400,000 years ago that several species of man began to hunt large game on a regular basis, and only in the last 100,000 years – with the rise of *Homo sapiens* – that man jumped to the top of the food chain.

That spectacular leap from the middle to the top had enormous consequences. Other animals at the top of the pyramid, such as lions and sharks, evolved into that position very gradually, over millions of years. This enabled the ecosystem to develop checks and balances that prevent lions and sharks from wreaking too much havoc. As lions became deadlier, so gazelles evolved to run faster, hyenas to cooperate better, and rhinoceroses to be more bad-tempered. In contrast, humankind ascended to the top so quickly that the ecosystem was not given time to adjust. Moreover, humans themselves failed to adjust. Most top predators of the planet are majestic creatures. Millions of years of dominion have filled them with self-confidence. Sapiens by contrast is more like a banana-republic dictator. Having so recently been one of the underdogs of the savannah, we are full of fears and anxieties over our position, which makes us doubly cruel and dangerous. Many historical calamities, from deadly wars to ecological catastrophes, have resulted from this over-hasty jump.

A Race of Cooks

A significant step on the way to the top was the domestication of fire. Some human species may have made occasional use of fire as early as 800,000 years ago. By about 300,000 years ago, *Homo erectus*, Neanderthals and the forefathers of *Homo sapiens* were using fire on a daily basis. Humans now had a dependable source of light and warmth, and a deadly weapon against prowling lions. Not long afterwards, humans may even have started deliberately to torch their neighbourhoods. A carefully managed fire could turn impassable barren thickets into prime grasslands teeming with game. In addition, once the fire died down, Stone Age entrepreneurs could walk through the smoking remains and harvest charcoaled animals, nuts and tubers.

But the best thing fire did was cook. Foods that humans cannot digest in their natural forms – such as wheat, rice and potatoes – became staples of our diet thanks to cooking. Fire not only changed food's chemistry, it changed its biology as well. Cooking killed germs and parasites that infested food. Humans also had a far easier time chewing and digesting old favourites such as fruits, nuts, insects and carrion if they were cooked. Whereas chimpanzees spend five hours a day chewing raw food, a single hour suffices for people eating cooked food.

The advent of cooking enabled humans to eat more kinds of food, to devote less time to eating, and to make do with smaller teeth and shorter intestines. Some scholars believe there is a direct link between the advent of cooking, the shortening of the human intestinal track, and the growth of the human brain. Since long intestines and large brains are both massive energy consumers, it's hard to have both. By shortening the intestines and decreasing their energy consumption, cooking inadvertently opened the way to the jumbo brains of Neanderthals and Sapiens.[1]

Fire also opened the first significant gulf between man and the other animals. The power of almost all animals depends on their bodies: the strength of their muscles, the size of their teeth, the breadth of their wings. Though they may harness winds and currents, they are unable to control these natural forces, and are always constrained by their physical design. Eagles, for example, identify thermal columns rising from the ground, spread their giant wings and allow the hot air to lift them upwards. Yet eagles cannot control the location of the columns, and their maximum carrying capacity is strictly proportional to their wingspan.

When humans domesticated fire, they gained control of an obedient and potentially limitless force. Unlike eagles, humans could choose when and where to ignite a flame, and they were able to exploit fire for any number of tasks. Most importantly, the power of fire was not limited by the form, structure or strength of the human body. A single woman with a flint or fire stick could burn down an entire forest in a matter of hours. The domestication of fire was a sign of things to come.

Our Brothers' Keepers

Despite the benefits of fire, 150,000 years ago humans were still marginal creatures. They could now scare away lions, warm themselves during cold nights, and burn down the occasional forest. Yet counting all species together, there were still no more than perhaps a million humans living between the Indonesian archipelago and the Iberian peninsula, a mere blip on the ecological radar.

Our own species, *Homo sapiens*, was already present on the world stage, but so far it was just minding its own business in a corner of Africa. We don't know exactly where and when animals that can be classified as *Homo sapiens* first evolved from some earlier type of humans, but most scientists agree that by 150,000 years ago, East Africa was populated by Sapiens that looked just like us. If one of them turned up in a modern morgue, the local pathologist would notice nothing peculiar. Thanks to the blessings of fire, they had smaller teeth and jaws than their ancestors, whereas they had massive brains, equal in size to ours.

Scientists also agree that about 70,000 years ago, Sapiens from East Africa spread into the Arabian peninsula, and from there they quickly overran the entire Eurasian landmass.

When *Homo sapiens* landed in Arabia, most of Eurasia was already settled by other humans. What happened to them? There are two conflicting theories. The 'Interbreeding Theory' tells a story of attraction, sex and mingling. As the African immigrants spread around the world, they bred with other human populations, and people today are the outcome of this interbreeding.

For example, when Sapiens reached the Middle East and Europe, they encountered the Neanderthals. These humans were more muscular than Sapiens, had larger brains, and were better adapted to cold climes. They used tools and fire, were good hunters, and apparently took care of their sick and infirm. (Archaeologists have discovered the bones of Neanderthals who lived for many years with severe physical handicaps, evidence that they were cared for by their relatives.) Neanderthals are often depicted in caricatures as the archetypical brutish and stupid 'cave people', but recent evidence has changed their image.

According to the Interbreeding Theory, when Sapiens spread into Neanderthal lands, Sapiens bred with Neanderthals until the two populations merged. If this is the case, then today's Eurasians are not pure Sapiens. They are a mixture of Sapiens and Neanderthals.

Similarly, when Sapiens reached East Asia, they interbred with the local Erectus, so the Chinese and Koreans are a mixture of Sapiens and Erectus.

The opposing view, called the 'Replacement Theory', tells a very different story – one of incompatibility, revulsion, and perhaps even genocide. According to this theory, Sapiens and other humans had different anatomies, and most likely different mating habits and even body odours. They would have had little sexual interest in one another. And even if a Neanderthal Romeo and a Sapiens Juliet fell in love, they could not produce fertile children, because the genetic gulf separating the two populations was already unbridgeable. The two populations remained completely distinct, and when the Neanderthals died out, or were killed off, their genes died with them. According to this view, Sapiens replaced all the previous human populations without merging with them. If that is the case, the lineages of all contemporary humans can be traced back, exclusively, to East Africa, 70,000 years ago. We are all 'pure Sapiens'.

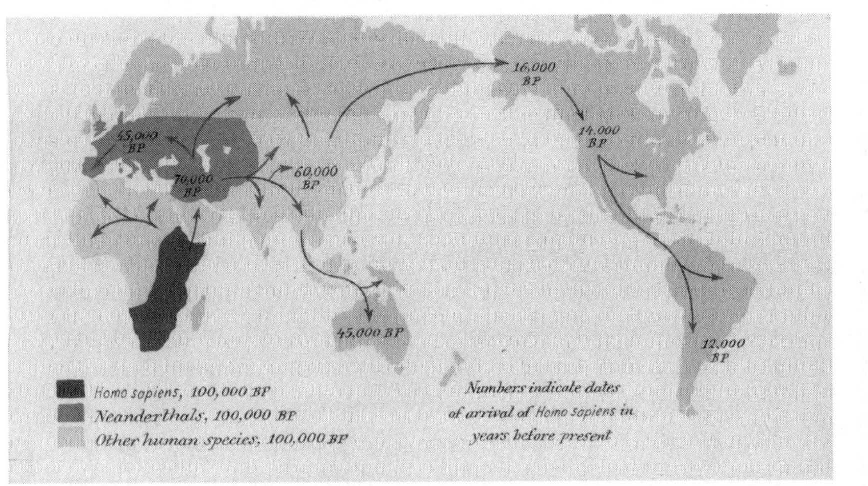

Map 1. *Homo sapiens* conquers the globe.

A lot hinges on this debate. From an evolutionary perspective, 70,000 years is a relatively short interval. If the Replacement Theory is correct, all living humans have roughly the same genetic baggage, and racial distinctions among them are negligible. But if the Interbreeding Theory is right, there might well be genetic differences between Africans, Europeans and Asians that go back hundreds of thousands of years. This is political dynamite, which could provide material for explosive racial theories.

In recent decades the Replacement Theory has been the common wisdom in the field. It had firmer archaeological backing, and was more politically correct (scientists had no desire to open up the Pandora's box of racism by claiming significant genetic diversity among modern human populations). But that ended in 2010, when the results of a four-year effort to map the Neanderthal genome were published. Geneticists were able to collect enough intact Neanderthal DNA from fossils to make a broad comparison between it and the DNA of contemporary humans. The results stunned the scientific community.

It turned out that 1–4 per cent of the unique human DNA of modern populations in the Middle East and Europe is Neanderthal DNA. That's not a huge amount, but it's significant. A second shock came several months later, when DNA extracted from the fossilised finger from Denisova was mapped. The results proved that up to 6 per cent of the unique human DNA of modern Melanesians and Aboriginal Australians is Denisovan DNA.

If these results are valid – and it's important to keep in mind that further research is under way and may either reinforce or modify these conclusions – the Interbreeders got at least some things right. But that doesn't mean that the Replacement Theory is completely wrong. Since Neanderthals and Denisovans contributed only a small amount of DNA to our present-day genome, it is impossible to speak of a 'merger' between Sapiens and other human species. Although differences between them were not large enough to completely prevent fertile intercourse, they were sufficient to make such contacts very rare.

How then should we understand the biological relatedness of Sapiens, Neanderthals and Denisovans? Clearly, they were not completely different species like horses and donkeys. On the other hand, they were not just different populations of the same species, like bulldogs and spaniels. Biological reality is not black and white. There are also important grey areas. Every two species that evolved from a common ancestor, such as horses and donkeys, were at one time just two populations of the same species, like bulldogs and spaniels. There must have been a point when the two populations were already quite different from one another, but still capable on rare occasions of having sex and producing fertile offspring. Then another mutation severed this last connecting thread, and they went their separate evolutionary ways.

It seems that about 50,000 years ago, Sapiens, Neanderthals and Denisovans were at that borderline point. They were almost, but not quite, entirely separate species. As we shall see in the next chapter, Sapiens were already very different from Neanderthals and Denisovans not only in their genetic code and physical traits, but

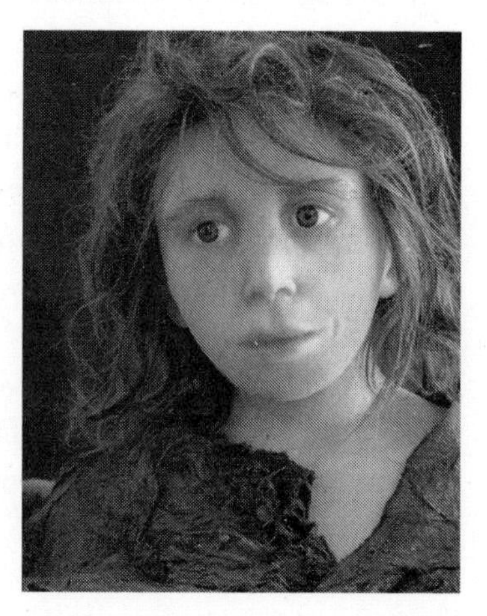

3. A speculative reconstruction of a Neanderthal child. Genetic evidence hints that at least some Neanderthals may have had fair skin and hair.

also in their cognitive and social abilities, yet it appears it was still just possible, on rare occasions, for a Sapiens and a Neanderthal to produce a fertile offspring. So the populations did not merge, but a few lucky Neanderthal genes did hitch a ride on the Sapiens Express. It is unsettling – and perhaps thrilling – to think that we Sapiens could at one time have sex with an animal from a different species, and produce children together.

But if the Neanderthals, Denisovans and other human species didn't merge with Sapiens, why did they vanish? One possibility is that *Homo sapiens* drove them to extinction. Imagine a Sapiens band reaching a Balkan valley where Neanderthals had lived for hundreds of thousands of years. The newcomers began to hunt the deer and gather the nuts and berries that were the Neanderthals' traditional staples. Sapiens were more proficient hunters and gatherers – thanks to better technology and superior social skills – so they multiplied and spread. The less resourceful Neanderthals found it increasingly difficult to feed themselves. Their population dwindled and they slowly died out, except perhaps for one or two members who joined their Sapiens neighbours.

Another possibility is that competition for resources flared up into violence and genocide. Tolerance is not a Sapiens trademark. In modern times, a small difference in skin colour, dialect or religion has been enough to prompt one group of Sapiens to set about exterminating another group. Would ancient Sapiens have been more tolerant towards an entirely different human species? It may well be that when Sapiens encountered Neanderthals, the result was the first and most significant ethnic-cleansing campaign in history.

Whichever way it happened, the Neanderthals (and the other human species) pose one of history's great what ifs. Imagine how things might have turned out had the Neanderthals or Denisovans survived alongside *Homo sapiens*. What kind of cultures, societies and political structures would have emerged in a world where several different human species coexisted? How, for example, would religious faiths have unfolded? Would the book of Genesis have declared that Neanderthals descend from Adam and Eve, would Jesus have died for the sins of the Denisovans, and would the Qur'an have reserved seats in heaven for all righteous humans, whatever their species? Would Neanderthals have been able to serve in the Roman legions, or in the sprawling bureaucracy of imperial China? Would the American Declaration of Independence hold as a self-evident truth that all members of the genus *Homo* are created equal? Would Karl Marx have urged workers of all species to unite?

Over the past 10,000 years, *Homo sapiens* has grown so accustomed to being the only human species that it's hard for us to conceive of any other possibility. Our lack of brothers and sisters makes it easier to imagine that we are the epitome of creation, and that a chasm separates us from the rest of the animal kingdom. When Charles Darwin indicated that *Homo sapiens* was just another kind of animal, people were outraged. Even today many refuse to believe it. Had the Neanderthals survived, would we still imagine ourselves to be a creature apart? Perhaps this is exactly why our ancestors wiped out the Neanderthals. They were too familiar to ignore, but too different to tolerate.

Whether Sapiens are to blame or not, no sooner had they arrived at a new location than the native population became extinct. The last remains of *Homo soloensis* are dated to about 50,000 years ago. *Homo denisova* disappeared shortly thereafter. Neanderthals made their exit roughly 30,000 years ago. The last dwarf-like humans vanished from Flores Island about 12,000 years ago. They left behind some bones, stone tools, a few genes in our DNA and a lot of unanswered questions. They also left behind us, *Homo sapiens*, the last human species.

What was the Sapiens' secret of success? How did we manage to settle so rapidly in so many distant and ecologically different habitats? How did we push all other human species into oblivion? Why couldn't even the strong, brainy, cold-proof Neanderthals survive our onslaught? The debate continues to rage. The most likely answer is the very thing that makes the debate possible: *Homo sapiens* conquered the world thanks above all to its unique language.

2

The Tree of Knowledge

IN THE PREVIOUS CHAPTER WE SAW THAT although Sapiens had already populated East Africa 150,000 years ago, they began to overrun the rest of planet Earth and drive the other human species to extinction only about 70,000 years ago. In the intervening millennia, even though these archaic Sapiens looked just like us and their brains were as big as ours, they did not enjoy any marked advantage over other human species, did not produce particularly sophisticated tools, and did not accomplish any other special feats.

In fact, in the first recorded encounter between Sapiens and Neanderthals, the Neanderthals won. About 100,000 years ago, some Sapiens groups migrated north to the Levant, which was Neanderthal territory, but failed to secure a firm footing. It might have been due to nasty natives, an inclement climate, or unfamiliar local parasites. Whatever the reason, the Sapiens eventually retreated, leaving the Neanderthals as masters of the Middle East.

This poor record of achievement has led scholars to speculate that the internal structure of the brains of these Sapiens was probably different from ours. They looked like us, but their cognitive abilities – learning, remembering, communicating – were far more limited. Teaching such ancient Sapiens to speak English, persuading them of the truth of Christian dogma, or getting them to understand the theory of evolution would probably have been hopeless undertakings.

Conversely, we would have had a very hard time learning their communication system and way of thinking.

But then, beginning about 70,000 years ago, *Homo sapiens* started doing very special things. Around that date Sapiens bands left Africa for a second time. This time they drove the Neanderthals and all other human species not only from the Middle East, but from the face of the earth. Within a remarkably short period, Sapiens reached Europe and East Asia. About 45,000 years ago, they somehow crossed the open sea and landed in Australia – a continent hitherto untouched by humans. The period from about 70,000 years ago to about 30,000 years ago witnessed the invention of boats, oil lamps, bows and arrows and needles (essential for sewing warm clothing). The first objects that can reliably be called art date from this era (see the Stadel lion-man on page 25), as does the first clear evidence for religion, commerce and social stratification.

Most researchers believe that these unprecedented accomplishments were the product of a revolution in Sapiens' cognitive abilities. They maintain that the people who drove the Neanderthals to extinction, settled Australia, and carved the Stadel lion-man were as intelligent, creative and sensitive as we are. If we were to come across the artists of the Stadel Cave, we could learn their language and they ours. We'd be able to explain to them everything we know – from the adventures of Alice in Wonderland to the paradoxes of quantum physics – and they could teach us how their people view the world.

The appearance of new ways of thinking and communicating, between 70,000 and 30,000 years ago, constitutes the Cognitive Revolution. What caused it? We're not sure. The most commonly believed theory argues that accidental genetic mutations changed the inner wiring of the brains of Sapiens, enabling them to think in unprecedented ways and to communicate using an altogether new type of language. We might call it the Tree of Knowledge mutation. Why did it occur in Sapiens DNA rather than in that of Neanderthals? It was a matter of pure chance, as far as we can tell. But it's more important to understand the consequences of the Tree of Knowledge mutation than its causes. What was so special about the new Sapiens language that it enabled us to conquer the world?*

It was not the first communication system. Every animal knows how to communicate. Even insects, such as bees and ants, know how to inform one another of the whereabouts of food. Neither was it the first vocal communication system. Many animals, including all ape and monkey species, use vocal signs. For example, green monkeys use calls of various kinds to warn one another of danger. Zoologists have identified one call that means 'Careful! An eagle!' A slightly different call warns 'Careful! A lion!' When researchers played a recording of the first call to a group of monkeys, the monkeys stopped what they were doing and looked upwards in fear. When the same group heard a recording of the second call, the lion warning, they quickly scrambled up a tree. Sapiens can produce many more distinct sounds than green monkeys, but whales and elephants have equally impressive abilities. A parrot can say anything Albert Einstein could say, as well as mimicking the sounds of phones ringing, doors slamming and sirens wailing. Whatever advantage Einstein had over a parrot, it wasn't vocal. What, then, is so special about our language?

The most common answer is that our language is amazingly supple. We can connect a limited number of sounds and signs to produce an infinite number of sentences, each with a distinct meaning. We can thereby ingest, store and communicate a prodigious amount of information about the surrounding world. A green monkey can yell to its comrades, 'Careful! A lion!' But a modern

* Here and in the following pages, when speaking about Sapiens language, I refer to the basic linguistic abilities of our species and not to a particular dialect. English, Hindi and Chinese are all variants of Sapiens language. Apparently, even at the time of the Cognitive Revolution, different Sapiens groups had different dialects.

human can tell her friends that this morning, near the bend in the river, she saw a lion tracking a herd of bison. She can then describe the exact location, including the different paths leading to the area. With this information, the members of her band can put their heads together and discuss whether they should approach the river, chase away the lion, and hunt the bison.

A second theory agrees that our unique language evolved as a means of sharing information about the world. But the most important information that needed to be conveyed was about humans, not about lions and bison. Our language evolved as a way of gossiping.

4. An ivory figurine of a 'lion-man' (or 'lioness-woman') from the Stadel Cave in Germany (*c*.32,000 years ago). The body is human, but the head is leonine. This is one of the first indisputable examples of art, and probably of religion, and of the ability of the human mind to imagine things that do not really exist.

According to this theory *Homo sapiens* is primarily a social animal. Social cooperation is our key for survival and reproduction. It is not enough for individual men and women to know the whereabouts of lions and bison. It's much more important for them to know who in their band hates whom, who is sleeping with whom, who is honest, and who is a cheat.

The amount of information that one must obtain and store in order to track the ever-changing relationships of even a few dozen individuals is staggering. (In a band of fifty individuals, there are 1,225 one-on-one relationships, and countless more complex social combinations.) All apes show a keen interest in such social information, but they have trouble gossiping effectively. Neanderthals and archaic *Homo sapiens* probably also had a hard time talking behind each other's backs – a much maligned ability which is in fact essential for cooperation in large numbers. The new linguistic skills that modern Sapiens acquired about seventy millennia ago enabled them to gossip for hours on end. Reliable information about who could be trusted meant that small bands could expand into larger bands, and Sapiens could develop tighter and more sophisticated types of cooperation.[1]

The gossip theory might sound like a joke, but numerous studies support it. Even today the vast majority of human communication – whether in the form of emails, phone calls or newspaper columns – is gossip. It comes so naturally to us that it seems as if our language evolved for this very purpose. Do you think that history professors chat about the reasons for the First World War when they meet for lunch, or that nuclear physicists spend their coffee breaks at scientific conferences talking about quarks? Sometimes. But more often, they gossip about the professor who caught her husband cheating, or the quarrel between the head of the department and the dean, or the rumours that a colleague used his research funds to buy a Lexus. Gossip usually focuses on wrongdoings. Rumour-mongers are the original fourth estate, journalists who inform society about and thus protect it from cheats and freeloaders.

Most likely, both the gossip theory and the there-is-a-lion-near-the-river theory are valid. Yet the truly unique feature of our language is not its ability to transmit information about men and lions. Rather, it's the ability to transmit information about things that do not exist at all. As far as we know, only Sapiens can talk about entire kinds of entities that they have never seen, touched or smelled.

Legends, myths, gods and religions appeared for the first time with the Cognitive Revolution. Many animals and human species could previously say, 'Careful! A lion!' Thanks to the Cognitive Revolution, *Homo sapiens* acquired the ability to say, 'The lion is the guardian spirit of our tribe.' This ability to speak about fictions is the most unique feature of Sapiens language.

It's relatively easy to agree that only *Homo sapiens* can speak about things that don't really exist, and believe six impossible things before breakfast. You could never convince a monkey to give you a banana by promising him limitless bananas after death in monkey heaven. But why is it important? After all, fiction can be dangerously misleading or distracting. People who go to the forest looking for fairies and unicorns would seem to have less chance of survival than people who go looking for mushrooms and deer. And if you spend hours praying to non-existing guardian spirits, aren't you wasting precious time, time better spent foraging, fighting and fornicating?

However, fiction has enabled us not merely to imagine things, but to do so *collectively*. We can weave common myths such as the biblical creation story, the Dreamtime myths of Aboriginal Australians, and the nationalist myths of modern states. Such myths give Sapiens the unprecedented ability to cooperate flexibly in large numbers. Ants and bees can also work together in huge numbers, but they do so in a very rigid manner and only with close relatives. Wolves and chimpanzees cooperate far more flexibly than ants, but they can do so only with small numbers of other individuals that they know intimately. Sapiens can cooperate in extremely flexible ways with countless numbers of strangers. That's why Sapiens rule the world, whereas ants eat our leftovers and chimps are locked up in zoos and research laboratories.

The Legend of Peugeot

Our chimpanzee cousins usually live in small troops of several dozen individuals. They form close friendships, hunt together and fight shoulder to shoulder against baboons, cheetahs and enemy chimpanzees. Their social structure tends to be hierarchical. The dominant member, who is almost always a male, is termed the 'alpha male'. Other males and females exhibit their submission to the alpha male by bowing before him while making grunting sounds, not unlike human subjects kowtowing before a king. The alpha male strives to maintain social harmony within his troop. When two individuals fight, he will intervene and stop the violence. Less benevolently, he might monopolise particularly coveted foods and prevent lower-ranking males from mating with the females.

When two males are contesting the alpha position, they usually do so by forming extensive coalitions of supporters, both male and female, from within the group. Ties between coalition members are based on intimate daily contact – hugging, touching, kissing, grooming and mutual favours. Just as human politicians on election campaigns go around shaking hands and kissing babies, so aspirants to the top position in a chimpanzee group spend much time hugging, back-slapping and kissing baby chimps. The alpha male usually wins his position not because he is physically stronger, but because he leads a large and stable coalition. These coalitions play a central part not only during overt struggles for the alpha position, but in almost all day-to-day activities. Members of a coalition spend more time together, share food, and help one another in times of trouble.

There are clear limits to the size of groups that can be formed and maintained in such a way. In order to function, all members of a group must know each other intimately. Two chimpanzees who have never met, never fought, and never engaged in mutual grooming will not know whether they can trust one another, whether it would be worthwhile to help one another, and which of them ranks higher. Under natural conditions, a typical chimpanzee troop consists of about twenty to fifty individuals. As the number of chimpanzees in a troop increases, the social order destabilises, eventually leading to a rupture and the formation of a new troop by some of the animals. Only in a handful of cases have zoologists observed groups larger than a hundred. Separate groups seldom cooperate, and tend to compete for territory and food. Researchers have documented prolonged warfare between groups, and even one case of 'genocidal' activity in which one troop systematically slaughtered most members of a neighbouring band.[2]

Similar patterns probably dominated the social lives of early humans, including archaic *Homo sapiens*. Humans, like chimps, have social instincts that enabled our ancestors to form friendships and hierarchies, and to hunt or fight together. However, like the social instincts of chimps, those of humans were adapted only for small intimate groups. When the group grew too large, its social order destabilised and the band split. Even if a particularly fertile valley could feed 500 archaic Sapiens, there was no way that so many strangers could live together. How could they agree who should be leader, who should hunt where, or who should mate with whom?

In the wake of the Cognitive Revolution, gossip helped *Homo sapiens* to form larger and more stable bands. But even gossip has its limits. Sociological research has shown that the maximum 'natural' size of a group bonded by gossip is about 150 individuals. Most people can neither intimately know, nor gossip effectively about, more than 150 human beings.

Even today, a critical threshold in human organisations falls somewhere around this magic number. Below this threshold, communities, businesses, social networks and military units can maintain themselves based mainly on intimate acquaintance and rumour-mongering. There is no need for formal ranks, titles and law books to keep order.[3] A platoon of thirty soldiers or even a company of a hundred soldiers can function well on the basis of intimate relations, with a minimum of formal discipline. A well-respected sergeant can become 'king of the company' and exercise authority even over commissioned officers. A small family business can survive and flourish without a board of directors, a CEO or an accounting department.

But once the threshold of 150 individuals is crossed, things can no longer work that way. You cannot run a division with thousands of soldiers the same way you run a platoon. Successful family businesses usually face a crisis when they grow larger and hire more personnel. If they cannot reinvent themselves, they go bust.

How did *Homo sapiens* manage to cross this critical threshold, eventually founding cities comprising tens of thousands of inhabitants and empires ruling hundreds of millions? The secret was probably the appearance of fiction. Large numbers of strangers can cooperate successfully by believing in common myths.

Any large-scale human cooperation – whether a modern state, a medieval church, an ancient city or an archaic tribe – is rooted in common myths that exist only in people's collective imagination. Churches are rooted in common religious myths. Two Catholics who have never met can nevertheless go together on crusade or pool funds to build a hospital because they both believe that God was incarnated in human flesh and allowed Himself to be crucified to redeem our sins. States are rooted in common national myths. Two Serbs who have never met might risk their lives to save one another because both believe in the existence of the Serbian nation, the Serbian homeland and the Serbian flag. Judicial systems are rooted in common legal myths. Two lawyers who have never met can nevertheless combine efforts to defend a complete stranger because they both believe in the existence of laws, justice, human rights – and the money paid out in fees.

Yet none of these things exists outside the stories that people invent and tell one another. There are no gods in the universe, no nations, no money, no human rights, no laws and no justice outside the common imagination of human beings.

People easily acknowledge that 'primitive tribes' cement their social order by believing in ghosts and spirits, and gathering each full moon to dance together around the campfire. What we fail to appreciate is that our modern institutions function on exactly the same basis. Take for example the world of business corporations. Modern business-people and lawyers are, in fact, powerful sorcerers. The principal difference between them and tribal shamans is that modern lawyers tell far stranger tales. The legend of Peugeot affords us a good example.

An icon that somewhat resembles the Stadel lion-man appears today on cars, trucks and motorcycles from Paris to Sydney. It's the hood ornament that adorns vehicles made by Peugeot, one of the oldest and largest of Europe's carmakers. Peugeot began as a small family business in the village of Valentigney, just 300 kilometres from the Stadel Cave. Today the company employs about 200,000 people worldwide, most of whom are complete strangers to each other. These strangers cooperate so effectively that in 2008 Peugeot produced more than 1.5 million automobiles, earning revenues of about €55 billion.

In what sense can we say that Peugeot SA (the company's official name) exists? There are many Peugeot vehicles, but these are obviously not the company. Even if every Peugeot in the world were simultaneously junked and sold for scrap metal, Peugeot SA would not disappear. It would continue to manufacture new cars and issue its annual report. The company owns factories, machinery and showrooms, and employs mechanics, accountants and secretaries, but all these together do not comprise Peugeot. A disaster might kill every single one of Peugeot's employees, and go on to destroy

5. **The Peugeot Lion.**

all of its assembly lines and executive offices. Even then, the company could borrow money, hire new employees, build new factories and buy new machinery. Peugeot has managers and shareholders, but neither do they constitute the company. All the managers could be dismissed and all its shares sold, but the company itself would remain intact.

It doesn't mean that Peugeot SA is invulnerable or immortal. If a judge were to mandate the dissolution of the company, its factories would remain standing and its workers, accountants, managers and shareholders would continue to live – but Peugeot SA would immediately vanish. In short, Peugeot SA seems to have no essential connection to the physical world. Does it really exist?

Peugeot is a figment of our collective imagination. Lawyers call this a 'legal fiction'. It can't be pointed at; it is not a physical object. But it exists as a legal entity. Just like you or me, it is bound by the laws of the countries in which it operates. It can open a bank account and own property. It pays taxes, and it can be sued and even prosecuted separately from any of the people who own or work for it.

Peugeot belongs to a particular genre of legal fictions called 'limited liability companies'. The idea behind such companies is among humanity's most ingenious inventions. *Homo sapiens* lived for untold millennia without them. During most of recorded history property could be owned only by flesh-and-blood humans, the kind that stood on two legs and had big brains. If in thirteenth-century France Jean set up a wagon-manufacturing workshop, he himself was the business. If a wagon he'd made broke down a week after purchase, the disgruntled buyer would have sued Jean personally. If Jean had borrowed 1,000 gold coins to set up his workshop and the business failed, he would have had to repay the loan by selling his private property – his house, his cow, his land. He might even have had to sell his children into servitude. If he couldn't cover the debt, he could be thrown in prison by the state or enslaved by his creditors. He was fully liable, without limit, for all obligations incurred by his workshop.

If you had lived back then, you would probably have thought twice before you opened an enterprise of your own. And indeed this legal situation discouraged entrepreneurship. People were afraid to start new businesses and take economic risks. It hardly seemed worth taking the chance that their families could end up utterly destitute.

This is why people began collectively to imagine the existence of limited liability companies. Such companies were legally independent of the people who set them up, or invested money in them, or managed them. Over the last few centuries such companies have become the main players in the economic arena, and we have grown so used to them that we forget they exist only in our imagination. In the US, the technical term for a limited liability company is a 'corporation', which is ironic, because the term derives from '*corpus*' ('body' in Latin) – the one thing these corporations lack. Despite their having no real bodies, the American legal system treats corporations as legal persons, as if they were flesh-and-blood human beings.

And so did the French legal system back in 1896, when Armand Peugeot, who had inherited from his parents a metalworking shop that produced springs, saws and bicycles, decided to go into the automobile business. To that end, he set up a limited liability company. He named the company after himself, but it was independent of him. If one of the cars broke down, the buyer could sue Peugeot, but not Armand Peugeot. If the company borrowed millions of francs and then went bust, Armand Peugeot did not owe its creditors a single franc. The loan, after all, had been given to Peugeot, the company, not to Armand Peugeot, the *Homo sapiens*. Armand Peugeot died in 1915. Peugeot, the company, is still alive and well.

How exactly did Armand Peugeot, the man, create Peugeot, the company? In much the same way that priests and sorcerers have created gods and demons throughout history, and in which thousands of French *curés* were still creating Christ's body every Sunday in the parish churches. It all revolved around telling stories, and convincing people to believe them. In the case of the French *curés*, the crucial story was that of Christ's life and death as told by the Catholic Church. According to this story, if a Catholic priest dressed in his sacred garments solemnly said the right words at the right moment, mundane bread and wine turned into God's flesh and blood. The priest exclaimed, '*Hoc est corpus meum!*' (Latin for 'This is my body!') and hocus pocus – the bread turned into Christ's flesh. Seeing that the priest had properly and assiduously observed all the procedures, millions of devout French Catholics behaved as if God really existed in the consecrated bread and wine.

In the case of Peugeot SA the crucial story was the French legal code, as written by the French parliament. According to the French legislators, if a certified lawyer followed all the proper liturgy and rituals, wrote all the required spells and oaths on a wonderfully decorated piece of paper, and affixed his ornate signature to the bottom of the document, then hocus pocus – a new company was incorporated. When in 1896 Armand Peugeot wanted to create his company, he paid a lawyer to go through all these sacred procedures. Once the lawyer had performed all the right rituals and pronounced all the necessary spells and oaths, millions of upright French citizens behaved as if the Peugeot company really existed.

Telling effective stories is not easy. The difficulty lies not in telling the story, but in convincing everyone else to believe it. Much of history revolves around this question: how does one convince millions of people to believe particular stories about gods, or nations, or limited liability companies? Yet when it succeeds, it gives Sapiens immense power, because it enables millions of strangers to cooperate and work towards common goals. Just try to imagine how difficult it would have been to create states, or churches, or legal systems if we could speak only about things that really exist, such as rivers, trees and lions.

Over the years, people have woven an incredibly complex network of stories. Within this network, fictions such as Peugeot not only exist, but also accumulate immense power. The kinds of things that people create through this network of stories are known in academic circles as 'fictions', 'social constructs' or 'imagined realities'. An imagined reality is not a lie. I lie when I say that there is a lion near the river when I know perfectly well that there is no lion there. There is nothing special about lies. Green monkeys and chimpanzees can lie. A green monkey, for example, has been observed calling 'Careful! A lion!' when there was no lion around. This alarm conveniently frightened away a fellow monkey who had just found a banana, leaving the liar all alone to steal the prize for itself.

Unlike lying, an imagined reality is something that everyone believes in, and as long as this communal belief persists, the imagined reality exerts force in the world. The sculptor from the Stadel Cave may sincerely have believed in the existence of the lion-man guardian spirit. Some sorcerers are charlatans, but most sincerely believe in the existence of gods and demons. Most millionaires sincerely believe in the existence of money and limited liability companies. Most human-rights activists sincerely believe in the existence of human rights. No one was lying when, in 2011, the UN demanded that the Libyan government respect the human rights of its citizens, even though the UN, Libya and human rights are all figments of our fertile imaginations.

Ever since the Cognitive Revolution, Sapiens have thus been living in a dual reality. On the one hand, the objective reality of rivers, trees and lions; and on the other hand, the imagined reality of gods, nations and corporations. As time went by, the imagined reality became ever more powerful, so that today the very survival of rivers, trees and lions depends on the grace of imagined entities such as the United States and Google.

Bypassing the Genome

The ability to create an imagined reality out of words enabled large numbers of strangers to cooperate effectively. But it also did something more. Since large-scale human cooperation is based on myths, the way people cooperate can be altered by changing the myths – by telling different stories. Under the right circumstances myths can change rapidly. In 1789 the French population switched almost overnight from believing in the myth of the divine right of kings to believing in the myth of the sovereignty of the people. Consequently, ever since the Cognitive Revolution *Homo sapiens* has been able to revise its behaviour rapidly in accordance with changing needs. This opened a fast lane of cultural evolution, bypassing the traffic jams of genetic evolution. Speeding down this fast lane, *Homo sapiens* soon far outstripped all other human and animal species in its ability to cooperate.

The behaviour of other social animals is determined to a large extent by their genes. DNA is not an autocrat. Animal behaviour is also influenced by environmental factors and individual quirks. Nevertheless, in a given environment, animals of the same species will tend to behave in a similar way. Significant changes in social behaviour cannot occur, in general, without genetic mutations. For example, common chimpanzees have a genetic tendency to live in hierarchical groups headed by an alpha male. Members of a closely related chimpanzee species, bonobos, usually live in more egalitarian groups dominated by female alliances. Female common chimpanzees cannot take lessons from their bonobo relatives and stage a feminist revolution. Male chimps cannot gather in a constitutional assembly to abolish the office of alpha male and declare that from here on out all chimps are to be treated as equals. Such dramatic changes in behaviour would occur only if something changed in the chimpanzees' DNA.

For similar reasons, archaic humans did not initiate any revolutions. As far as we can tell, changes in social patterns, the invention of new technologies and the settlement of alien habitats resulted from genetic mutations and environmental pressures more than from cultural initiatives. This is why it took humans hundreds of thousands of years to make these steps. Two million years ago, genetic mutations resulted in the appearance of a new human species called *Homo erectus*. Its emergence was accompanied by the development of a new stone tool technology, now recognised as a defining feature of this species. As long as *Homo erectus* did not undergo further genetic alterations, its stone tools remained roughly the same – for close to 2 million years!

In contrast, ever since the Cognitive Revolution, Sapiens have been able to change their behaviour quickly, transmitting new behaviours to future generations without any need of genetic or environmental change. As a prime example, consider the repeated appearance of childless elites, such as the Catholic priesthood, Buddhist monastic orders and Chinese eunuch bureaucracies. The existence of such elites goes against the most fundamental principles of natural selection, since these dominant members of society willingly give up procreation. Whereas chimpanzee alpha males use their power to have sex with as many females as possible – and consequently sire a large proportion of their troop's young – the Catholic alpha male abstains completely from sexual intercourse and childcare. This abstinence does not result from unique environmental conditions such as a severe lack of food or want of potential mates. Nor is it the result of some quirky genetic mutation. The Catholic Church has survived for centuries, not by passing on a 'celibacy gene' from one pope to the next, but by passing on the stories of the New Testament and of Catholic canon law.

In other words, while the behaviour patterns of archaic humans remained fixed for tens of thousands of years, Sapiens could transform their social structures, the nature of their interpersonal relations, their economic activities and a host of other behaviours within a decade or two. Consider a resident of Berlin, born in 1900 and living to the ripe age of one hundred. She spent her childhood in the Hohenzollern Empire of Wilhelm II; her adult years in the Weimar Republic, the Nazi Third Reich and Communist East Germany; and she died a citizen of a democratic and reunified Germany. She had managed to be a part of five very different sociopolitical systems, though her DNA remained exactly the same.

This was the key to Sapiens' success. In a one-on-one brawl, a Neanderthal would probably have beaten a Sapiens. But in a conflict of hundreds, Neanderthals wouldn't stand a chance. Neanderthals could share information about the whereabouts of lions, but they probably could not tell – and revise – stories about tribal spirits. Without an ability to compose fiction, Neanderthals were unable to cooperate effectively in large numbers, nor could they adapt their social behaviour to rapidly changing challenges.

While we can't get inside a Neanderthal mind to understand how they thought, we have indirect evidence of the limits to their cognition compared with their Sapiens rivals. Archaeologists excavating 30,000-year-old Sapiens sites in the European heartland occasionally find there seashells from the Mediterranean and Atlantic coasts. In

6. The Catholic alpha male abstains from sexual intercourse and childcare, even though there is no genetic or ecological reason for him to do so.

all likelihood, these shells got to the continental interior through long-distance trade between different Sapiens bands. Neanderthal sites lack any evidence of such trade. Each group manufactured its own tools from local materials.[4]

Another example comes from the South Pacific. Sapiens bands that lived on the island of New Ireland, north of New Guinea, used a volcanic glass called obsidian to manufacture particularly strong and sharp tools. New Ireland, however, has no natural deposits of obsidian. Laboratory tests revealed that the obsidian they used was brought from deposits on New Britain, an island 400 kilometres away. Some of the inhabitants of these islands must have been skilled navigators who traded from island to island over long distances.[5]

Trade may seem a very pragmatic activity, one that needs no fictive basis. Yet the fact is that no animal other than Sapiens engages in trade, and all Sapiens trade-networks were based on fictions. Trade cannot exist without trust, and it is very difficult to trust strangers. The global trade network of today is based on our trust in such fictional entities as currencies, banks and corporations. When two strangers in a tribal society want to trade, they establish trust by appealing to a common god, mythical ancestor or totem animal. In modern society, currency notes usually display religious images, revered ancestors and corporate totems.

If archaic Sapiens believing in such fictions traded shells and obsidian, it stands to reason that they could also have traded information, thus creating a much denser and wider knowledge network than the one that served Neanderthals and other archaic humans.

Hunting techniques provide another illustration of these differences. Neanderthals usually hunted alone or in small groups. Sapiens, on the other hand, developed techniques that relied on cooperation between many dozens of individuals, and perhaps even between different bands. One particularly effective method was to surround an entire herd of animals, such as wild horses, then chase them into a narrow gorge, where it was easy to slaughter them en masse. If all went according to plan, the bands could harvest tons of meat, fat and animal skins in a single afternoon of collective effort, and either consume these riches in a giant potlatch, or dry, smoke or (in Arctic areas) freeze them for later usage. Archaeologists have discovered sites where entire herds were butchered annually in such ways. There are even sites where fences and obstacles were erected in order to create artificial traps and slaughtering grounds.

We may presume that Neanderthals were not pleased to see their traditional hunting grounds turned into Sapiens-controlled slaughterhouses. However, if violence broke out between the two species, Neanderthals were not much better off than wild horses. Fifty Neanderthals cooperating in traditional and static patterns were no match for 500 versatile and innovative Sapiens. And even if the Sapiens lost the first round, they could quickly invent new stratagems that would enable them to win the next time.

What happened in the Cognitive Revolution?

New ability	Wider consequences
The ability to transmit larger quantities of information about the world surrounding *Homo sapiens*	Planning and carrying out complex actions, such as avoiding lions and hunting bison
The ability to transmit larger quantities of information about Sapiens social relationships	Larger and more cohesive groups, numbering up to 150 individuals
The ability to transmit information about things that do not really exist, such as tribal spirits, nations, limited liability companies and human rights	a. Cooperation between very large numbers of strangers b. Rapid innovation of social behaviour

History and Biology

The immense diversity of imagined realities that Sapiens invented, and the resulting diversity of behaviour patterns, are the main components of what we call 'cultures'. Once cultures appeared, they never ceased to change and develop, and these unstoppable alterations are what we call 'history'.

The Cognitive Revolution is accordingly the point when history declared its independence from biology. Until the Cognitive Revolution, the doings of all human species belonged to the realm of biology, or, if you so prefer, prehistory (I tend to avoid the term 'prehistory', because it wrongly implies that even before the Cognitive Revolution, humans were in a category of their own). From the Cognitive Revolution onwards, historical narratives replace biological theories as our primary means of explaining the development of *Homo sapiens.* To understand the rise of Christianity or the French Revolution, it is not enough to comprehend the interaction of genes, hormones and organisms. It is necessary to take into account the interaction of ideas, images and fantasies as well.

This does not mean that *Homo sapiens* and human culture became exempt from biological laws. We are still animals, and our physical, emotional and cognitive abilities are still shaped by our DNA. Our societies are built from the same building blocks as Neanderthal or chimpanzee societies, and the more we examine these building blocks – sensations, emotions, family ties – the less difference we find between us and other apes.

It is, however, a mistake to look for the differences at the level of the individual or the family. One on one, even ten on ten, we are embarrassingly similar to chimpanzees. Significant differences begin to appear only when we cross the threshold of 150 individuals, and when we reach 1,000–2,000 individuals, the differences are astounding. If you tried to bunch together thousands of chimpanzees into Tiananmen Square, Wall Street, the Vatican or the headquarters of the United Nations, the result would be pandemonium. By contrast, Sapiens regularly gather by the thousands in such places. Together, they create orderly patterns – such as trade networks, mass celebrations and political institutions – that they could never have created in isolation. The real difference between us and chimpanzees is the mythical glue that binds together large numbers of individuals, families and groups. This glue has made us the masters of creation.

Of course, we also needed other skills, such as the ability to make and use tools. Yet tool-making is of little consequence unless it is coupled with the ability to cooperate with many others. How is it that we now have intercontinental missiles with nuclear warheads, whereas 30,000 years ago we had only sticks with flint spearheads? Physiologically, there has been no significant improvement in our tool-making capacity over the last 30,000 years. Albert Einstein was far less dexterous with his hands than was an ancient hunter-gatherer. However, our capacity to cooperate with large numbers of strangers has improved dramatically. The ancient flint spearhead was manufactured in minutes by a single person, who relied on the advice and help of a few intimate friends. The production of a modern nuclear warhead requires the cooperation of millions of strangers all over the world – from the workers who mine the uranium ore in the depths of the earth to theoretical physicists who write long mathematical formulae to describe the interactions of subatomic particles.

To summarise the relationship between biology and history after the Cognitive Revolution:

a. Biology sets the basic parameters for the behaviour and capacities of *Homo sapiens*. The whole of history takes place within the bounds of this biological arena.

b. However, this arena is extraordinarily large, allowing Sapiens to play an astounding variety of games. Thanks to their ability to invent fiction, Sapiens create more and more complex games, which each generation develops and elaborates even further.

c. Consequently, in order to understand how Sapiens behave, we must describe the historical evolution of their actions. Referring only to our biological constraints would be like a radio sportscaster who, attending the World Cup football championship, offers his listeners a detailed description of the playing field rather than an account of what the players are doing.

What games did our Stone Age ancestors play in the arena of history? As far as we know, the people who carved the Stadel lion-man some 30,000 years ago had the same physical, emotional and intellectual abilities we have. What did they do when they woke up in the morning? What did they eat for breakfast – and lunch? What were their societies like? Did they have monogamous relationships and nuclear families? Did they have ceremonies, moral codes, sports contests and religious rituals? Did they fight wars? The next chapter takes a peek behind the curtain of the ages, examining what life was like in the millennia separating the Cognitive Revolution from the Agricultural Revolution.

3

A Day in the Life of Adam and Eve

TO UNDERSTAND OUR NATURE, HISTORY and psychology, we must get inside the heads of our hunter-gatherer ancestors. For nearly the entire history of our species, Sapiens lived as foragers. The past 200 years, during which ever-increasing numbers of Sapiens have obtained their daily bread as urban labourers and office workers, and the preceding 10,000 years, during which most Sapiens lived as farmers and herders, are the blink of an eye compared to the tens of thousands of years during which our ancestors hunted and gathered.

The flourishing field of evolutionary psychology argues that many of our present-day social and psychological characteristics were shaped during this long pre-agricultural era. Even today, scholars in this field claim, our brains and minds are adapted to a life of hunting and gathering. Our eating habits, our conflicts and our sexuality are all the result of the way our hunter-gatherer minds interact with our current post-industrial environment, with its mega-cities, aeroplanes, telephones and computers. This environment gives us more material resources and longer lives than those enjoyed by any previous generation, but it often makes us feel alienated, depressed and pressured. To understand why, evolutionary psychologists argue, we need to delve into the hunter-gatherer world that shaped us, the world that we subconsciously still inhabit.

Why, for example, do people gorge on high-calorie food that is doing little good to their bodies? Today's affluent societies are in the throes of a plague of obesity, which is rapidly spreading to developing countries. It's a puzzle why we binge on the sweetest and greasiest food we can find, until we consider the eating habits of our forager forebears. In the savannahs and forests they inhabited, high-calorie sweets were extremely rare and food in general was in short supply. A typical forager 30,000 years ago had access to only one type of sweet food – ripe fruit. If a Stone Age woman came across a tree groaning with figs, the most sensible thing to do was to eat as many of them as she could on the spot, before the local baboon band picked the tree bare. The instinct to gorge on high-calorie food was hard-wired into our genes. Today we may be living in high-rise apartments with over-stuffed refrigerators, but our DNA still thinks we are in the savannah. That's what makes us spoon down an entire tub of Ben & Jerry's when we find one in the freezer and wash it down with a jumbo Coke.

This 'gorging gene' theory is widely accepted. Other theories are far more contentious. For example, some evolutionary psychologists argue that ancient foraging bands were not composed of nuclear families centred on monogamous couples. Rather, foragers lived in communes devoid of private property, monogamous relationships and even fatherhood. In such a band, a woman could have sex and form intimate bonds with several men (and women) simultaneously, and all of the band's adults cooperated in parenting its children. Since no man knew definitively which of the children were his, men showed equal concern for all youngsters.

Such a social structure is not an Aquarian utopia. It's well documented among animals, notably our closest relatives, the chimpanzees and bonobos. There are even a number of present-day human cultures in which collective fatherhood is practised, as for example among the Barí Indians. According to the beliefs of such societies, a child is not born from the sperm of a single man, but from the accumulation of sperm in a woman's womb. A good mother will make a point of having sex with several different men, especially when she is pregnant, so that her child will enjoy the qualities (and paternal care) not merely of the best hunter, but also of the best storyteller, the strongest warrior and the most considerate lover. If this sounds silly, bear in mind that before the development of modern embryological studies, people had no solid evidence that babies are always sired by a single father rather than by many.

The proponents of this 'ancient commune' theory argue that the frequent infidelities that characterise modern marriages, and the high rates of divorce, not to mention the cornucopia of psychological complexes from which both children and adults suffer, all result from forcing humans to live in nuclear families and monogamous relationships that are incompatible with our biological software.[1]

Many scholars vehemently reject this theory, insisting that both monogamy and the forming of nuclear families are core human behaviours. Though ancient hunter-gatherer societies tended to be more communal and egalitarian than modern societies, these researchers argue, they were nevertheless comprised of separate cells, each containing a jealous couple and the children they held in common. This is why today monogamous relationships and nuclear families are the norm in the vast majority of cultures, why men and women tend to be very possessive of their partners and children, and why even in modern states such as North Korea and Syria political authority passes from father to son.

In order to resolve this controversy and understand our sexuality, society and politics, we need to learn something about the living conditions of our ancestors, to examine how Sapiens lived between the Cognitive Revolution of 70,000 years ago, and the start of the Agricultural Revolution about 12,000 years ago.

Unfortunately, there are few certainties regarding the lives of our forager ancestors. The debate between the 'ancient commune' and 'eternal monogamy' schools is based on flimsy evidence. We obviously have no written records from the age of foragers, and the archaeological evidence consists mainly of fossilised bones and stone tools. Artefacts made of more perishable materials – such as wood, bamboo or leather – survive only under unique conditions. The common impression that pre-agricultural humans lived in an age of stone is a misconception based on this archaeological bias. The Stone Age should more accurately be called the Wood Age, because most of the tools used by ancient hunter-gatherers were made of wood.

Any reconstruction of the lives of ancient hunter-gatherers from the surviving artefacts is extremely problematic. One of the most glaring differences between the ancient foragers and their agricultural and industrial descendants is that foragers had very few artefacts to begin with, and these played a comparatively modest role in their lives. Over the course of his or her life, a typical member of a modern affluent society will own several million artefacts – from cars and houses to disposable nappies and milk cartons. There's hardly an activity, a belief, or even an emotion that is not mediated by objects of our own devising. Our eating habits are mediated by a mind-boggling collection of such items, from spoons and glasses to genetic engineering labs and gigantic ocean-going ships. In play, we use a plethora of toys, from plastic cards to 100,000-seater stadiums. Our romantic and sexual relations are accoutred by rings, beds, nice clothes, sexy underwear, condoms, fashionable restaurants, cheap motels, airport lounges, wedding halls and catering companies. Religions bring the sacred into our lives with Gothic churches, Muslim mosques, Hindu ashrams, Torah scrolls, Tibetan prayer wheels, priestly cassocks, candles, incense, Christmas trees, matzah balls, tombstones and icons.

We hardly notice how ubiquitous our stuff is until we have to move it to a new house. Foragers moved house every month, every week, and sometimes even every day, toting whatever they had on their backs. There were no moving companies, wagons or even pack animals to share the burden. They consequently had to make do with only the most essential possessions. It's reasonable to presume, then, that the greater part of their mental, religious and emotional lives was conducted without the help of artefacts. An archaeologist working 100,000 years from now could piece together a reasonable picture of Muslim belief and practice from the myriad objects he unearthed in a ruined mosque. But we are largely at a loss in trying to comprehend the beliefs and rituals of ancient hunter-gatherers. It's much the same dilemma that a future historian would face if he had to depict the social world of twenty-first-century teenagers solely on the basis of their surviving snail mail – since no records will remain of their phone conversations, emails, blogs and text messages.

A reliance on artefacts will thus bias an account of ancient hunter-gatherer life. One way to remedy this is to look at modern forager societies. These can be studied directly, by anthropological observation. But there are good reasons to be very careful in extrapolating from modern forager societies to ancient ones.

Firstly, all forager societies that have survived into the modern era have been influenced by neighbouring agricultural and industrial societies. Consequently, it's risky to assume that what is true of them was also true tens of thousands of years ago.

Secondly, modern forager societies have survived mainly in areas with difficult climatic conditions and inhospitable terrain, ill-suited for agriculture. Societies that have adapted to the extreme conditions of places such as the Kalahari Desert in southern Africa may well provide a very misleading model for understanding ancient societies in fertile areas such as the Yangtze River Valley. In particular, population density in an area like the Kalahari Desert is far lower than it was around the ancient Yangtze, and this has far-reaching implications for key questions about the size and structure of human bands and the relations between them.

Thirdly, the most notable characteristic of hunter-gatherer societies is how different they are one from the other. They differ not only from one part of the world to another but even in the same region.

One good example is the huge variety the first European settlers found among the Aboriginal peoples of Australia. Just before the British conquest, between 300,000 and 700,000 hunter-gatherers lived on the continent in 200–600 tribes, each of which was further divided into several bands.[2] Each tribe had its own language, religion, norms and customs. Living around what is now Adelaide in southern Australia were several patrilineal clans that reckoned descent from the father's side. These clans bonded together into tribes on a strictly territorial basis. In contrast, some tribes in northern Australia gave more importance to a person's maternal ancestry, and a person's tribal identity depended on his or her totem rather than his or her territory.

It stands to reason that the ethnic and cultural variety among ancient hunter-gatherers was equally impressive, and that the 5 million to 8 million foragers who populated the world on the eve of the Agricultural Revolution were divided into thousands of separate tribes with thousands of different languages and cultures.[3] This, after all, was one of the main legacies of the Cognitive Revolution. Thanks to the appearance of fiction, even people with the same genetic make-up who lived under similar ecological conditions were able to create very different imagined realities, which manifested themselves in different norms and values.

For example, there's every reason to believe that a forager band that lived 30,000 years ago on the spot where Oxford University now stands would have spoken a different language from one living where Cambridge is now situated. One band might have been belligerent and the other peaceful. Perhaps the Cambridge band was communal while the one at Oxford was based on nuclear families. The Cantabrigians might have spent long hours carving wooden statues of their guardian spirits, whereas the Oxonians may have worshipped through dance. The former perhaps believed in reincarnation, while the latter thought this was nonsense. In one society, homosexual relationships might have been accepted, while in the other they were taboo.

In other words, while anthropological observations of modern foragers can help us understand some of the possibilities available to ancient foragers, the ancient horizon of possibilities was much broader, and most of it is hidden from our view.* The heated debates about *Homo sapiens*' 'natural way of life' miss the main point. Ever since the Cognitive Revolution, there hasn't been a single natural way of life for Sapiens. There are only cultural choices, from among a bewildering palette of possibilities.

The Original Affluent Society

What generalisations can we make about life in the pre-agricultural world nevertheless? It seems safe to say that the vast majority of people lived in small bands numbering several dozen or at most several hundred individuals, and that all these individuals were humans. It is important to note this last point, because it is far from obvious. Most members of agricultural and industrial societies are domesticated animals. They are not equal to their masters, of course, but they are members all the same. Today, the society called New Zealand is composed of 4.5 million Sapiens and 50 million sheep.

There was just one exception to this general rule: the dog. The dog was the first animal domesticated by *Homo sapiens*, and this occurred *before* the Agricultural Revolution. Experts disagree about the exact date, but we have incontrovertible evidence of domesticated dogs from about 15,000 years ago. They may have joined the human pack thousands of years earlier.

Dogs were used for hunting and fighting, and as an alarm system against wild beasts and human intruders. With the passing of

* A 'horizon of possibilities' means the entire spectrum of beliefs, practices and experiences that are open before a particular society, given its ecological, technological and cultural limitations. Each society and each individual usually explores only a tiny fraction of their horizon of possibilities.

generations, the two species co-evolved to communicate well with each other. Dogs that were most attentive to the needs and feelings of their human companions got extra care and food, and were more likely to survive. Simultaneously, dogs learned to manipulate people for their own needs. A 15,000-year bond has yielded a much deeper understanding and affection between humans and dogs than between humans and any other animal.[4] In some cases dead dogs were even buried ceremoniously, much like humans.

Members of a band knew each other very intimately, and were surrounded throughout their lives by friends and relatives. Loneliness and privacy were rare. Neighbouring bands probably competed for resources and even fought one another, but they also had friendly contacts. They exchanged members, hunted together, traded rare luxuries, celebrated religious festivals, and joined forces against foreigners. Such cooperation was one of the important trademarks of *Homo sapiens*, and gave it a crucial edge over other human species. Sometimes relations with neighbouring bands were tight enough that together they constituted a single tribe, sharing a common language, common myths, and common norms and values.

Yet we should not overestimate the intensity of such external relations. Even if in times of crisis the tribe acted as one, and even if the tribe periodically gathered to hunt, fight or feast together, most people still spent most of their time in a small band. Trade was mostly limited to prestige items such as shells, amber, and pigments. There is no evidence that people traded staple goods like fruits and meat, or that the existence of one band depended on the importing of goods from another. Sociopolitical relations, too, tended to be sporadic. The tribe did not serve as a permanent political framework, and even if it had seasonal meeting places, there were no permanent towns or institutions. The average person might live many months without seeing or hearing a human from outside of her own band, and she encountered throughout her life no more

The Upper Galilee Museum of Prehistory

7. First pet? A 12,000-year-old tomb found in northern Israel. It contains the skeleton of a fifty-year-old woman next to that of a puppy (bottom left corner). The puppy was buried close to the woman's head. Her left hand is resting on the dog in a way that might indicate an emotional connection. There are, of course, other possible explanations. Perhaps, for example, the puppy was a gift to the gatekeeper of the next world.

than a few thousand humans. The Sapiens population was thinly spread over vast territories. Before the Agricultural Revolution, the human population of the entire planet was smaller than that of today's Cairo.

Most Sapiens bands lived on the road, roaming from place to place in search of food. Their movements were influenced by the changing seasons, the annual migrations of animals and the growth cycles of plants. They usually travelled back and forth across the same home territory, an area of between several dozen and many hundreds of square kilometres.

Occasionally, bands wandered outside their turf and explored new lands, whether due to natural calamities, violent conflicts, demographic pressures or the initiative of a charismatic leader. These wanderings were the engine of human worldwide expansion. If a forager band split once every forty years and its splinter group migrated to a new territory a hundred kilometres to the east, the distance from East Africa to China would have been covered in about 10,000 years.

In some exceptional cases, when food sources were particularly rich, bands settled down in seasonal and even permanent camps. Techniques for drying, smoking and freezing food also made it possible to stay put for longer periods. Most importantly, alongside seas and rivers rich in seafood and waterfowl, humans set up permanent fishing villages – the first permanent settlements in history, long predating the Agricultural Revolution. Fishing villages might have appeared on the coasts of Indonesian islands as early as 45,000 years ago. These may have been the base from which *Homo sapiens* launched its first transoceanic enterprise: the invasion of Australia.

In most habitats, Sapiens bands fed themselves in an elastic and opportunistic fashion. They scrounged for termites, picked berries, dug for roots, stalked rabbits and hunted bison and mammoth. Notwithstanding the popular image of 'man the hunter', gathering was Sapiens' main activity, and it provided most of their calories, as well as raw materials such as flint, wood and bamboo.

Sapiens did not forage only for food and materials. They foraged for knowledge as well. To survive, they needed a detailed mental map of their territory. To maximise the efficiency of their daily search for food, they required information about the growth patterns of each plant and the habits of each animal. They needed to know which foods were nourishing, which made you sick, and how to use others as cures. They needed to know the progress of the seasons and what warning signs preceded a thunderstorm or a dry spell. They studied every stream, every walnut tree, every bear cave, and every flint-stone deposit in their vicinity. Each individual had to understand how to make a stone knife, how to mend a torn cloak, how to lay a rabbit trap, and how to face avalanches, snakebites or hungry lions. Mastery of each of these many skills required years of apprenticeship and practice. The average ancient forager could turn a flint stone into a spear point within minutes. When we try to imitate this feat, we usually fail miserably. Most of us lack expert knowledge of the flaking properties of flint and basalt and the fine motor skills needed to work them precisely.

In other words, the average forager had wider, deeper and more varied knowledge of her immediate surroundings than most of her modern descendants. Today, most people in industrial societies don't need to know much about the natural world in order to survive. What do you really need to know in order to get by as a computer engineer, an insurance agent, a history teacher or a factory worker? You need to know a lot about your own tiny field of expertise, but for the vast majority of life's necessities you rely blindly on the help of other experts, whose own knowledge is also limited to a tiny field of expertise. The human collective knows far more today than did the ancient bands. But at the individual level, ancient foragers were the most knowledgeable and skilful people in history.

There is some evidence that the size of the average Sapiens brain has actually *decreased* since the age of foraging.[5] Survival in that era required superb mental abilities from everyone. When agriculture and industry came along people could increasingly rely on the skills of others for survival, and new 'niches for imbeciles' were opened up. You could survive and pass your unremarkable genes to the next generation by working as a water carrier or an assembly-line worker.

Foragers mastered not only the surrounding world of animals, plants and objects, but also the internal world of their own bodies and senses. They listened to the slightest movement in the grass to learn whether a snake might be lurking there. They carefully observed the foliage of trees in order to discover fruits, beehives and birds' nests. They moved with a minimum of effort and noise, and knew how to sit, walk and run in the most agile and efficient manner.

Varied and constant use of their bodies made them as fit as marathon runners. They had physical dexterity that people today are unable to achieve even after years of practising yoga or t'ai chi.

The hunter-gatherer way of life differed significantly from region to region and from season to season, but on the whole foragers seem to have enjoyed a more comfortable and rewarding lifestyle than most of the peasants, shepherds, labourers and office clerks who followed in their footsteps.

While people in today's affluent societies work an average of forty to forty-five hours a week, and people in the developing world work sixty and even eighty hours a week, hunter-gatherers living today in the most inhospitable of habitats – such as the Kalahari Desert – work on average for just thirty-five to forty-five hours a week. They hunt only one day out of three, and gathering takes up just three to six hours daily. In normal times, this is enough to feed the band. It may well be that ancient hunter-gatherers living in zones more fertile than the Kalahari spent even less time obtaining food and raw materials. On top of that, foragers enjoyed a lighter load of household chores. They had no dishes to wash, no carpets to vacuum, no floors to polish, no nappies to change and no bills to pay.

The forager economy provided most people with more interesting lives than agriculture or industry do. Today, a Chinese factory hand leaves home around seven in the morning, makes her way through polluted streets to a sweatshop, and there operates the same machine, in the same way, day in, day out, for ten long and mind-numbing hours, returning home around seven in the evening in order to wash dishes and do the laundry. Thirty thousand years ago, a Chinese forager might leave camp with her companions at, say, eight in the morning. They'd roam the nearby forests and meadows, gathering mushrooms, digging up edible roots, catching frogs and occasionally running away from tigers. By early afternoon, they were back at the camp to make lunch. That left them plenty of time to gossip, tell stories, play with the children and just hang out. Of course the tigers sometimes caught them, or a snake bit them, but on the other hand they didn't have to deal with automobile accidents and industrial pollution.

In most places and at most times, foraging provided ideal nutrition. That is hardly surprising – this had been the human diet for hundreds of thousands of years, and the human body was well adapted to it. Evidence from fossilised skeletons indicates that ancient foragers were less likely to suffer from starvation or malnutrition, and were generally taller and healthier than their peasant descendants. Average life expectancy was apparently just thirty to forty years, but this was due largely to the high incidence of child mortality. Children who made it through the perilous first years had a good chance of reaching the age of sixty, and some even made it to their eighties. Among modern foragers, forty-five-year-old women can expect to live another twenty years, and about 5–8 per cent of the population is over sixty.[6]

The foragers' secret of success, which protected them from starvation and malnutrition, was their varied diet. Farmers tend to eat a very limited and unbalanced diet. Especially in premodern times, most of the calories feeding an agricultural population came from a single crop – such as wheat, potatoes or rice – that lacks some of the vitamins, minerals and other nutritional materials humans need. The typical peasant in traditional China ate rice for breakfast, rice for lunch and rice for dinner. If she was lucky, she could expect to eat the same on the following day. By contrast, ancient foragers regularly ate dozens of different foodstuffs. The peasant's ancient ancestor, the forager, may have eaten berries and mushrooms for breakfast; fruits, snails and turtle for lunch; and rabbit steak with wild onions for dinner. Tomorrow's menu might have been completely different. This variety ensured that the ancient foragers received all the necessary nutrients.

Furthermore, by not being dependent on any single kind of food, they were less liable to suffer when one particular food source failed. Agricultural societies are ravaged by famine when drought, fire or earthquake devastates the annual rice or potato crop. Forager societies were hardly immune to natural disasters, and suffered from periods of want and hunger, but they were usually able to deal with such calamities more easily. If they lost some of their staple foodstuffs, they could gather or hunt other species, or move to a less affected area.

Ancient foragers also suffered less from infectious diseases. Most of the infectious diseases that have plagued agricultural and industrial societies (such as smallpox, measles and tuberculosis) originated in domesticated animals and were transferred to humans only after the Agricultural Revolution. Ancient foragers, who had domesticated only dogs, were free of these scourges. Moreover, most people in agricultural and industrial societies lived in dense, unhygienic permanent settlements – ideal hotbeds for disease. Foragers roamed the land in small bands that could not sustain epidemics.

The wholesome and varied diet, the relatively short working week, and the rarity of infectious diseases have led many experts to define pre-agricultural forager societies as 'the original affluent societies'. It would be a mistake, however, to idealise the lives of these ancients. Though they lived better lives than most people in agricultural and industrial societies, their world could still be harsh and unforgiving. Periods of want and hardship were not uncommon, child mortality was high, and an accident which would be minor today could easily become a death sentence. Most people probably enjoyed the close intimacy of the roaming band, but those unfortunates who incurred the hostility or mockery of their fellow band members probably suffered terribly. Modern foragers occasionally abandon and even kill old or disabled people who cannot keep up with the band. Unwanted babies and children may be slain, and there are even cases of religiously inspired human sacrifice.

The Aché people, hunter-gatherers who lived in the jungles of Paraguay until the 1960s, offer a glimpse into the darker side of foraging. When a valued band member died, the Aché customarily killed a little girl and buried the two together. Anthropologists who interviewed the Aché recorded a case in which a band abandoned a middle-aged man who fell sick and was unable to keep up with the others. He was left under a tree. Vultures perched above him, expecting a hearty meal. But the man recuperated, and, walking briskly, he managed to rejoin the band. His body was covered with the birds' faeces, so he was henceforth nicknamed 'Vulture Droppings'.

When an old Aché woman became a burden to the rest of the band, one of the younger men would sneak behind her and kill her with an axe-blow to the head. An Aché man told the inquisitive anthropologists stories of his prime years in the jungle. 'I customarily killed old women. I used to kill my aunts . . . The women were afraid of me . . . Now, here with the whites, I have become weak.' Babies born without hair, who were considered underdeveloped, were killed immediately. One woman recalled that her first baby girl was killed because the men in the band did not want another girl. On another occasion a man killed a small boy because he was 'in a bad mood and the child was crying'. Another child was buried alive because 'it was funny-looking and the other children laughed at it'.[7]

We should be careful, though, not to judge the Aché too quickly. Anthropologists who lived with them for years report that violence between adults was very rare. Both women and men were free to change partners at will. They smiled and laughed constantly, had no leadership hierarchy, and generally shunned domineering people. They were extremely generous with their few possessions, and were not obsessed with success or wealth. The things they valued most in life were good social interactions and high-quality friendships.[8] They viewed the killing of children, sick people and the elderly as many people today view abortion and euthanasia. It should also be noted that the Aché were hunted and killed without mercy by Paraguayan farmers. The need to evade their enemies probably caused the Aché to adopt an exceptionally harsh attitude towards anyone who might become a liability to the band.

The truth is that Aché society, like every human society, was very complex. We should beware of demonising or idealising it on the basis of a superficial acquaintance. The Aché were neither angels nor fiends – they were humans. So, too, were the ancient hunter-gatherers.

Talking Ghosts

What can we say about the spiritual and mental life of the ancient hunter-gatherers? The basics of the forager economy can be reconstructed with some confidence based on quantifiable and objective factors. For example, we can calculate how many calories per day a person needed in order to survive, how many calories were obtained from a kilogram of walnuts, and how many walnuts could be gathered from a square kilometre of forest. With this data, we can make an educated guess about the relative importance of walnuts in their diet.

But did they consider walnuts a delicacy or a humdrum staple? Did they believe that walnut trees were inhabited by spirits? Did they find walnut leaves pretty? If a forager boy wanted to take a forager girl to a romantic spot, did the shade of a walnut tree suffice? The world of thought, belief and feeling is by definition far more difficult to decipher.

Most scholars agree that animistic beliefs were common among ancient foragers. Animism (from '*anima*', 'soul' or 'spirit' in Latin) is the belief that almost every place, every animal, every plant and every natural phenomenon has awareness and feelings, and can communicate directly with humans. Thus, animists may believe that the big rock at the top of the hill has desires and needs. The rock might be angry about something that people did and rejoice over some other action. The rock might admonish people or ask for favours. Humans, for their part, can address the rock, to mollify or threaten it. Not only the rock, but also the oak tree at the bottom of the hill is an animated being, and so is the stream flowing below the hill, the spring in the forest clearing, the bushes growing around it, the path to the clearing, and the field mice, wolves and crows that drink there. In the animist world, objects and living things are not the only animated beings. There are also immaterial entities – the spirits of the dead, and friendly and malevolent beings, the kind that we today call demons, fairies and angels.

Animists believe that there is no barrier between humans and other beings. They can all communicate directly through speech, song, dance and ceremony. A hunter may address a herd of deer and ask that one of them sacrifice itself. If the hunt succeeds, the hunter may ask the dead animal to forgive him. When someone falls sick, a shaman can contact the spirit that caused the sickness and try to pacify it or scare it away. If need be, the shaman may ask for help from other spirits. What characterises all these acts of communication is that the entities being addressed are local beings. They are not universal gods, but rather a particular deer, a particular tree, a particular stream, a particular ghost.

Just as there is no barrier between humans and other beings, neither is there a strict hierarchy. Non-human entities do not exist merely to provide for the needs of man. Nor are they all-powerful gods who run the world as they wish. The world does not revolve around humans or around any other particular group of beings.

Animism is not a specific religion. It is a generic name for thousands of very different religions, cults and beliefs. What makes all of them 'animist' is this common approach to the world and to man's place in it. Saying that ancient foragers were probably animists is like saying that premodern agriculturists were mostly theists.

Theism (from '*theos*', 'god' in Greek) is the view that the universal order is based on a hierarchical relationship between humans and a small group of ethereal entities called gods. It is certainly true to say that premodern agriculturists tended to be theists, but it does not teach us much about the particulars. The generic rubric 'theists' covers Jewish rabbis from eighteenth-century Poland, witch-burning Puritans from seventeenth-century Massachusetts, Aztec priests from fifteenth-century Mexico, Sufi mystics from twelfth-century Iran, tenth-century Viking warriors, second-century Roman legionaries, and first-century Chinese bureaucrats. Each of these viewed the others' beliefs and practices as weird and heretical. The differences between the beliefs and practices of groups of 'animistic' foragers were probably just as big. Their religious experience may have been turbulent and filled with controversies, reforms and revolutions.

But these cautious generalisations are about as far as we can go. Any attempt to describe the specifics of archaic spirituality is highly speculative, as there is next to no evidence to go by and the little evidence we have – a handful of artefacts and cave paintings – can be interpreted in myriad ways. The theories of scholars who claim to know what the foragers felt shed much more light on the prejudices of their authors than on Stone Age religions.

Instead of erecting mountains of theory over a molehill of tomb relics, cave paintings and bone statuettes, it is better to be frank and admit that we have only the haziest notions about the religions of ancient foragers. We assume that they were animists, but that's not very informative. We don't know which spirits they prayed to, which festivals they celebrated, or which taboos they observed. Most importantly, we don't know what stories they told. It's one of the biggest holes in our understanding of human history.

The sociopolitical world of the foragers is another area about which we know next to nothing. As explained above, scholars cannot even agree on the basics, such as the existence of private property, nuclear families and monogamous relationships. It's likely that different bands had different structures. Some may have been as hierarchical, tense and violent as the nastiest chimpanzee group, while others were as laid-back, peaceful and lascivious as a bunch of bonobos.

In Sungir, Russia, archaeologists discovered in 1955 a 30,000-year-old burial site belonging to a mammoth-hunting culture. In one grave they found the skeleton of a fifty-year-old man, covered with strings of mammoth ivory beads, containing about 3,000 beads in total. On the dead man's head was a hat decorated with fox teeth, and on his wrists twenty-five ivory bracelets. Other graves from the same site contained far fewer goods. Scholars deduced that the Sungir mammoth-hunters lived in a hierarchical society, and that the dead man was perhaps the leader of a band or of an entire tribe comprising several bands. It is unlikely that a few dozen members of a single band could have produced so many grave goods by themselves.

Archaeologists then discovered an even more interesting tomb. It contained two skeletons, buried head to head. One belonged to a boy aged about twelve or thirteen, and the other to a girl of about nine or ten. The boy was covered with 5,000 ivory beads. He wore a fox-tooth hat and a belt with 250 fox teeth (at least sixty foxes had to have their teeth pulled to get that many). The girl was adorned with 5,250 ivory beads. Both children were surrounded by statuettes and various ivory objects. A skilled craftsman (or craftswoman) probably needed about forty-five minutes to prepare a single ivory bead. In other words, fashioning the 10,000 ivory beads that covered the two children, not to mention the other objects, required some 7,500 hours of delicate work, well over three years of labour by an experienced artisan!

It is highly unlikely that at such a young age the Sungir children had proved themselves as leaders or mammoth-hunters. Only cultural beliefs can explain why they received such an extravagant burial. One theory is that they owed their rank to their parents. Perhaps

8. A painting from Lascaux Cave, *c.*15,000–20,000 years ago. What exactly do we see, and what is the painting's meaning? Some argue that we see a man with the head of a bird and an erect penis, being killed by a bison. Beneath the man is another bird which might symbolise the soul, released from the body at the moment of death. If so, the picture depicts not a prosaic hunting accident, but rather the passage from this world to the next. But we have no way of knowing whether any of these speculations are true. It's a Rorschach test that reveals much about the preconceptions of modern scholars, and little about the beliefs of ancient foragers.

they were the children of the leader, in a culture that believed in either family charisma or strict rules of succession. According to a second theory, the children had been identified at birth as the incarnations of some long-dead spirits. A third theory argues that the children's burial reflects the way they died rather than their status in life. They were ritually sacrificed – perhaps as part of the burial rites of the leader – and then entombed with pomp and circumstance.[9]

9. Hunter-gatherers made these handprints about 9,000 years ago in the 'Hands Cave', in Argentina. It looks as if these long-dead hands are reaching towards us from within the rock. This is one of the most moving relics of the ancient forager world – but nobody knows what it means.

Whatever the correct answer, the Sungir children are among the best pieces of evidence that 30,000 years ago Sapiens could invent sociopolitical codes that went far beyond the dictates of our DNA and the behaviour patterns of other human and animal species.

Peace or War?

Finally, there's the thorny question of the role of war in forager societies. Some scholars imagine ancient hunter-gatherer societies as peaceful paradises, and argue that war and violence began only with the Agricultural Revolution, when people started to accumulate private property. Other scholars maintain that the world of the ancient foragers was exceptionally cruel and violent. Both schools of thought are castles in the air, connected to the ground by the thin strings of meagre archaeological remains and anthropological observations of present-day foragers.

The anthropological evidence is intriguing but very problematic. Foragers today live mainly in isolated and inhospitable areas such as the Arctic or the Kalahari, where population density is very low and opportunities to fight other people are limited. Moreover, in recent generations, foragers have been increasingly subject to the authority of modern states, which prevent the eruption of large-scale conflicts. European scholars have had only two opportunities to observe large and relatively dense populations of independent foragers: in north-western North America in the nineteenth century, and in northern Australia during the nineteenth and early twentieth centuries. Both Amerindian and Aboriginal Australian cultures witnessed frequent armed conflicts. It is debatable, however, whether this represents a 'timeless' condition or the impact of European imperialism.

The archaeological findings are both scarce and opaque. What telltale clues might remain of any war that took place tens of thousands of years ago? There were no fortifications and walls back then, no artillery shells or even swords and shields. An ancient spear point might have been used in war, but it could have been used in a hunt as well. Fossilised human bones are no less hard to interpret. A fracture might indicate a war wound or an accident. Nor is the absence of fractures and cuts on an ancient skeleton conclusive proof that the person to whom the skeleton belonged did not die a violent death. Death can be caused by trauma to soft tissues that leaves no marks on bone. Even more importantly, during pre-industrial warfare more than 90 per cent of war dead were killed by starvation, cold and disease rather than by weapons. Imagine that 30,000 years ago one tribe defeated its neighbour and expelled it from coveted foraging grounds. In the decisive battle, ten members of the defeated tribe were killed. In the following year, another hundred members of the losing tribe died from starvation, cold and disease. Archaeologists who come across these 110 skeletons may too easily conclude that most fell victim to some natural disaster. How would we be able to tell that they were all victims of a merciless war?

Duly warned, we can now turn to the archaeological findings. In Portugal, a survey was made of 400 skeletons from the period immediately predating the Agricultural Revolution. Only two skeletons showed clear marks of violence. A similar survey of 400 skeletons from the same period in Israel discovered a single crack in a single skull that could be attributed to human violence. A third survey of 400 skeletons from various pre-agricultural sites in the Danube Valley found evidence of violence on eighteen skeletons. Eighteen out of 400 may not sound like a lot, but it's actually a very high percentage. If all eighteen indeed died violently, it means that about 4.5 per cent of deaths in the ancient Danube Valley were caused by human violence. Today, the global average is only 1.5 per cent, taking war and crime together. During the twentieth century, only 5 per cent of human deaths resulted from human violence – and this in a century that saw the bloodiest wars and most massive genocides in history. If this revelation is typical, the ancient Danube Valley was as violent as the twentieth century.*

The depressing findings from the Danube Valley are supported by a string of equally depressing findings from other areas. At Jabl Sahaba in Sudan, a 12,000-year-old cemetery containing fifty-nine skeletons was discovered. Arrowheads and spear points were found embedded in or lying near the bones of twenty-four skeletons, 40 per cent of the find. The skeleton of one woman revealed twelve

* It might be argued that not all eighteen ancient Danubians actually died from the violence whose marks can be seen on their remains. Some were only injured. However, this is probably counterbalanced by deaths from trauma to soft tissues and from the invisible deprivations that accompany war.

injuries. In Ofnet Cave in Bavaria, archaeologists discovered the remains of thirty-eight foragers, mainly women and children, who had been thrown into two burial pits. Half the skeletons, including those of children and babies, bore clear signs of damage by human weapons such as clubs and knives. The few skeletons belonging to mature males bore the worst marks of violence. In all probability, an entire forager band was massacred at Ofnet.

Which better represents the world of the ancient foragers: the peaceful skeletons from Israel and Portugal, or the abattoirs of Jabl Sahaba and Ofnet? The answer is neither. Just as foragers exhibited a wide array of religions and social structures, so, too, did they probably demonstrate a variety of violence rates. While some areas and some periods of time may have enjoyed peace and tranquillity, others were riven by ferocious conflicts.[10]

The Curtain of Silence

If the larger picture of ancient forager life is hard to reconstruct, particular events are largely irretrievable. When a Sapiens band first entered a valley inhabited by Neanderthals, the following years might have witnessed a breathtaking historical drama. Unfortunately, nothing would have survived from such an encounter except, at best, a few fossilised bones and a handful of stone tools that remain mute under the most intense scholarly inquisitions. We may extract from them information about human anatomy, human technology, human diet, and perhaps even human social structure. But they reveal nothing about the political alliance forged between neighbouring Sapiens bands, about the spirits of the dead that blessed this alliance, or about the ivory beads secretly given to the local witch doctor in order to secure the blessing of the spirits.

This curtain of silence shrouds tens of thousands of years of history. These long millennia may well have witnessed wars and revolutions, ecstatic religious movements, profound philosophical theories, incomparable artistic masterpieces. The foragers may have had their all-conquering Napoleons, who ruled empires half the size of Luxembourg; gifted Beethovens who lacked symphony orchestras but brought people to tears with the sound of their bamboo flutes; and charismatic prophets who revealed the words of a local oak tree rather than those of a universal creator god. But these are all mere guesses. The curtain of silence is so thick that we cannot even be sure such things occurred – let alone describe them in detail.

Scholars tend to ask only those questions that they can reasonably expect to answer. Without the discovery of as yet unavailable research tools, we will probably never know what the ancient foragers believed or what political dramas they experienced. Yet it is vital to ask questions for which no answers are available, otherwise we might be tempted to dismiss 60,000 of 70,000 years of human history with the excuse that 'the people who lived back then did nothing of importance'.

The truth is that they did a lot of important things. In particular, they shaped the world around us to a much larger degree than most people realise. Trekkers visiting the Siberian tundra, the deserts of central Australia and the Amazonian rainforest believe that they have entered pristine landscapes, virtually untouched by human hands. But that's an illusion. The foragers were there before us and they brought about dramatic changes even in the densest jungles and the most desolate wildernesses. The next chapter explains how the foragers completely reshaped the ecology of our planet long before the first agricultural village was built. The wandering bands of storytelling Sapiens were the most important and most destructive force the animal kingdom had ever produced.

4

The Flood

PRIOR TO THE COGNITIVE REVOLUTION, humans of all species lived exclusively on the Afro-Asian landmass. True, they had settled a few islands by swimming short stretches of water or crossing them on improvised rafts. Flores, for example, was colonised as far back as 850,000 years ago. Yet they were unable to venture into the open sea, and none reached America, Australia, or remote islands such as Madagascar, New Zealand and Hawaii.

The sea barrier prevented not just humans but also many other Afro-Asian animals and plants from reaching this 'Outer World'. As a result, the organisms of distant lands like Australia and Madagascar evolved in isolation for millions upon millions of years, taking on shapes and natures very different from those of their distant Afro-Asian relatives. Planet Earth was separated into several distinct ecosystems, each made up of a unique assembly of animals and plants. *Homo sapiens* was about to put an end to this biological exuberance.

Following the Cognitive Revolution, Sapiens acquired the technology, the organisational skills, and perhaps even the vision necessary to break out of Afro-Asia and settle the Outer World. Their first achievement was the colonisation of Australia some 45,000 years ago. Experts are hard-pressed to explain this feat. In order to reach Australia, humans had to cross a number of sea channels, some more than a hundred kilometres wide, and upon arrival they had to adapt nearly overnight to a completely new ecosystem.

The most reasonable theory suggests that, about 45,000 years ago, the Sapiens living in the Indonesian archipelago (a group of islands separated from Asia and from each other by only narrow straits) developed the first seafaring societies. They learned how to build and manoeuvre ocean-going vessels and became long-distance fishermen, traders and explorers. This would have brought about an unprecedented transformation in human capabilities and life-styles. Every other mammal that went to sea – seals, sea cows, dolphins – had to evolve for aeons to develop specialised organs and a hydrodynamic body. The Sapiens in Indonesia, descendants of apes who lived on the African savannah, became Pacific seafarers without growing flippers and without having to wait for their noses to migrate to the top of their heads as whales did. Instead, they built boats and learned how to steer them. And these skills enabled them to reach and settle Australia.

True, archaeologists have yet to unearth rafts, oars or fishing villages that date back as far as 45,000 years ago (they would be difficult to discover, because rising sea levels have buried the ancient Indonesian shoreline under a hundred metres of ocean). Nevertheless, there is strong circumstantial evidence to support this theory, especially the fact that in the thousands of years following the settlement of Australia, Sapiens colonised a large number of small and isolated islands to its north. Some, such as Buka and Manus, were separated from the closest land by 200 kilometres of open water. It's hard to believe that anyone could have reached and colonised Manus without sophisticated vessels and sailing skills. As mentioned earlier, there is also firm evidence for regular sea trade between some of these islands, such as New Ireland and New Britain.[1]

The journey of the first humans to Australia is one of the most important events in history, at least as important as Columbus' journey to America or the Apollo 11 expedition to the moon. It was the first time any human had managed to leave the Afro-Asian ecological system – indeed, the first time any large terrestrial mammal had managed to cross from Afro-Asia to Australia. Of even greater importance was what the human pioneers did in this new world. The moment the first hunter-gatherer set foot on an Australian beach was the moment that *Homo sapiens* climbed to the top rung in the food chain, and became the deadliest species ever in the 4-billion-year history of life on earth.

Up until then humans had displayed some innovative adaptations and behaviours, but their effect on their environment had been negligible. They had demonstrated remarkable success in moving into and adjusting to various habitats, but they did so without drastically changing those habitats. The settlers of Australia, or more accurately, its conquerors, didn't just adapt. They transformed the Australian ecosystem beyond recognition.

The first human footprint on a sandy Australian beach was immediately washed away by the waves. Yet when the invaders advanced inland, they left behind a different footprint, one that would never be expunged. As they pushed on, they encountered a strange universe of unknown creatures that included a 200-kilogram, two-metre kangaroo, and a marsupial lion, as massive as a modern tiger, that was the continent's largest predator. Koalas far too big to be cuddly and cute rustled in the trees and flightless birds twice the size of ostriches sprinted on the plains. Dragon-like lizards and snakes five metres long slithered through the undergrowth. The giant diprotodon, a two-and-a-half-ton wombat, roamed the forests. Except for the birds and reptiles, all these animals were marsupials – like kangaroos, they gave birth to tiny, helpless, fetus-like young which they then nurtured with milk in abdominal pouches. Marsupial mammals were almost unknown in Africa and Asia, but in Australia they reigned supreme.

Within a few thousand years, virtually all of these giants vanished. Of the twenty-four Australian animal species weighing fifty kilograms or more, twenty-three became extinct.[2] A large number of smaller species also disappeared. Food chains throughout the entire

Australian ecosystem were broken and rearranged. It was the most important transformation of the Australian ecosystem for millions of years. Was it all the fault of *Homo sapiens*?

Guilty as Charged

Some scholars try to exonerate our species, placing the blame on the vagaries of the climate (the usual scapegoat in such cases). Yet it is hard to believe that *Homo sapiens* was completely innocent. There are three pieces of evidence that weaken the climate alibi, and implicate our ancestors in the extinction of the Australian megafauna.

Firstly, even though Australia's climate changed some 45,000 years ago, it wasn't a very remarkable upheaval. It's hard to see how the new weather patterns alone could have caused such a massive extinction. It's common today to explain anything and everything as the result of climate change, but the truth is that earth's climate never rests. It is in constant flux. Every event in history occurred against the background of some climate change.

In particular, our planet has experienced numerous cycles of cooling and warming. During the last million years, there has been an ice age on average every 100,000 years. The last one ran from about 75,000 to 15,000 years ago. Not unusually severe for an ice age, it had twin peaks, the first about 70,000 years ago and the second at about 20,000 years ago. The giant diprotodon appeared in Australia more than 1.5 million years ago and successfully weathered at least ten previous ice ages. It also survived the first peak of the last ice age, around 70,000 years ago. Why, then, did it disappear 45,000 years ago? Of course, if diprotodons had been the only large animal to disappear at this time, it might have been just a fluke. But more than 90 per cent of Australia's megafauna disappeared along with the diprotodon. The evidence is circumstantial, but it's hard to imagine that Sapiens, just by coincidence, arrived in Australia at the precise point that all these animals were dropping dead of the chills.[3]

Secondly, when climate change causes mass extinctions, sea creatures are usually hit as hard as land dwellers. Yet there is no evidence of any significant disappearance of oceanic fauna 45,000 years ago. Human involvement can easily explain why the wave of extinction obliterated the terrestrial megafauna of Australia while sparing that of the nearby oceans. Despite its burgeoning navigational abilities, *Homo sapiens* was still overwhelmingly a terrestrial menace.

Thirdly, mass extinctions akin to the archetypal Australian decimation occurred again and again in the ensuing millennia – whenever people settled another part of the Outer World. In these cases Sapiens' guilt is irrefutable. For example, the megafauna of New Zealand – which had weathered the alleged 'climate change' of *c.*45,000 years ago without a scratch – suffered devastating blows immediately after the first humans set foot on the islands. The Maori, New Zealand's first Sapiens colonisers, reached the islands about 800 years ago. Within a couple of centuries, the majority of the local megafauna was extinct, along with 60 per cent of all bird species.

A similar fate befell the mammoth population of Wrangel Island in the Arctic Ocean (200 kilometres north of the Siberian coast). Mammoths had flourished for millions of years over most of the northern hemisphere, but as *Homo sapiens* spread – first over Eurasia and then over North America – the mammoths retreated. By 10,000 years ago there was not a single mammoth to be found in the world, except on a few remote Arctic islands, most conspicuously Wrangel. The mammoths of Wrangel continued to prosper for a few more millennia, then suddenly disappeared about 4,000 years ago, just when the first humans reached the island.

Were the Australian extinction an isolated event, we could grant humans the benefit of the doubt. But the historical record makes *Homo sapiens* look like an ecological serial killer.

*

All the settlers of Australia had at their disposal was Stone Age technology. How could they cause an ecological disaster? There are three explanations that mesh quite nicely.

Large animals – the primary victims of the Australian extinction – breed slowly. Pregnancy is long, offspring per pregnancy are few, and there are long breaks between pregnancies. Consequently, if humans cut down even one diprotodon every few months, it would be enough to cause diprotodon deaths to outnumber births. Within a few thousand years the last lonesome diprotodon would pass away, and with her the entire species.[4]

In fact, for all their size, diprotodons and Australia's other giants probably wouldn't have been that hard to hunt because they would have been taken totally by surprise by their two-legged assailants. Various human species had been prowling and evolving in Afro-Asia for 2 million years. They slowly honed their hunting skills, and began going after large animals around 400,000 years ago. The big beasts of Africa and Asia learned to avoid humans, so when the new mega-predator – *Homo sapiens* – appeared on the Afro-Asian scene, the large animals already knew to keep their distance from creatures that looked like it. In contrast, the Australian giants had no time to learn to run away. Humans don't come across as particularly dangerous. They don't have long, sharp teeth or muscular, lithe bodies. So when a diprotodon, the largest marsupial ever to walk the earth, set eyes for the first time on this frail-looking ape, he gave it one glance and then went back to chewing leaves. These animals had to evolve a fear of humankind, but before they could do so they were gone.

The second explanation is that by the time Sapiens reached Australia, they had already mastered fire agriculture. Faced with an alien and threatening environment, they deliberately burned vast areas of impassable thickets and dense forests to create open grasslands, which attracted more easily hunted game, and were better suited to their needs. They thereby completely changed the ecology of large parts of Australia within a few short millennia.

One body of evidence supporting this view is the fossil plant record. Eucalyptus trees were rare in Australia 45,000 years ago. But the arrival of *Homo sapiens* inaugurated a golden age for the species. Since eucalyptuses regenerate after fire particularly well, they spread far and wide while other trees disappeared.

These changes in vegetation influenced the animals that ate the plants and the carnivores that ate the vegetarians. Koalas, which subsist exclusively on eucalyptus leaves, happily munched their way into new territories. Most other animals suffered greatly. Many Australian food chains collapsed, driving the weakest links into extinction.[5]

A third explanation agrees that hunting and fire agriculture played a significant role in the extinction, but emphasises that we can't completely ignore the role of climate. The climate changes that beset Australia about 45,000 years ago destabilised the ecosystem and made it particularly vulnerable. Under normal circumstances the system would probably have recuperated, as had happened many times previously. However, humans appeared on the stage at just this critical juncture and pushed the brittle ecosystem into the abyss. The combination of climate change and human hunting is particularly devastating for large animals, since it attacks them from different angles. It is hard to find a good survival strategy that will work simultaneously against multiple threats.

Without further evidence, there's no way of deciding between the three scenarios. But there are certainly good reasons to believe that if *Homo sapiens* had never gone Down Under, it would still be home to marsupial lions, diprotodons and giant kangaroos.

The End of Sloth

The extinction of the Australian megafauna was probably the first significant mark *Homo sapiens* left on our planet. It was followed by an even larger ecological disaster, this time in America. *Homo sapiens* was the first and only human species to reach the western hemisphere landmass, arriving about 16,000 years ago, that is in or around 14,000 BC. The first Americans arrived on foot, which they could do because, at the time, sea levels were low enough that a land bridge connected north-eastern Siberia with north-western Alaska. Not that it was easy – the journey was an arduous one, perhaps harder than the sea passage to Australia. To make the crossing, Sapiens first had to learn how to withstand the extreme Arctic conditions of northern Siberia, an area on which the sun never shines in winter, and where temperatures can drop to minus fifty degrees Celsius.

No previous human species had managed to penetrate places like northern Siberia. Even the cold-adapted Neanderthals restricted themselves to relatively warmer regions further south. But *Homo sapiens*, whose body was adapted to living in the African savannah rather than in the lands of snow and ice, devised ingenious solutions. When roaming bands of Sapiens foragers migrated into colder climates, they learned to make snowshoes and effective thermal clothing composed of layers of furs and skins, sewn together tightly with the help of needles. They developed new weapons and sophisticated hunting techniques that enabled them to track and kill mammoths and the other big game of the far north. As their thermal clothing and hunting techniques improved, Sapiens dared to venture deeper and deeper into the frozen regions. And as they moved north, their clothes, hunting strategies and other survival skills continued to improve.

But why did they bother? Why banish oneself to Siberia by choice? Perhaps some bands were driven north by wars, demographic pressures or natural disasters. Others might have been lured northwards by more positive reasons, such as animal protein. The Arctic lands were full of large, juicy animals such as reindeer and mammoths. Every mammoth was a source of a vast quantity of meat (which, given the frosty temperatures, could even be frozen for later use), tasty fat, warm fur and valuable ivory. As the findings from Sungir testify, mammoth-hunters did not just survive in the frozen north – they thrived. As time passed, the bands spread far and wide, pursuing mammoths, mastodons, rhinoceroses and reindeer. Around 14,000 BC, the chase took some of them from north-eastern Siberia to Alaska. Of course, they didn't know they were discovering a new world. For mammoth and man alike, Alaska was a mere extension of Siberia.

At first, glaciers blocked the way from Alaska to the rest of America, though some pioneers might have bypassed these obstacles by sailing along the coast. Around 12,000 BC global warming melted the ice and opened an easier land passage. Making use of the new corridor, people moved south en masse, spreading over the entire continent. Though originally adapted to hunting large game in the Arctic, they soon adjusted to an amazing variety of climates and ecosystems. Descendants of the Siberians settled the thick forests of the eastern United States, the swamps of the Mississippi Delta, the deserts of Mexico and steaming jungles of Central America. Some made their homes in the river world of the Amazon basin, others struck roots in Andean mountain valleys or the open pampas of Argentina. And all this happened in a mere millennium or two! By 10,000 BC, humans already inhabited the most southern point in America, the island of Tierra del Fuego at the continent's southern tip. The human blitzkrieg across America testifies to the incomparable ingenuity and the unsurpassed adaptability of *Homo sapiens*. No other animal had ever moved into such a huge variety of radically different habitats so quickly, everywhere using virtually the same genes.[6]

The settling of America was hardly bloodless. It left behind a long trail of victims. American fauna 14,000 years ago was far richer than it is today. When the first Americans marched south from Alaska into the plains of Canada and the western United States, they encountered mammoths and mastodons, rodents the size of bears, herds of horses and camels, oversized lions and dozens of large species the likes of which are completely unknown today, among them fearsome sabre-tooth cats and giant ground sloths that weighed up to eight tons and reached a height of six metres. South America hosted an even more exotic menagerie of large mammals, reptiles and birds. The Americas were a great laboratory of evolutionary experimentation, a place where animals and plants unknown in Africa and Asia had evolved and thrived.

But no longer. Within 2,000 years of the Sapiens' arrival, most of these unique species were gone. According to current estimates, within that short interval, North America lost thirty-four out of its forty-seven genera of large mammals. South America lost fifty out of sixty. The sabre-tooth cats, after flourishing for more than 30 million years, disappeared, and so did the giant ground sloths, the oversized lions, native American horses, native American camels, the giant rodents and the mammoths. Thousands of species of smaller mammals, reptiles, birds and even insects and parasites also became extinct (when the mammoths died out, all species of mammoth ticks followed them to oblivion).

For decades, palaeontologists and zooarchaeologists – people who search for and study animal remains – have been combing the plains and mountains of the Americas in search of the fossilised bones of ancient camels and the petrified faeces of giant ground sloths. When they find what they seek, the treasures are carefully packed up and sent to laboratories, where every bone and every coprolite (the technical name for fossilised turds) is meticulously studied and dated. Time and again, these analyses yield the same results: the freshest dung balls and the most recent camel bones date to the period when humans flooded America, that is, between approximately 12,000 and 9000 BC. Only in one area have scientists discovered younger dung balls: on several Caribbean islands, in particular Cuba and Hispaniola, they found petrified ground-sloth scat dating to about 5000 BC. This is exactly the time when the first humans managed to cross the Caribbean Sea and settle these two large islands.

Again, some scholars try to exonerate *Homo sapiens* and blame climate change (which requires them to posit that, for some mysterious reason, the climate in the Caribbean islands remained static for 7,000 years while the rest of the western hemisphere warmed). But in America, the dung ball cannot be dodged. We are the culprits. There is no way around that truth. Even if climate change abetted us, the human contribution was decisive.[7]

Noah's Ark

If we combine the mass extinctions in Australia and America, and add the smaller-scale extinctions that took place as *Homo sapiens* spread over Afro-Asia – such as the extinction of all other human species – and the extinctions that occurred when ancient foragers settled remote islands such as Cuba, the inevitable conclusion is that the first wave of Sapiens colonisation was one of the biggest and swiftest ecological disasters to befall the animal kingdom. Hardest hit were the large furry creatures. At the time of the Cognitive Revolution, the planet was home to about 200 genera of large terrestrial mammals weighing over fifty kilograms. At the time of the Agricultural Revolution, only about a hundred remained. *Homo sapiens* drove to extinction about half of the planet's big beasts long before humans invented the wheel, writing or iron tools.

This ecological tragedy was restaged in miniature countless times after the Agricultural Revolution. The archaeological record of island after island tells the same sad story. The tragedy opens with a scene showing a rich and varied population of large animals, without any trace of humans. In scene two, Sapiens appear, evidenced by a human bone, a spear point, or perhaps a potsherd. Scene three quickly follows, in which men and women occupy centre stage and most large animals, along with many smaller ones, are gone.

The large island of Madagascar, about 400 kilometres east of the African mainland, offers a famous example. Through millions of

10. Reconstructions of two giant ground sloths (Megatherium) and behind them two giant armadillos (Glyptodon). Now extinct, giant armadillos measured over three metres in length and weighed up to two tons, whereas giant ground sloths reached heights of up to six metres, and weighed up to eight tons.

years of isolation, a unique collection of animals evolved there. These included the elephant bird, a flightless creature three metres tall and weighing almost half a ton – the largest bird in the world – and the giant lemurs, the globe's largest primates. The elephant birds and the giant lemurs, along with most of the other large animals of Madagascar, suddenly vanished about 1,500 years ago – precisely when the first humans set foot on the island.

In the Pacific Ocean, the main wave of extinction began in about 1500 BC, when Polynesian farmers settled the Solomon Islands, Fiji and New Caledonia. They killed off, directly or indirectly, hundreds of species of birds, insects, snails and other local inhabitants. From there, the wave of extinction moved gradually to the east, the south and the north, into the heart of the Pacific Ocean, obliterating on its way the unique fauna of Samoa and Tonga (1200 BC); the Marquis Islands (AD 1); Easter Island, the Cook Islands and Hawaii (AD 500); and finally New Zealand (AD 1200).

Similar ecological disasters occurred on almost every one of the thousands of islands that pepper the Atlantic Ocean, Indian Ocean, Arctic Ocean and Mediterranean Sea. Archaeologists have discovered on even the tiniest islands evidence of the existence of birds, insects and snails that lived there for countless generations, only to vanish when the first human farmers arrived. None but a few extremely remote islands escaped man's notice until the modern age, and these islands kept their fauna intact. The Galapagos Islands, to give one famous example, remained uninhabited by humans until the nineteenth century, thus preserving their unique menagerie, including their giant tortoises, which, like the ancient diprotodons, show no fear of humans.

The First Wave Extinction, which accompanied the spread of the foragers, was followed by the Second Wave Extinction, which accompanied the spread of the farmers, and gives us an important perspective on the Third Wave Extinction, which industrial activity is causing today. Don't believe tree-huggers who claim that our ancestors lived in harmony with nature. Long before the Industrial Revolution, *Homo sapiens* held the record among all organisms for driving the most plant and animal species to their extinctions. We have the dubious distinction of being the deadliest species in the annals of biology.

Perhaps if more people were aware of the First Wave and Second Wave extinctions, they'd be less nonchalant about the Third Wave they are part of. If we knew how many species we've already eradicated, we might be more motivated to protect those that still survive. This is especially relevant to the large animals of the oceans. Unlike their terrestrial counterparts, the large sea animals suffered relatively little from the Cognitive and Agricultural Revolutions. But many of them are on the brink of extinction now as a result of industrial pollution and human overuse of oceanic resources. If things continue at the present pace, it is likely that whales, sharks, tuna and dolphins will follow the diprotodons, ground sloths and mammoths to oblivion. Among all the world's large creatures, the only survivors of the human flood will be humans themselves, and the farmyard animals that serve as galley slaves in Noah's Ark.

Part Two

The Agricultural Revolution

11. A wall painting from an Egyptian grave, dated to about 3,500 years ago, depicting typical agricultural scenes.

5

History's Biggest Fraud

FOR 2.5 MILLION YEARS HUMANS FED themselves by gathering plants and hunting animals that lived and bred without their intervention. *Homo erectus*, *Homo ergaster* and the Neanderthals plucked wild figs and hunted wild sheep without deciding where fig trees would take root, in which meadow a herd of sheep should graze, or which billy goat would inseminate which nanny goat. *Homo sapiens* spread from East Africa to the Middle East, to Europe and Asia, and finally to Australia and America – but everywhere they went, Sapiens too continued to live by gathering wild plants and hunting wild animals. Why do anything else when your lifestyle feeds you amply and supports a rich world of social structures, religious beliefs and political dynamics?

All this changed about 10,000 years ago, when Sapiens began to devote almost all their time and effort to manipulating the lives of a few animal and plant species. From sunrise to sunset humans sowed seeds, watered plants, plucked weeds from the ground and led sheep to prime pastures. This work, they thought, would provide them with more fruit, grain and meat. It was a revolution in the way humans lived – the Agricultural Revolution.

The transition to agriculture began around 9500–8500 BC in the hill country of south-eastern Turkey, western Iran and the Levant. It began slowly and in a restricted geographical area. Wheat and goats were domesticated by approximately 9000 BC; peas and lentils around 8000 BC; olive trees by 5000 BC; horses by 4000 BC; and grapevines by 3500 BC. Some animals and plants, such as camels and cashew nuts, were domesticated even later, but by 3500 BC the main wave of domestication was over. Even today, with all our advanced technologies, more than 90 per cent of the calories that feed humanity come from the handful of plants that our ancestors domesticated between 9500 and 3500 BC – wheat, rice, maize (called 'corn' in the US), potatoes, millet and barley. No noteworthy plant or animal has been domesticated in the last 2,000 years. If our minds are those of hunter-gatherers, our cuisine is that of ancient farmers.

Scholars once believed that agriculture spread from a single Middle Eastern point of origin to the four corners of the world. Today, scholars agree that agriculture sprang up in other parts of the world not by the action of Middle Eastern farmers exporting their revolution but entirely independently. People in Central America domesticated maize and beans without knowing anything about wheat and pea cultivation in the Middle East. South Americans learned how to raise potatoes and llamas, unaware of what was going on in either Mexico or the Levant. China's first revolutionaries domesticated rice, millet and pigs. North America's first gardeners were those who got tired of combing the undergrowth for edible gourds and decided to cultivate pumpkins. New Guineans tamed sugar cane and bananas, while the first West African farmers made African millet, African rice, sorghum and wheat conform to their needs. From these initial focal points, agriculture spread far and wide. By the first century AD the vast majority of people throughout most of the world were agriculturists.

Why did agricultural revolutions erupt in the Middle East, China and Central America but not in Australia, Alaska or South Africa? The reason is simple: most species of plants and animals can't be domesticated. Sapiens could dig up delicious truffles and hunt down woolly mammoths, but domesticating either species was out of the question. The fungi were far too elusive, the giant beasts too ferocious. Of the thousands of species that our ancestors hunted and gathered, only a few were suitable candidates for farming and herding. Those few species lived in particular places, and those are the places where agricultural revolutions occurred.

Scholars once proclaimed that the agricultural revolution was a great leap forward for humanity. They told a tale of progress fuelled by human brain power. Evolution gradually produced ever more intelligent people. Eventually, people were so smart that they were able to decipher nature's secrets, enabling them to tame sheep and cultivate wheat. As soon as this happened, they cheerfully abandoned the gruelling, dangerous and often spartan life of hunter-gatherers, settling down to enjoy the pleasant, satiated life of farmers.

That tale is a fantasy. There is no evidence that people became more intelligent with time. Foragers knew the secrets of nature long before the Agricultural Revolution, since their survival depended on an intimate knowledge of the animals they hunted and the plants they gathered. Rather than heralding a new era of easy living, the

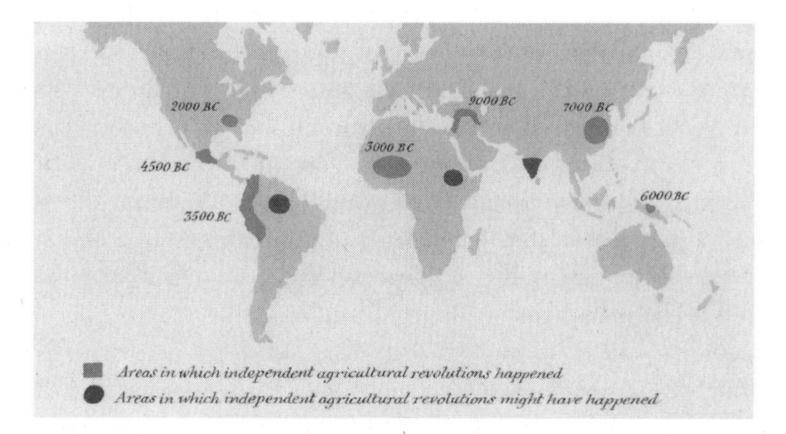

Map 2. Locations and dates of agricultural revolutions. The data is contentious, and the map is constantly being redrawn to incorporate the latest archaeological discoveries.[1]

Agricultural Revolution left farmers with lives generally more difficult and less satisfying than those of foragers. Hunter-gatherers spent their time in more stimulating and varied ways, and were less in danger of starvation and disease. The Agricultural Revolution certainly enlarged the sum total of food at the disposal of humankind, but the extra food did not translate into a better diet or more leisure. Rather, it translated into population explosions and pampered elites. The average farmer worked harder than the average forager, and got a worse diet in return. The Agricultural Revolution was history's biggest fraud.[2]

Who was responsible? Neither kings, nor priests, nor merchants. The culprits were a handful of plant species, including wheat, rice and potatoes. These plants domesticated *Homo sapiens*, rather than vice versa.

Think for a moment about the Agricultural Revolution from the viewpoint of wheat. Ten thousand years ago wheat was just a wild grass, one of many, confined to a small range in the Middle East. Suddenly, within just a few short millennia, it was growing all over the world. According to the basic evolutionary criteria of survival and reproduction, wheat has become one of the most successful plants in the history of the earth. In areas such as the Great Plains of North America, where not a single wheat stalk grew 10,000 years ago, you can today walk for hundreds upon hundreds of kilometres without encountering any other plant. Worldwide, wheat covers about 2.25 million square kilometres of the globe's surface, almost ten times the size of Britain. How did this grass turn from insignificant to ubiquitous?

Wheat did it by manipulating *Homo sapiens* to its advantage. This ape had been living a fairly comfortable life hunting and gathering until about 10,000 years ago, but then began to invest more and more effort in cultivating wheat. Within a couple of millennia, humans in many parts of the world were doing little from dawn to dusk other than taking care of wheat plants. It wasn't easy. Wheat demanded a lot of them. Wheat didn't like rocks and pebbles, so Sapiens broke their backs clearing fields. Wheat didn't like sharing its space, water and nutrients with other plants, so men and women laboured long days weeding under the scorching sun. Wheat got sick, so Sapiens had to keep a watch out for worms and blight. Wheat was attacked by rabbits and locust swarms, so the farmers built fences and stood guard over the fields. Wheat was thirsty, so humans dug irrigation canals or lugged heavy buckets from the well to water it. Its hunger even impelled Sapiens to collect animal faeces to nourish the ground in which wheat grew.

The body of *Homo sapiens* had not evolved for such tasks. It was adapted to climbing apple trees and running after gazelles, not to clearing rocks and carrying water buckets. Human spines, knees, necks and arches paid the price. Studies of ancient skeletons indicate that the transition to agriculture brought about a plethora of ailments, such as slipped discs, arthritis and hernias. Moreover, the new agricultural tasks demanded so much time that people were forced to settle permanently next to their wheat fields. This completely changed their way of life. We did not domesticate wheat. It domesticated us. The word 'domesticate' comes from the Latin '*domus*', which means 'house'. Who's the one living in a house? Not the wheat. It's the Sapiens.

How did wheat convince *Homo sapiens* to exchange a rather good life for a more miserable existence? What did it offer in return? It did not offer a better diet. Remember, humans are omnivorous apes who thrive on a wide variety of foods. Grains made up only a small fraction of the human diet before the Agricultural Revolution. A diet based on cereals is poor in minerals and vitamins, hard to digest, and really bad for your teeth and gums.

Wheat did not give people economic security. The life of a peasant is less secure than that of a hunter-gatherer. Foragers relied on dozens of species to survive, and could therefore weather difficult years even without stocks of preserved food. If the availability of one species was reduced, they could gather and hunt more of other species. Farming societies have, until very recently, relied for the great bulk of their calorie intake on a small variety of domesticated plants. In many areas, they relied on just a single staple, such as wheat, potatoes or rice. If the rains failed or clouds of locusts arrived or if a fungus infected that staple species, peasants died by the thousands and millions.

Nor could wheat offer security against human violence. The early farmers were at least as violent as their forager ancestors, if not more so. Farmers had more possessions and needed land for planting. The loss of pasture land to raiding neighbours could mean the difference between subsistence and starvation, so there was much less room for compromise. When a foraging band was hard-pressed by a stronger rival, it could usually move on. It was difficult and dangerous, but it was feasible. When a strong enemy threatened an agricultural village, retreat meant giving up fields, houses and granaries. In many

12. Tribal warfare in New Guinea between two farming communities (1960). Such scenes were probably widespread in the thousands of years following the Agricultural Revolution.

cases, this doomed the refugees to starvation. Farmers, therefore, tended to stay put and fight to the bitter end.

Many anthropological and archaeological studies indicate that in simple agricultural societies with no political frameworks beyond village and tribe, human violence was responsible for about 15 per cent of deaths, including 25 per cent of male deaths. In contemporary New Guinea, violence accounts for 30 per cent of male deaths in one agricultural tribal society, the Dani, and 35 per cent in another, the Enga. In Ecuador, perhaps 50 per cent of adult Waoranis meet a violent death at the hands of another human![3] In time, human violence was brought under control through the development of larger social frameworks – cities, kingdoms and states. But it took thousands of years to build such huge and effective political structures.

Village life certainly brought the first farmers some immediate benefits, such as better protection against wild animals, rain and cold. Yet for the average person, the disadvantages probably outweighed the advantages. This is hard for people in today's prosperous societies to appreciate. Since we enjoy affluence and security, and since our affluence and security are built on foundations laid by the Agricultural Revolution, we assume that the Agricultural Revolution was a wonderful improvement. Yet it is wrong to judge thousands of years of history from the perspective of today. A much more representative viewpoint is that of a three-year-old girl dying from malnutrition in first-century China because her father's crops have failed. Would she say, 'I am dying from malnutrition, but in 2,000 years, people will have plenty to eat and live in big air-conditioned houses, so my suffering is a worthwhile sacrifice'?

What then did wheat offer agriculturists, including that malnourished Chinese girl? It offered nothing for people as individuals. Yet it did bestow something on *Homo sapiens* as a species. Cultivating wheat provided much more food per unit of territory, and thereby enabled *Homo sapiens* to multiply exponentially. Around 13,000 BC, when people fed themselves by gathering wild plants and hunting wild animals, the area around the oasis of Jericho, in Palestine, could support at most one roaming band of about a hundred relatively healthy and well-nourished people. Around 8500 BC, when wild plants gave way to wheat fields, the oasis supported a large but cramped village of 1,000 people, who suffered far more from disease and malnourishment.

The currency of evolution is neither hunger nor pain, but rather copies of DNA helixes. Just as the economic success of a company is measured only by the number of dollars in its bank account, not by the happiness of its employees, so the evolutionary success of a species is measured by the number of copies of its DNA. If no more DNA copies remain, the species is extinct, just as a company without money is bankrupt. If a species boasts many DNA copies, it is a success, and the species flourishes. From such a perspective, 1,000 copies are always better than a hundred copies. This is the essence of the Agricultural Revolution: the ability to keep more people alive under worse conditions.

Yet why should individuals care about this evolutionary calculus? Why would any sane person lower his or her standard of living just to multiply the number of copies of the *Homo sapiens* genome? Nobody agreed to this deal: the Agricultural Revolution was a trap.

The Luxury Trap

The rise of farming was a very gradual affair spread over centuries and millennia. A band of *Homo sapiens* gathering mushrooms and nuts and hunting deer and rabbit did not all of a sudden settle in a permanent village, ploughing fields, sowing wheat and carrying water from the river. The change proceeded by stages, each of which involved just a small alteration in daily life.

Homo sapiens reached the Middle East around 70,000 years ago.

For the next 50,000 years our ancestors flourished there without agriculture. The natural resources of the area were enough to support its human population. In times of plenty people had a few more children, and in times of need a few less. Humans, like many mammals, have hormonal and genetic mechanisms that help control procreation. In good times females reach puberty earlier, and their chances of getting pregnant are a bit higher. In bad times puberty is late and fertility decreases.

To these natural population controls were added cultural mechanisms. Babies and small children, who move slowly and demand much attention, were a burden on nomadic foragers. People tried to space their children three to four years apart. Women did so by nursing their children around the clock and until a late age (around-the-clock suckling significantly decreases the chances of getting pregnant). Other methods included full or partial sexual abstinence (backed perhaps by cultural taboos), abortions and occasionally infanticide.[4]

During these long millennia people occasionally ate wheat grain, but this was a marginal part of their diet. About 18,000 years ago, the last ice age gave way to a period of global warming. As temperatures rose, so did rainfall. The new climate was ideal for Middle Eastern wheat and other cereals, which multiplied and spread. People began eating more wheat, and in exchange they inadvertently spread its growth. Since it was impossible to eat wild grains without first winnowing, grinding and cooking them, people who gathered these grains carried them back to their temporary campsites for processing. Wheat grains are small and numerous, so some of them inevitably fell on the way to the campsite and were lost. Over time, more and more wheat grew along favourite human trails and near campsites.

When humans burned down forests and thickets, this also helped wheat. Fire cleared away trees and shrubs, allowing wheat and other grasses to monopolise the sunlight, water and nutrients. Where wheat became particularly abundant, and game and other food sources were also plentiful, human bands could gradually give up their nomadic lifestyle and settle down in seasonal and even permanent camps.

At first they might have camped for four weeks during the harvest. A generation later, as wheat plants multiplied and spread, the harvest camp might have lasted for five weeks, then six, and finally it became a permanent village. Evidence of such settlements has been discovered throughout the Middle East, particularly in the Levant, where the Natufian culture flourished from 12,500 BC to 9500 BC. The Natufians were hunter-gatherers who subsisted on dozens of wild species, but they lived in permanent villages and devoted much of their time to the intensive gathering and processing of wild cereals. They built stone houses and granaries. They stored grain for times of need. They invented new tools such as stone scythes for harvesting wild wheat, and stone pestles and mortars to grind it.

In the years following 9500 BC, the descendants of the Natufians continued to gather and process cereals, but they also began to cultivate them in more and more elaborate ways. When gathering wild grains, they took care to lay aside part of the harvest to sow the fields next season. They discovered that they could achieve much better results by sowing the grains deep in the ground rather than haphazardly scattering them on the surface. So they began to hoe and plough. Gradually they also started to weed the fields, to guard them against parasites, and to water and fertilise them. As more effort was directed towards cereal cultivation, there was less time to gather and hunt wild species. The foragers became farmers.

No single step separated the woman gathering wild wheat from the woman farming domesticated wheat, so it's hard to say exactly when the decisive transition to agriculture took place. But, by 8500 BC, the Middle East was peppered with permanent villages such as Jericho, whose inhabitants spent most of their time cultivating a few domesticated species.

With the move to permanent villages and the increase in food supply, the population began to grow. Giving up the nomadic lifestyle enabled women to have a child every year. Babies were weaned at an earlier age – they could be fed on porridge and gruel. The extra hands were sorely needed in the fields. But the extra mouths quickly wiped out the food surpluses, so even more fields had to be planted. As people began living in disease-ridden settlements, as children fed more on cereals and less on mother's milk, and as each child competed for his or her porridge with more and more siblings, child mortality soared. In most agricultural societies at least one out of every three children died before reaching twenty.[5] Yet the increase in births still outpaced the increase in deaths; humans kept having larger numbers of children.

With time, the 'wheat bargain' became more and more burdensome. Children died in droves, and adults ate bread by the sweat of their brows. The average person in Jericho of 8500 BC lived a harder life than the average person in Jericho of 9500 BC or 13,000 BC. But nobody realised what was happening. Every generation continued to live like the previous generation, making only small improvements here and there in the way things were done. Paradoxically, a series of 'improvements', each of which was meant to make life easier, added up to a millstone around the necks of these farmers.

Why did people make such a fateful miscalculation? For the same reason that people throughout history have miscalculated. People were unable to fathom the full consequences of their decisions. Whenever they decided to do a bit of extra work – say, to hoe the fields instead of scattering seeds on the surface – people thought, 'Yes, we will have to work harder. But the harvest will be so bountiful! We won't have to worry any more about lean years. Our children will never go to sleep hungry.' It made sense. If you worked harder, you would have a better life. That was the plan.

The first part of the plan went smoothly. People indeed worked harder. But people did not foresee that the number of children would increase, meaning that the extra wheat would have to be shared between more children. Neither did the early farmers understand that feeding children with more porridge and less breast milk would weaken their immune system, and that permanent settlements would be hotbeds for infectious diseases. They did not foresee that by increasing their dependence on a single source of food, they were actually exposing themselves even more to the depredations of drought. Nor did the farmers foresee that in good years their bulging granaries would tempt thieves and enemies, compelling them to start building walls and doing guard duty.

Then why didn't humans abandon farming when the plan backfired? Partly because it took generations for the small changes to accumulate and transform society and, by then, nobody remembered that they had ever lived differently. And partly because population growth burned humanity's boats. If the adoption of ploughing increased a village's population from 100 to 110, which ten people would have volunteered to starve so that the others could go back to the good old times? There was no going back. The trap snapped shut.

The pursuit of an easier life resulted in much hardship, and not for the last time. It happens to us today. How many young college graduates have taken demanding jobs in high-powered firms, vowing that they will work hard to earn money that will enable them to retire and pursue their real interests when they are thirty-five? But by the time they reach that age, they have large mortgages, children to school, houses in the suburbs that necessitate at least two cars per family, and a sense that life is not worth living without really good wine and expensive holidays abroad. What are they supposed to do, go back to digging up roots? No, they double their efforts and keep slaving away.

One of history's few iron laws is that luxuries tend to become necessities and to spawn new obligations. Once people get used to a certain luxury, they take it for granted. Then they begin to count on it. Finally they reach a point where they can't live without it. Let's take another familiar example from our own time. Over the last few decades, we have invented countless time-saving devices that are supposed to make life more relaxed – washing machines, vacuum cleaners, dishwashers, telephones, mobile phones, computers, email. Previously it took a lot of work to write a letter, address and stamp an envelope, and take it to the mailbox. It took days or weeks, maybe even months, to get a reply. Nowadays I can dash off an email, send it halfway around the globe, and (if my addressee is online) receive a reply a minute later. I've saved all that trouble and time, but do I live a more relaxed life?

Sadly not. Back in the snail-mail era, people usually only wrote letters when they had something important to relate. Rather than writing the first thing that came into their heads, they considered carefully what they wanted to say and how to phrase it. They expected to receive a similarly considered answer. Most people wrote and received no more than a handful of letters a month and seldom felt compelled to reply immediately. Today I receive dozens of emails each day, all from people who expect a prompt reply. We thought we were saving time; instead we revved up the treadmill of life to ten times its former speed and made our days more anxious and agitated.

Here and there a Luddite holdout refuses to open an email account, just as thousands of years ago some human bands refused to take up farming and so escaped the luxury trap. But the Agricultural Revolution didn't need every band in a given region to join up. It only took one. Once one band settled down and started tilling, whether in the Middle East or Central America, agriculture was irresistible. Since farming created the conditions for swift demographic growth, farmers could usually overcome foragers by sheer weight of numbers. The foragers could either run away, abandoning their hunting grounds to field and pasture, or take up the ploughshare themselves. Either way, the old life was doomed.

The story of the luxury trap carries with it an important lesson. Humanity's search for an easier life released immense forces of change that transformed the world in ways nobody envisioned or wanted.

Nobody plotted the Agricultural Revolution or sought human dependence on cereal cultivation. A series of trivial decisions aimed mostly at filling a few stomachs and gaining a little security had the cumulative effect of forcing ancient foragers to spend their days carrying water buckets under a scorching sun.

Divine Intervention

The above scenario explains the Agricultural Revolution as a miscalculation. It's very plausible. History is full of far more idiotic miscalculations. But there's another possibility. Maybe it wasn't the search for an easier life that brought about the transformation. Maybe Sapiens had other aspirations, and were consciously willing to make their lives harder in order to achieve them.

Scientists usually seek to attribute historical developments to cold economic and demographic factors. It sits better with their rational and mathematical methods. In the case of modern history, scholars cannot avoid taking into account non-material factors such as ideology and culture. The written evidence forces their hand. We have enough documents, letters and memoirs to prove that the Second World War was not caused by food shortages or demographic pressures. But we have no documents from the Natufian culture, so when dealing with ancient periods the materialist school reigns supreme. It is difficult to prove that preliterate people were motivated by faith rather than economic necessity.

Yet, in some rare cases, we are lucky enough to find telltale clues. In 1995 archaeologists began to excavate a site in south-east Turkey called Göbekli Tepe. In the oldest stratum they discovered no signs of a settlement, houses or daily activities. They did, however, find monumental pillared structures decorated with spectacular engravings. Each stone pillar weighed up to seven tons and reached a height of five metres. In a nearby quarry they found a half-chiselled pillar weighing fifty tons. Altogether, they uncovered more than ten monumental structures, the largest of them nearly thirty metres across.

Archaeologists are familiar with such monumental structures from sites around the world – the best-known example is Stonehenge in Britain. Yet as they studied Göbekli Tepe, they discovered an amazing fact. Stonehenge dates to 2500 BC, and was built by a developed agricultural society. The structures at Göbekli Tepe are dated to about 9500 BC, and all available evidence indicates that they were built by hunter-gatherers. The archaeological community initially found it difficult to credit these findings, but one test after another confirmed both the early date of the structures and the pre-agricultural society of their builders. The capabilities of ancient foragers, and the complexity of their cultures, seem to be far more impressive than was previously suspected.

Why would a foraging society build such structures? They had no obvious utilitarian purpose. They were neither mammoth slaughterhouses nor places to shelter from rain or hide from lions. That leaves us with the theory that they were built for some mysterious cultural purpose that archaeologists have a hard time deciphering. Whatever it was, the foragers thought it worth a huge amount of effort and time. The only way to build Göbekli Tepe was for thousands of foragers belonging to different bands and tribes to cooperate over an extended period of time. Only a sophisticated religious or ideological system could sustain such efforts.

Göbekli Tepe held another sensational secret. For many years, geneticists have been tracing the origins of domesticated wheat. Recent discoveries indicate that at least one domesticated variant, einkorn wheat, originated in the Karaçadag Hills – about thirty kilometres from Göbekli Tepe.[6]

This can hardly be a coincidence. It's likely that the cultural centre of Göbekli Tepe was somehow connected to the initial domestication of wheat by humankind and of humankind by wheat. In

13. Opposite: The remains of a monumental structure from Göbekli Tepe. Right: One of the decorated stone pillars (about five metres high).

order to feed the people who built and used the monumental structures, particularly large quantities of food were required. It may well be that foragers switched from gathering wild wheat to intense wheat cultivation, not to increase their normal food supply, but rather to support the building and running of a temple. In the conventional picture, pioneers first built a village, and when it prospered, they set up a temple in the middle. But Göbekli Tepe suggests that the temple may have been built first, and that a village later grew up around it.

Victims of the Revolution

The Faustian bargain between humans and grains was not the only deal our species made. Another deal was struck concerning the fate of animals such as sheep, goats, pigs and chickens. Nomadic bands that stalked wild sheep gradually altered the constitutions of the

herds on which they preyed. This process probably began with selective hunting. Humans learned that it was to their advantage to hunt only adult rams and old or sick sheep. They spared fertile females and young lambs in order to safeguard the long-term vitality of the local herd. The second step might have been to actively defend the herd against predators, driving away lions, wolves and rival human bands. The band might next have corralled the herd into a narrow gorge in order to better control and defend it. Finally, people began to make a more careful selection among the sheep in order to tailor them to human needs. The most aggressive rams, those that showed the greatest resistance to human control, were slaughtered first. So were the skinniest and most inquisitive females. (Shepherds are not fond of sheep whose curiosity takes them far from the herd.) With each passing generation, the sheep became fatter, more submissive and less curious. *Voilà!* Mary had a little lamb and everywhere that Mary went the lamb was sure to go.

Alternatively, hunters may have caught and 'adopted' a lamb, fattening it during the months of plenty and slaughtering it in the leaner season. At some stage they began keeping a greater number of such lambs. Some of these reached puberty and began to procreate. The most aggressive and unruly lambs were first to the slaughter. The most submissive, most appealing lambs were allowed to live longer and procreate. The result was a herd of domesticated and submissive sheep.

Such domesticated animals – sheep, chickens, donkeys and others – supplied food (meat, milk, eggs), raw materials (skins, wool) and muscle power. Transportation, ploughing, grinding and other tasks, hitherto performed by human sinew, were increasingly carried out by animals. In most farming societies people focused on plant cultivation; raising animals was a secondary activity. But a new kind of society also appeared in some places, based primarily on the exploitation of animals: tribes of pastoralist herders.

As humans spread around the world, so did their domesticated animals. Ten thousand years ago, not more than a few million sheep, cattle, goats, boars and chickens lived in restricted Afro-Asian niches. Today the world contains about a billion sheep, a billion pigs, more than a billion cattle, and more than 25 billion chickens. And they are all over the globe. The domesticated chicken is the most widespread fowl ever. Following *Homo sapiens*, domesticated cattle, pigs and sheep are the second, third and fourth most widespread large mammals in the world. From a narrow evolutionary perspective, which measures success by the number of DNA copies, the Agricultural Revolution was a wonderful boon for chickens, cattle, pigs and sheep.

Unfortunately, the evolutionary perspective is an incomplete measure of success. It judges everything by the criteria of survival and reproduction, with no regard for individual suffering and happiness. Domesticated chickens and cattle may well be an evolutionary success story, but they are also among the most miserable creatures that ever lived. The domestication of animals was founded on a series of brutal practices that only became crueller with the passing of the centuries.

The natural lifespan of wild chickens is about seven to twelve years, and of cattle about twenty to twenty-five years. In the wild, most chickens and cattle died long before that, but they still had a fair chance of living for a respectable number of years. In contrast, the vast majority of domesticated chickens and cattle are slaughtered at the age of between a few weeks and a few months, because this has always been the optimal slaughtering age from an economic perspective. (Why keep feeding a cock for three years if it has already reached its maximum weight after three months?)

Egg-laying hens, dairy cows and draught animals are sometimes allowed to live for many years. But the price is subjugation to a way of life completely alien to their urges and desires. It's reasonable to assume, for example, that bulls prefer to spend their days wandering over open prairies in the company of other bulls and cows rather than pulling carts and ploughshares under the yoke of a whip-wielding ape.

In order to turn bulls, horses, donkeys and camels into obedient draught animals, their natural instincts and social ties had to be broken, their aggression and sexuality contained, and their freedom of movement curtailed. Farmers developed techniques such as locking animals inside pens and cages, bridling them in harnesses and leashes, training them with whips and cattle prods, and mutilating them. The process of taming almost always involves the castration of males. This restrains male aggression and enables humans selectively to control the herd's procreation.

In many New Guinean societies, the wealth of a person has traditionally been determined by the number of pigs he or she owns. To ensure that the pigs can't run away, farmers in northern New Guinea slice off a chunk of each pig's nose. This causes severe pain whenever the pig tries to sniff. Since the pigs cannot find food or

14. A painting from an Egyptian grave, *c.*1200 BC: a pair of oxen ploughing a field. In the wild, cattle roamed as they pleased in herds with a complex social structure. The castrated and domesticated ox wasted away his life under the lash and in a narrow pen, labouring alone or in pairs in a way that suited neither its body nor its social and emotional needs. When an ox could no longer pull the plough, it was slaughtered. (Note the hunched position of the Egyptian farmer who, much like the ox, spent his life in hard labour oppressive to his body, his mind and his social relationships.)

even find their way around without sniffing, this mutilation makes them completely dependent on their human owners. In another area of New Guinea, it has been customary to gouge out pigs' eyes, so that they cannot even see where they're going.[7]

The dairy industry has its own ways of forcing animals to do its will. Cows, goats and sheep produce milk only after giving birth to calves, kids and lambs, and only as long as the youngsters are suckling. To continue a supply of animal milk, a farmer needs to have calves, kids or lambs for suckling, but must prevent them from monopolising the milk. One common method throughout history was to simply slaughter the calves and kids shortly after birth, milk the mother for all she was worth, and then get her pregnant again. This is still a very widespread technique. In many modern dairy farms a milk cow usually lives for about five years before being slaughtered. During these five years she is almost constantly pregnant, and is fertilised within 60 to 120 days after giving birth in order to preserve maximum milk production. Her calves are separated from her shortly after birth. The females are reared to become the next generation of dairy cows, whereas the males are handed over to the care of the meat industry.[8]

Another method is to keep the calves and kids near their mothers, but prevent them by various stratagems from suckling too much milk. The simplest way to do that is to allow the kid or calf to start suckling, but drive it away once the milk starts flowing. This method usually encounters resistance from both kid and mother. Some shepherd tribes used to kill the offspring, eat its flesh, and then stuff the skin. The stuffed offspring was then presented to the mother so that its presence would encourage her milk production. The Nuer tribe in the Sudan went so far as to smear stuffed animals with their mother's urine, to give the counterfeit calves a familiar, live scent. Another Nuer technique was to tie a ring of thorns around a calf's mouth, so that it pricks the mother and causes her to resist suckling.[9] Tuareg camel breeders in the Sahara used to puncture or cut off parts of the nose and upper lip of young camels in order to make suckling painful, thereby discouraging them from consuming too much milk.[10]

Not all agricultural societies were this cruel to their farm animals. The lives of some domesticated animals could be quite good. Sheep raised for wool, pet dogs and cats, war horses and race horses often enjoyed comfortable conditions. The Roman emperor Caligula allegedly planned to appoint his favourite horse, Incitatus, to the consulship. Shepherds and farmers throughout history showed affection

15. A modern calf in an industrial meat farm. Immediately after birth the calf is separated from its mother and locked inside a tiny cage not much bigger than the calf's own body. There the calf spends its entire life – about four months on average. It never leaves its cage, nor is it allowed to play with other calves or even walk – all so that its muscles will not grow strong. Soft muscles mean a soft and juicy steak. The first time the calf has a chance to walk, stretch its muscles and touch other calves is on its way to the slaughterhouse. In evolutionary terms, cattle represent one of the most successful animal species ever to exist. At the same time, they are some of the most miserable animals on the planet.

for their animals and have taken great care of them, just as many slaveholders felt affection and concern for their slaves. It was no accident that kings and prophets styled themselves as shepherds and likened the way they and the gods cared for their people to a shepherd's care for his flock.

Yet from the viewpoint of the herd, rather than that of the shepherd, it's hard to avoid the impression that for the vast majority of domesticated animals, the Agricultural Revolution was a terrible catastrophe. Their evolutionary 'success' is meaningless. A rare wild rhinoceros on the brink of extinction is probably more satisfied than a calf who spends its short life inside a tiny box, fattened to produce juicy steaks. The contented rhinoceros is no less content for being among the last of its kind. The numerical success of the calf's species is little consolation for the suffering the individual endures.

This discrepancy between evolutionary success and individual suffering is perhaps the most important lesson we can draw from the Agricultural Revolution. When we study the narrative of plants such as wheat and maize, maybe the purely evolutionary perspective makes sense. Yet in the case of animals such as cattle, sheep and Sapiens, each with a complex world of sensations and emotions, we have to consider how evolutionary success translates into individual experience. In the following chapters we will see time and again how a dramatic increase in the collective power and ostensible success of our species went hand in hand with much individual suffering.

6

Building Pyramids

THE AGRICULTURAL REVOLUTION IS ONE of the most controversial events in history. Some partisans proclaim that it set humankind on the road to prosperity and progress. Others insist that it led to perdition. This was the turning point, they say, where Sapiens cast off its intimate symbiosis with nature and sprinted towards greed and alienation. Whichever direction the road led, there was no going back. Farming enabled populations to increase so radically and rapidly that no complex agricultural society could ever again sustain itself if it returned to hunting and gathering. Around 10,000 BC, before the transition to agriculture, earth was home to about 5–8 million nomadic foragers. By the first century AD, only 1–2 million foragers remained (mainly in Australia, America and Africa), but their numbers were dwarfed by the world's 250 million farmers.[1]

The vast majority of farmers lived in permanent settlements; only a few were nomadic shepherds. Settling down caused most people's turf to shrink dramatically. Ancient hunter-gatherers usually lived in territories covering many dozens and even hundreds of square kilometres. 'Home' was the entire territory, with its hills, streams, woods and open sky. Peasants, on the other hand, spent most of their days working a small field or orchard, and their domestic lives centred on a cramped structure of wood, stone or mud, measuring no more than a few dozen metres – the house. The typical peasant developed a very strong attachment to this structure. This was a far-reaching revolution, whose impact was psychological as much as architectural. Henceforth, attachment to 'my house' and separation from the neighbours became the psychological hallmark of a much more self-centred creature.

The new agricultural territories were not only far smaller than those of ancient foragers, but also far more artificial. Aside from the use of fire, hunter-gatherers made few deliberate changes to the lands in which they roamed. Farmers, on the other hand, lived in artificial human islands that they laboriously carved out of the surrounding wilds. They cut down forests, dug canals, cleared fields, built houses, ploughed furrows, and planted fruit trees in tidy rows. The resulting artificial habitat was meant only for humans and 'their' plants and animals, and was often fenced off by walls and hedges. Farmer families did all they could to keep out wayward weeds and wild animals. If such interlopers made their way in, they were driven out. If they persisted, their human antagonists sought ways to exterminate them. Particularly strong defences were erected around the home. From the dawn of agriculture until this very day, billions of humans armed with branches, swatters, shoes and poison sprays have waged relentless war against the diligent ants, furtive roaches, adventurous spiders and misguided beetles that constantly infiltrate the human domicile.

For most of history these man-made enclaves remained very small, surrounded by expanses of untamed nature. The earth's surface measures about 510 million square kilometres, of which 155 million is land. As late as AD 1400, the vast majority of farmers, along with their plants and animals, clustered together in an area of just 11 million square kilometres – 2 per cent of the planet's surface.[2] Everywhere else was too cold, too hot, too dry, too wet, or otherwise unsuited for cultivation. This minuscule 2 per cent of the earth's surface constituted the stage on which history unfolded.

People found it difficult to leave their artificial islands. They could not abandon their houses, fields and granaries without grave risk of loss. Furthermore, as time went on they accumulated more and more things – objects, not easily transportable, that tied them down. Ancient farmers might seem to us dirt poor, but a typical family possessed more artefacts than an entire forager tribe.

The Coming of the Future

While agricultural space shrank, agricultural time expanded. Foragers usually didn't waste much time thinking about next month or next summer. Farmers sailed in their imagination years and decades into the future.

Foragers discounted the future because they lived from hand to mouth and could only preserve food or accumulate possessions with difficulty. Of course, they clearly engaged in some advanced planning. The creators of the cave paintings of Chauvet, Lascaux and Altamira almost certainly intended them to last for generations. Social alliances and political rivalries were long-term affairs. It often took years to repay a favour or to avenge a wrong. Nevertheless, in the subsistence economy of hunting and gathering, there was an obvious limit to such long-term planning. Paradoxically, it saved foragers a lot of anxieties. There was no sense in worrying about things that they could not influence.

The Agricultural Revolution made the future far more important than it had ever been before. Farmers must always keep the future in mind and must work in its service. The agricultural economy was based on a seasonal cycle of production, comprising long months of cultivation followed by short peak periods of harvest. On the night following the end of a plentiful harvest the peasants might celebrate for all they were worth, but within a week or so they were again up at dawn for a long day in the field. Although there was enough food for today, next week and even next month, they had to worry about next year and the year after that.

Concern about the future was rooted not only in seasonal cycles of production, but also in the fundamental uncertainty of agriculture. Since most villages lived by cultivating a very limited variety of domesticated plants and animals, they were at the mercy of droughts, floods and pestilence. Peasants were obliged to produce more than they consumed so that they could build up reserves. Without grain in the silo, jars of olive oil in the cellar, cheese in the pantry and sausages hanging from the rafters, they would starve in bad years. And bad years were bound to come, sooner or later. A peasant living on the assumption that bad years would not come didn't live long.

Consequently, from the very advent of agriculture, worries about the future became major players in the theatre of the human mind. Where farmers depended on rains to water their fields, the onset of the rainy season meant that each morning the farmers gazed towards the horizon, sniffing the wind and straining their eyes. Is that a cloud? Would the rains come on time? Would there be enough? Would violent storms wash the seeds from the fields and batter down seedlings? Meanwhile, in the valleys of the Euphrates, Indus and Yellow rivers, other peasants monitored, with no less trepidation, the height of the water. They needed the rivers to rise in order to spread the fertile topsoil washed down from the highlands, and to enable their vast irrigation systems to fill with water. But floods that surged too high or came at the wrong time could destroy their fields as much as a drought.

Peasants were worried about the future not just because they had more cause for worry, but also because they could do something about it. They could clear another field, dig another irrigation canal, sow more crops. The anxious peasant was as frenetic and hardworking as a harvester ant in the summer, sweating to plant olive trees whose oil would be pressed by his children and grandchildren, putting off until the winter or the following year the eating of the food he craved today.

The stress of farming had far-reaching consequences. It was the foundation of large-scale political and social systems. Sadly, the diligent peasants almost never achieved the future economic security they so craved through their hard work in the present. Everywhere, rulers and elites sprang up, living off the peasants' surplus food and leaving them with only a bare subsistence.

These forfeited food surpluses fuelled politics, wars, art and philosophy. They built palaces, forts, monuments and temples. Until the late modern era, more than 90 per cent of humans were peasants who rose each morning to till the land by the sweat of their brows. The extra they produced fed the tiny minority of elites – kings, government officials, soldiers, priests, artists and thinkers – who fill the history books. History is something that very few people have been doing while everyone else was ploughing fields and carrying water buckets.

An Imagined Order

The food surpluses produced by peasants, coupled with new transportation technology, eventually enabled more and more people to cram together first into large villages, then into towns, and finally into cities, all of them joined together by new kingdoms and commercial networks.

Yet in order to take advantage of these new opportunities, food surpluses and improved transportation were not enough. The mere fact that one can feed a thousand people in the same town or a million people in the same kingdom does not guarantee that they can agree how to divide the land and water, how to settle disputes and conflicts, and how to act in times of drought or war. And if no agreement can be reached, strife spreads, even if the storehouses are bulging. It was not food shortages that caused most of history's wars and revolutions. The French Revolution was spearheaded by affluent lawyers, not by famished peasants. The Roman Republic reached the height of its power in the first century BC, when treasure fleets from throughout the Mediterranean enriched the Romans beyond their ancestors' wildest dreams. Yet it was at that moment of maximum affluence that the Roman political order collapsed into a series of deadly civil wars. Yugoslavia in 1991 had more than enough resources to feed all its inhabitants, and still disintegrated into a terrible bloodbath.

The problem at the root of such calamities is that humans evolved for millions of years in small bands of a few dozen individuals. The handful of millennia separating the Agricultural Revolution from the appearance of cities, kingdoms and empires was not enough time to allow an instinct for mass cooperation to evolve.

Despite the lack of such biological instincts, during the foraging era, hundreds of people were able to cooperate thanks to their shared myths. However, this cooperation was loose and limited. Every Sapiens band continued to run its life independently and to provide for most of its own needs. An archaic sociologist living 20,000 years ago, who had no knowledge of events following the Agricultural Revolution, might well have concluded that mythology had a fairly limited scope. Stories about ancestral spirits and tribal totems were strong enough to enable 500 people to trade seashells, celebrate the odd festival, and join forces to wipe out a Neanderthal band, but no more than that. Mythology, the ancient sociologist would have thought, could not possibly enable millions of strangers to cooperate on a daily basis.

But that turned out to be wrong. Myths, it transpired, *are* stronger than anyone could have imagined. When the Agricultural Revolution opened opportunities for the creation of crowded cities and mighty empires, people invented stories about great gods, motherlands and joint stock companies to provide the needed social links. While human evolution was crawling at its usual snail's pace, the human imagination was building astounding networks of mass cooperation, unlike any other ever seen on earth.

Around 8500 BC the largest settlements in the world were villages such as Jericho, which contained a few hundred individuals. By 7000 BC the town of Çatalhöyük in Anatolia numbered between 5,000 and 10,000 individuals. It may well have been the world's biggest settlement at the time. During the fifth and fourth millennia BC, cities with tens of thousands of inhabitants sprouted in the Fertile Crescent, and each of these held sway over many nearby villages. In 3100 BC the entire lower Nile Valley was united into the first Egyptian kingdom. Its pharaohs ruled thousands of square kilometres and hundreds of thousands of people. Around 2250 BC Sargon the Great forged the first empire, the Akkadian. It boasted over a million subjects and a standing army of 5,400 soldiers. Between 1000 BC and 500 BC, the first mega-empires appeared in the Middle East: the Late Assyrian Empire, the Babylonian Empire, and the Persian Empire. They ruled over many millions of subjects and commanded tens of thousands of soldiers.

In 221 BC the Qin dynasty united China, and shortly afterwards Rome united the Mediterranean basin. Taxes levied on 40 million Qin subjects paid for a standing army of hundreds of thousands of soldiers and a complex bureaucracy that employed more than 100,000 officials. The Roman Empire at its zenith collected taxes from up to 100 million subjects. This revenue financed a standing army of 250,000–500,000 soldiers, a road network still in use 1,500 years later, and theatres and amphitheatres that host spectacles to this day.

Impressive, no doubt, but we mustn't harbour rosy illusions about 'mass cooperation networks' operating in pharaonic Egypt or the Roman Empire. 'Cooperation' sounds very altruistic, but is not always voluntary and seldom egalitarian. Most human cooperation networks have been geared towards oppression and exploitation. The peasants paid for the burgeoning cooperation networks with their precious food surpluses, despairing when the tax collector wiped out an entire year of hard labour with a single stroke of his imperial pen. The famed Roman amphitheatres were often built by

16. A stone stela inscribed with the Code of Hammurabi, *c.*1776 BC.

slaves so that wealthy and idle Romans could watch other slaves engage in vicious gladiatorial combat. Even prisons and concentration camps are cooperation networks, and can function only because thousands of strangers somehow manage to coordinate their actions.

All these cooperation networks – from the cities of ancient Mesopotamia to the Qin and Roman empires – were 'imagined orders'. The social norms that sustained them were based neither on ingrained instincts nor on personal acquaintances, but rather on belief in shared myths.

How can myths sustain entire empires? We have already discussed one such example: Peugeot. Now let's examine two of the best-known myths of history: the Code of Hammurabi of *c.*1776 BC, which served as a cooperation manual for hundreds of thousands of ancient Babylonians; and the American Declaration of Independence of

IN CONGRESS, JULY 4, 1776.

The unanimous Declaration of the thirteen united States of America,

[illegible]

17. The Declaration of Independence of the United States, signed 4 July 1776.

AD 1776, which today still serves as a cooperation manual for hundreds of millions of modern Americans.

In 1776 BC Babylon was the world's biggest city. The Babylonian Empire was probably the world's largest, with more than a million subjects. It ruled most of Mesopotamia, including the bulk of modern Iraq and parts of present-day Syria and Iran. The Babylonian king most famous today is Hammurabi. His fame is due primarily to the text that bears his name, the Code of Hammurabi. This was a collection of laws and judicial decisions whose aim was to present Hammurabi as a role model of a just king, serve as a basis for a more uniform legal system across the Babylonian Empire, and teach future generations what justice is and how a just king acts.

Future generations took notice. The intellectual and bureaucratic elite of ancient Mesopotamia canonised the text, and apprentice scribes continued to copy it long after Hammurabi died and his empire lay in ruins. Hammurabi's Code is therefore a good source for understanding the ancient Mesopotamians' ideal of social order.[3]

The text begins by saying that the gods Anu, Enlil and Marduk – the leading deities of the Mesopotamian pantheon – appointed Hammurabi 'to make justice prevail in the land, to abolish the wicked and the evil, to prevent the strong from oppressing the weak'.[4] It then lists about 300 judgments, given in the set formula: 'If such and such a thing happens, such is the judgment.' For example, judgments 196–9 and 209–14 read:

196. If a superior man should blind the eye of another superior man, they shall blind his eye.

197. If he should break the bone of another superior man, they shall break his bone.

198. If he should blind the eye of a commoner or break the bone of a commoner, he shall weigh and deliver sixty shekels of silver.

199. If he should blind the eye of a slave of a superior man or break the bone of a slave of a superior man, he shall weigh and deliver one-half of the slave's value [in silver].[5]

209. If a superior man strikes a woman of superior class and thereby causes her to miscarry her fetus, he shall weigh and deliver ten shekels of silver for her fetus.

210. If that woman should die, they shall kill his daughter.

211. If he should cause a woman of commoner class to miscarry her fetus by the beating, he shall weigh and deliver five shekels of silver.

212. If that woman should die, he shall weigh and deliver thirty shekels of silver.

213. If he strikes a slave-woman of a superior man and thereby causes her to miscarry her fetus, he shall weigh and deliver two shekels of silver.
214. If that slave-woman should die, he shall weigh and deliver twenty shekels of silver.[6]

After listing his judgments, Hammurabi again declares that:

These are the just decisions which Hammurabi, the able king, has established and thereby has directed the land along the course of truth and the correct way of life . . . I am Hammurabi, noble king. I have not been careless or negligent towards humankind, granted to my care by the god Enlil, and with whose shepherding the god Marduk charged me.[7]

Hammurabi's Code asserts that Babylonian social order is rooted in universal and eternal principles of justice, dictated by the gods. The principle of hierarchy is of paramount importance. According to the code, people are divided into two genders and three classes: superior people, commoners and slaves. Members of each gender and class have different values. The life of a female commoner is worth thirty silver shekels and that of a slave-woman twenty silver shekels, whereas the eye of a male commoner is worth sixty silver shekels.

The code also establishes a strict hierarchy within families, according to which children are not independent persons, but rather the property of their parents. Hence, if one superior man kills the daughter of another superior man, the killer's daughter is executed in punishment. To us it may seem strange that the killer remains unharmed whereas his innocent daughter is killed, but to Hammurabi and the Babylonians this seemed perfectly just. Hammurabi's Code was based on the premise that if the king's subjects all accepted their positions in the hierarchy and acted accordingly, the empire's million inhabitants would be able to cooperate effectively. Their society could then produce enough food for its members, distribute it efficiently, protect itself against its enemies, and expand its territory so as to acquire more wealth and better security.

About 3,500 years after Hammurabi's death, the inhabitants of thirteen British colonies in North America felt that the king of Great Britain was treating them unjustly. Their representatives gathered in the city of Philadelphia, and on 4 July 1776 the colonies declared that their inhabitants were no longer subjects of the British Crown. Their Declaration of Independence proclaimed universal and eternal principles of justice, which, like those of Hammurabi, were inspired by a divine power. However, the most important principle dictated by the American god was somewhat different from the principle dictated by the gods of Babylon. The American Declaration of Independence asserts that:

> We hold these truths to be self-evident, that all men are created equal, that they are endowed by their Creator with certain unalienable rights, that among these are life, liberty, and the pursuit of happiness.

Like Hammurabi's Code, the American founding document promises that if humans act according to its sacred principles, millions of them would be able to cooperate effectively, living safely and peacefully in a just and prosperous society. Like the Code of Hammurabi, the American Declaration of Independence was not just a document of its time and place – it was accepted by future generations as well. For more than 200 years, American schoolchildren have been copying and learning it by heart.

The two texts present us with an obvious dilemma. Both the Code of Hammurabi and the American Declaration of Independence claim to outline universal and eternal principles of justice, but according to the Americans all people are equal, whereas according to the Babylonians people are decidedly unequal. The Americans would, of course, say that they are right, and that Hammurabi is wrong. Hammurabi, naturally, would retort that he is right, and that the Americans are wrong. In fact, they are both wrong. Hammurabi and the American Founding Fathers alike imagined a reality governed by universal and immutable principles of justice, such as equality or hierarchy. Yet the only place where such universal principles exist is in the fertile imagination of Sapiens, and in the myths they invent and tell one another. These principles have no objective validity.

It is easy for us to accept that the division of people into 'superiors' and 'commoners' is a figment of the imagination. Yet the idea that all humans are equal is also a myth. In what sense do all humans equal one another? Is there any objective reality, outside the human imagination, in which we are truly equal? Are all humans equal to one another biologically? Let us try to translate the most famous line of the American Declaration of Independence into biological terms:

We hold these truths to be self-evident, that all men are **created equal**, that they are **endowed** by their **Creator** with certain **unalienable rights**, that among these are life, **liberty**, and the pursuit of **happiness**.

According to the science of biology, people were not 'created'. They have evolved. And they certainly did not evolve to be 'equal'. The idea of equality is inextricably intertwined with the idea of creation. The Americans got the idea of equality from Christianity, which argues that every person has a divinely created soul, and that all souls are equal before God. However, if we do not believe in the Christian myths about God, creation and souls, what does it mean that all people are 'equal'? Evolution is based on difference, not on equality. Every person carries a somewhat different genetic code, and is exposed from birth to different environmental influences. This leads to the development of different qualities that carry with them different chances of survival. 'Created equal' should therefore be translated into 'evolved differently'.

Just as people were never created, neither, according to the science of biology, is there a 'Creator' who 'endows' them with anything. There is only a blind evolutionary process, devoid of any purpose, leading to the birth of individuals. 'Endowed by their Creator' should be translated simply into 'born'.

Similarly, there are no such things as rights in biology. There are only organs, abilities and characteristics. Birds fly not because they have a right to fly, but because they have wings. And it's not true that these organs, abilities and characteristics are 'unalienable'. Many of them undergo constant mutations, and may well be completely lost over time. The ostrich is a bird that lost its ability to fly. So 'unalienable rights' should be translated into 'mutable characteristics'.

And what are the characteristics that evolved in humans? 'Life', certainly. But 'liberty'? There is no such thing in biology. Just like equality, rights, and limited liability companies, liberty too is a political ideal rather than a biological phenomenon. From a purely biological viewpoint, there is little difference between the citizens of a republic and the subjects of a king. And what about 'happiness'? So far biological research has failed to come up with a clear definition of happiness or a way to measure it objectively. Most biological studies acknowledge only the existence of pleasure, which is more easily defined and measured. So 'life, liberty, and the pursuit of happiness' should be translated into 'life and the pursuit of pleasure'.

So here is that line from the American Declaration of Independence translated into biological terms:

> We hold these truths to be self-evident, that all men evolved differently, that they are born with certain mutable characteristics, and that among these are life and the pursuit of pleasure.

Advocates of equality and human rights may be outraged by this line of reasoning. Their response is likely to be 'We know that people are not equal biologically! But if we believe that we are all equal in essence, it will enable us to create a stable and prosperous society.' I have no argument with that. This is exactly what I mean by 'imagined order'. We believe in a particular order not because it is objectively true, but because believing in it enables us to cooperate effectively and forge a better society. Imagined orders are not evil conspiracies or useless mirages. Rather, they are the only way large numbers of humans can cooperate effectively. Bear in mind, though, that Hammurabi might have defended his principle of hierarchy using the same logic: 'I know that superiors, commoners and slaves are not inherently different kinds of people. But if we believe that they are, it will enable us to create a stable and prosperous society.'

True Believers

It's likely that more than a few readers squirmed in their chairs while reading the preceding paragraphs. Most of us today are educated to react in such a way. It is easy to accept that Hammurabi's Code was a myth, but we do not want to hear that human rights are also a myth. If people realise that human rights exist only in the imagination, isn't there a danger that our society will collapse? Voltaire said about God that 'There is no God, but don't tell that to my servant, lest he murder me at night'. Hammurabi would have said the same about his principle of hierarchy, and Thomas Jefferson about human rights. *Homo sapiens* has no natural rights, just as spiders, hyenas and chimpanzees have no natural rights. But don't tell that to our servants, lest they murder us at night.

Such fears are well justified. A natural order is a stable order. There is no chance that gravity will cease to function tomorrow, even if people stop believing in it. In contrast, an imagined order is always in danger of collapse, because it depends upon myths, and myths vanish once people stop believing in them. In order to safeguard an imagined order, continuous and strenuous efforts are imperative.

Some of these efforts take the shape of violence and coercion. Armies, police forces, courts and prisons are ceaselessly at work forcing people to act in accordance with the imagined order. If an ancient Babylonian blinded his neighbour, some violence was usually necessary in order to enforce the law of 'an eye for an eye'. When, in 1860, a majority of American citizens concluded that African slaves are human beings and must therefore enjoy the right of liberty, it took a bloody civil war to make the southern states acquiesce.

However, an imagined order cannot be sustained by violence alone. It requires some true believers as well. Prince Talleyrand, who began his chameleon-like career under Louis XVI, later served the revolutionary and Napoleonic regimes, and switched loyalties in time to end his days working for the restored monarchy, summed up decades of governmental experience by saying that 'You can do many things with bayonets, but it is rather uncomfortable to sit on them.' A single priest often does the work of a hundred soldiers – far more cheaply and effectively. Moreover, no matter how efficient bayonets are, somebody must wield them. Why should the soldiers, jailors, judges and police maintain an imagined order in which they do not believe? Of all human collective activities, the one most difficult to organise is violence. To say that a social order is maintained by military force immediately raises the question: what maintains the military order? It is impossible to organise an army solely by coercion. At least some of the commanders and soldiers must truly believe in something, be it God, honour, motherland, manhood or money.

An even more interesting question concerns those standing at the top of the social pyramid. Why should they wish to enforce an imagined order if they themselves don't believe in it? It is quite common to argue that the elite may do so out of cynical greed. Yet a cynic who believes in nothing is unlikely to be greedy. It does not take much to provide the objective biological needs of *Homo sapiens*. After those needs are met, more money can be spent on building pyramids, taking holidays around the world, financing election campaigns, funding your favourite terrorist organisation, or investing in the stock market and making yet more money – all of which are activities that a true cynic would find utterly meaningless. Diogenes, the Greek philosopher who founded the Cynical school, lived in a barrel. When Alexander the Great once visited Diogenes as he was relaxing in the sun, and asked if there were anything he might do for him, the Cynic answered the all-powerful conqueror, 'Yes, there is something you can do for me. Please move a little to the side. You are blocking the sunlight.'

This is why cynics don't build empires and why an imagined order can be maintained only if large segments of the population – and in particular large segments of the elite and the security forces – truly believe in it. Christianity would not have lasted 2,000 years if the majority of bishops and priests failed to believe in Christ. American democracy would not have lasted 250 years if the majority of presidents and congressmen failed to believe in human rights. The modern economic system would not have lasted a single day if the majority of investors and bankers failed to believe in capitalism.

The Prison Walls

How do you cause people to believe in an imagined order such as Christianity, democracy or capitalism? First, you never admit that the order is imagined. You always insist that the order sustaining society is an objective reality created by the great gods or by the laws of nature. People are unequal, not because Hammurabi said so, but because Enlil and Marduk decreed it. People are equal, not because Thomas Jefferson said so, but because God created them that way. Free markets are the best economic system, not because Adam Smith said so, but because these are the immutable laws of nature.

You also educate people thoroughly. From the moment they are born, you constantly remind them of the principles of the imagined order, which are incorporated into anything and everything. They are incorporated into fairy tales, dramas, paintings, songs, etiquette, political propaganda, architecture, recipes and fashions. For example, today people believe in equality, so it's fashionable for rich kids to wear jeans, which were originally working-class attire. In medieval Europe people believed in class divisions, so no young nobleman would have worn a peasant's smock. Back then, to be addressed as 'Sir' or 'Madam' was a rare privilege reserved for the nobility, and often purchased with blood. Today all polite correspondence, regardless of the recipient, begins with 'Dear Sir or Madam'.

The humanities and social sciences devote most of their energies to explaining exactly how the imagined order is woven into the tapestry of life. In the limited space at our disposal we can only scratch the surface. Three main factors prevent people from realising that the order organising their lives exists only in their imagination:

a. The imagined order is embedded in the material world. Though the imagined order exists only in our minds, it can be woven into the material reality around us, and even set in stone. Most Westerners today believe in individualism. They believe that every human is an individual, whose worth does not depend on what other people think of him or her. Each of us has within ourselves a brilliant ray of light that gives value and meaning to our lives. In modern Western schools teachers and parents tell children that if their classmates make fun of them, they should ignore it. Only they themselves, not others, know their true worth.

In modern architecture, this myth leaps out of the imagination to take shape in stone and mortar. The ideal modern house is divided into many small rooms so that each child can have a private space, hidden from view, providing maximum autonomy. This private room almost invariably has a door, and in many households it is accepted practice for the child to close, and perhaps lock, the door. Even parents are forbidden to enter without knocking and asking permission. The room is decorated as the child sees fit, with rock-star posters on the wall and dirty socks on the floor. Somebody growing up in such a space cannot help but imagine himself 'an individual', his true worth emanating from within rather than from without.

Medieval noblemen did not believe in individualism. Someone's worth was determined by their place in the social hierarchy, and by what other people said about them. Being laughed at was a horrible indignity. Noblemen taught their children to protect their good name whatever the cost. Like modern individualism, the medieval value system left the imagination and was manifested in the stone of medieval castles. The castle rarely contained private rooms for children (or anyone else, for that matter). The teenage son of a medieval baron did not have a private room on the castle's second floor, with posters of Richard the Lionheart and King Arthur on the walls and a locked door that his parents were not allowed to open. He slept alongside many other youths in a large hall. He was always on display and always had to take into account what others saw and said. Someone growing up in such conditions naturally concluded that a man's true worth was determined by his place in the social hierarchy and by what other people said of him.[8]

b. The imagined order shapes our desires. Most people do not wish to accept that the order governing their lives is imaginary, but in fact every person is born into a pre-existing imagined order, and his or her desires are shaped from birth by its dominant myths. Our personal desires thereby become the imagined order's most important defences.

For instance, the most cherished desires of present-day Westerners are shaped by romantic, nationalist, capitalist and humanist myths that have been around for centuries. Friends giving advice often tell each other, 'Follow your heart.' But the heart is a double agent that usually takes its instructions from the dominant myths of the day, and the very recommendation to 'follow your heart' was implanted in our minds by a combination of nineteenth-century Romantic myths and twentieth-century consumerist myths. The Coca-Cola Company, for example, has marketed Diet Coke around the world under the slogan 'Diet Coke. Do what feels good.'

Even what people take to be their most personal desires are usually programmed by the imagined order. Let's consider, for example, the popular desire to take a holiday abroad. There is nothing natural or obvious about this. A chimpanzee alpha male would never think of using his power in order to go on holiday into the territory of a neighbouring chimpanzee band. The elite of ancient Egypt spent their fortunes building pyramids and having their corpses mummified, but none of them thought of going shopping in Babylon or taking a skiing holiday in Phoenicia. People today spend a great deal of money on holidays abroad because they are true believers in the myths of romantic consumerism.

Romanticism tells us that in order to make the most of our human potential we must have as many different experiences as we can. We must open ourselves to a wide spectrum of emotions; we must sample various kinds of relationships; we must try different cuisines; we must learn to appreciate different styles of music. One of the best ways to do all that is to break free from our daily routine, leave behind our familiar setting, and go travelling in distant lands, where we can 'experience' the culture, the smells, the tastes and the norms of other people. We hear again and again the romantic myths about 'how a new experience opened my eyes and changed my life'.

Consumerism tells us that in order to be happy we must consume as many products and services as possible. If we feel that something is missing or not quite right, then we probably need to buy a product (a car, new clothes, organic food) or a service (housekeeping, relationship therapy, yoga classes). Every television commercial is another

18. The Great Pyramid of Giza. The kind of thing rich people in ancient Egypt did with their money.

little legend about how consuming some product or service will make life better.

Romanticism, which encourages variety, meshes perfectly with consumerism. Their marriage has given birth to the infinite 'market of experiences', on which the modern tourism industry is founded. The tourism industry does not sell flight tickets and hotel bedrooms. It sells experiences. Paris is not a city, nor India a country – they are both experiences, the consumption of which is supposed to widen our horizons, fulfil our human potential, and make us happier. Consequently, when the relationship between a millionaire and his wife is going through a rocky patch, he takes her on an expensive trip to Paris. The trip is not a reflection of some independent desire, but rather of an ardent belief in the myths of romantic consumerism. A wealthy man in ancient Egypt would never have dreamed of solving a relationship crisis by taking his wife on holiday to Babylon. Instead, he might have built for her the sumptuous tomb she had always wanted.

Like the elite of ancient Egypt, most people in most cultures dedicate their lives to building pyramids. Only the names, shapes and sizes of these pyramids change from one culture to the other. They may take the form, for example, of a suburban cottage with a swimming pool and an evergreen lawn, or a gleaming penthouse with an enviable view. Few question the myths that cause us to desire the pyramid in the first place.

c. The imagined order is inter-subjective. Even if by some superhuman effort I succeed in freeing my personal desires from the grip of the imagined order, I am just one person. In order to change the imagined order I must convince millions of strangers to cooperate with me. For the imagined order is not a subjective order existing in my own imagination – it is rather an inter-subjective order, existing in the shared imagination of thousands and millions of people.

In order to understand this, we need to understand the difference between 'objective', 'subjective', and 'inter-subjective'.

An **objective** phenomenon exists independently of human consciousness and human beliefs. Radioactivity, for example, is not a myth. Radioactive emissions occurred long before people discovered them, and they are dangerous even when people do not believe in them. Marie Curie, one of the discoverers of radioactivity, did not know, during her long years of studying radioactive materials, that they could harm her body. While she did not believe that radioactivity could kill her, she nevertheless died of aplastic anaemia, a disease caused by overexposure to radioactive materials.

The **subjective** is something that exists depending on the consciousness and beliefs of a single individual. It disappears or changes if that particular individual changes his or her beliefs. Many a child believes in the existence of an imaginary friend who is invisible and inaudible to the rest of the world. The imaginary friend exists solely in the child's subjective consciousness, and when the child grows up and ceases to believe in it, the imaginary friend fades away.

The **inter-subjective** is something that exists within the communication network linking the subjective consciousness of many individuals. If a single individual changes his or her beliefs, or even dies, it is of little importance. However, if most individuals in the network die or change their beliefs, the inter-subjective phenomenon will mutate or disappear. Inter-subjective phenomena are neither malevolent frauds nor insignificant charades. They exist in a different way from physical phenomena such as radioactivity, but their impact on the world may still be enormous. Many of history's most important drivers are inter-subjective: law, money, gods, nations.

Peugeot, for example, is not the imaginary friend of Peugeot's CEO. The company exists in the shared imagination of millions of people. The CEO believes in the company's existence because the board of directors also believes in it, as do the company's lawyers, the secretaries in the nearby office, the tellers in the bank, the brokers on the stock exchange, and car dealers from France to Australia. If the CEO alone were suddenly to stop believing in Peugeot's existence, he'd quickly land in the nearest mental hospital and someone else would occupy his office.

Similarly, the dollar, human rights and the United States of America exist in the shared imagination of billions, and no single individual can threaten their existence. If I alone were to stop believing in the dollar, in human rights or in the United States, it wouldn't much matter. These imagined orders are inter-subjective, so in order to change them we must simultaneously change the consciousness of billions of people, which is not easy. A change of such magnitude can be accomplished only with the help of a complex organisation, such as a political party, an ideological movement, or a religious cult. However, in order to establish such complex organisations, it's necessary to convince many strangers to cooperate with one another. And this will happen only if these strangers believe in some shared myths. It follows that in order to change an existing imagined order, we must first believe in an alternative imagined order.

In order to dismantle Peugeot, for example, we need to imagine something more powerful, such as the French legal system. In order to dismantle the French legal system we need to imagine something even more powerful, such as the French state. And if we would like to dismantle that too, we will have to imagine something yet more powerful.

There is no way out of the imagined order. When we break down our prison walls and run towards freedom, we are in fact running into the more spacious exercise yard of a bigger prison.

7

Memory Overload

EVOLUTION DID NOT ENDOW HUMANS with the ability to play football. True, it produced legs for kicking, elbows for fouling and mouths for cursing, but all that this enables us to do is perhaps practise penalty kicks by ourselves. To get into a game with the strangers we find in the schoolyard on any given afternoon, we not only have to work in concert with ten teammates we may never have met before, we also need to know that the eleven players on the opposing team are playing by the same rules. Other animals that engage strangers in ritualised aggression do so largely by instinct – puppies throughout the world have the rules for rough-and-tumble play hard-wired into their genes. But human teenagers have no genes for football. They can nevertheless play the game with complete strangers because they have all learned an identical set of ideas about football. These ideas are entirely imaginary, but if everyone shares them, we can all play the game.

The same applies, on a larger scale, to kingdoms, churches and trade networks, with one important difference. The rules of football are relatively simple and concise, much like those necessary for cooperation in a forager band or small village. Each player can easily store them in his brain and still have room for songs, images and shopping lists. But large systems of cooperation that involve not twenty-two but thousands or even millions of humans require the handling and storage of huge amounts of information, much more than any single human brain can contain and process.

The large societies found in some other species, such as ants and bees, are stable and resilient because most of the information needed to sustain them is encoded in the genome. A female honeybee larva can, for example, grow up to be either a queen or a worker, depending on what food it is fed. Its DNA programmes the necessary behaviours for whatever role it will fulfil in life. Hives can be very complex social structures, containing many different kinds of workers, such as harvesters, nurses and cleaners. But so far researchers have failed to locate lawyer bees. Bees don't need lawyers, because there is no danger that they might forget or violate the hive constitution. The queen does not cheat the cleaner bees of their food, and they never go on strike demanding higher wages.

But humans do such things all the time. Because the Sapiens social order is imagined, humans cannot preserve the critical information for running it simply by making copies of their DNA and passing these on to their progeny. A conscious effort has to be made to sustain laws, customs, procedures and manners, otherwise the social order would quickly collapse. For example, King Hammurabi decreed that people are divided into superiors, commoners and slaves. Unlike the beehive class system, this is not a natural division – there is no trace of it in the human genome. If the Babylonians could not keep this 'truth' in mind, their society would have ceased to function. Similarly, when Hammurabi passed his DNA to his offspring, it did not encode his ruling that a superior man who killed a commoner woman must pay thirty silver shekels. Hammurabi deliberately had to instruct his sons in the laws of his empire, and his sons and grandsons had to do the same.

Empires generate huge amounts of information. Beyond laws, empires have to keep accounts of transactions and taxes, inventories of military supplies and merchant vessels, and calendars of festivals and victories. For millions of years people stored information in a single place – their brains. Unfortunately, the human brain is not a good storage device for empire-sized databases, for three main reasons.

Firstly, its capacity is limited. True, some people have astonishing memories, and in ancient times there were memory professionals who could store in their heads the topographies of whole provinces and the law codes of entire states. Nevertheless, there is a limit that even master mnemonists cannot transcend. A lawyer might know by heart the entire law code of the Commonwealth of Massachusetts, but not the details of every legal proceeding that took place in Massachusetts from the Salem witch trials onward.

Secondly, humans die, and their brains die with them. Any information stored in a brain will be erased in less than a century. It is, of course, possible to pass memories from one brain to another, but after a few transmissions, the information tends to get garbled or lost.

Thirdly and most importantly, the human brain has been adapted to store and process only particular types of information. In order to survive, ancient hunter-gatherers had to remember the shapes, qualities and behaviour patterns of thousands of plant and animal species. They had to remember that a wrinkled yellow mushroom growing in autumn under an elm tree is most probably poisonous, whereas a similar-looking mushroom growing in winter under an oak tree is a good stomach-ache remedy. Hunter-gatherers also had to bear in mind the opinions and relations of several dozen band members. If Lucy needed a band member's help to get John to stop harassing her, it was important for her to remember that John had fallen out last week with Mary, who would thus be a likely and enthusiastic ally. Consequently, evolutionary pressures have adapted the human brain to store immense quantities of botanical, zoological, topographical and social information.

But when particularly complex societies began to appear in the wake of the Agricultural Revolution, a completely new type of information became vital – numbers. Foragers were never obliged to handle large amounts of mathematical data. No forager needed to remember, say, the number of fruit on each tree in the forest. So human brains did not adapt to storing and processing numbers. Yet in order to maintain a large kingdom, mathematical data was vital. It was never enough to legislate laws and tell stories about guardian gods. One also had to collect taxes. In order to tax hundreds of thousands of people, it was imperative to collect data about people's incomes and possessions; data about payments made; data about arrears, debts and fines; data about discounts and exemptions. This added up to millions of data bits, which had to be stored and processed. Without this capacity, the state would never know what resources it had and what further resources it could tap. When confronted with the need to memorise, recall and handle all these numbers, most human brains overdosed or fell asleep.

This mental limitation severely constrained the size and complexity of human collectives. When the amount of people and property in a particular society crossed a critical threshold, it became necessary to store and process large amounts of mathematical data. Since the human brain could not do it, the system collapsed. For thousands of years after the Agricultural Revolution, human social networks remained relatively small and simple.

The first to overcome the problem were the ancient Sumerians, who lived in southern Mesopotamia. There, a scorching sun beating upon rich muddy plains produced plentiful harvests and prosperous towns. As the number of inhabitants grew, so did the amount of information required to coordinate their affairs. Between the years 3500 BC and 3000 BC, some unknown Sumerian geniuses invented a system for storing and processing information outside their brains, one that was custom-built to handle large amounts of mathematical data. The Sumerians thereby released their social order from the limitations of the human brain, opening the way for the appearance of cities, kingdoms and empires. The data-processing system invented by the Sumerians is called 'writing'.

Signed, Kushim

Writing is a method for storing information through material signs. The Sumerian writing system did so by combining two types of signs, which were pressed in clay tablets. One type of signs represented numbers. There were signs for 1, 10, 60, 600, 3,600 and 36,000. (The Sumerians used a combination of base-6 and base-10 numeral systems. Their base-6 system bestowed on us several important legacies, such as the division of the day into twenty-four hours and of the circle into 360 degrees.) The other type of signs represented people, animals, merchandise, territories, dates and so forth. By combining both types of signs the Sumerians were able to preserve far more data than any human brain could remember or any DNA chain could encode.

At this early stage, writing was limited to facts and figures. The great Sumerian novel, if there ever was one, was never committed to clay tablets. Writing was time-consuming and the reading public tiny, so no one saw any reason to use it for anything other than essential record-keeping. If we look for the first words of wisdom reaching us from our ancestors, 5,000 years ago, we're in for a big disappointment. The earliest messages our ancestors have left us read, for example, '29,086 measures barley 37 months Kushim.' The most probable reading of this sentence is: 'A total of 29,086 measures of barley were received over the course of 37 months. Signed, Kushim.' Alas, the first texts of history contain no philosophical insights, no poetry, legends, laws, or even royal triumphs. They are humdrum economic documents, recording the payment of taxes, the accumulation of debts and the ownership of property.

Only one other type of text survived from these ancient days, and it is even less exciting: lists of words, copied over and over again by apprentice scribes as training exercises. Even had a bored student wanted to write out some of his poems instead of copy a bill of

19. A clay tablet with an administrative text from the city of Uruk, *c.*3400–3000 BC. 'Kushim' may be the generic title of an office-holder, or the name of a particular individual. If Kushim was indeed a person, he may be the first individual in history whose name is known to us! All the names applied earlier in human history – the Neanderthals, the Natufians, Chauvet Cave, Göbekli Tepe – are modern inventions. We have no idea what the builders of Göbekli Tepe actually called the place. With the appearance of writing, we are beginning to hear history through the ears of its protagonists. When Kushim's neighbours called out to him, they might really have shouted, 'Kushim!' It is telling that the first recorded name in history belongs to an accountant, rather than a prophet, a poet or a great conqueror.[1]

sale, he could not have done so. The earliest Sumerian writing was a partial rather than a full script. Full script is a system of material signs that can represent spoken language more or less completely. It can therefore express everything people can say, including poetry. Partial script, on the other hand, is a system of material signs that can represent only particular types of information, belonging to a

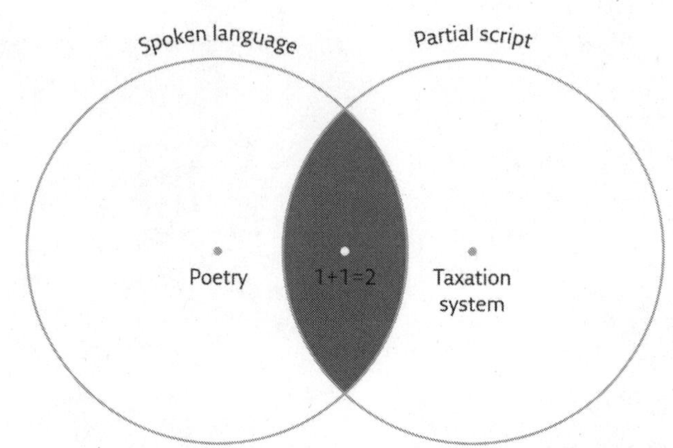

Partial script cannot express the entire spectrum of a spoken language, but it can express things that fall outside the scope of spoken language. Partial scripts such as the Sumerian and mathematical scripts cannot be used to write poetry, but they can keep tax accounts very effectively.

limited field of activity. Latin script, ancient Egyptian hieroglyphics and Braille are full scripts. You can use them to write tax registers, love poems, history books, food recipes and business law. In contrast, the earliest Sumerian script, like modern mathematical symbols and musical notation, are partial scripts. You can use mathematical script to make calculations, but you cannot use it to write love poems.

It didn't disturb the Sumerians that their script was ill-suited for writing poetry. They didn't invent it in order to copy spoken language, but rather to do things that spoken language failed at. There were some cultures, such as those of the pre-Columbian Andes, which used only partial scripts throughout their entire histories, unfazed by their scripts' limitations and feeling no need for a full version. Andean script was very different from its Sumerian counterpart. In fact, it was so different that many people would argue it wasn't a script at all. It was not written on clay tablets or pieces of paper. Rather, it was written by tying knots on colourful cords called quipus. Each

20. A man holding a quipu, as depicted in a Spanish manuscript following the fall of the Inca Empire.

quipu consisted of many cords of different colours, made of wool or cotton. On each cord, several knots were tied in different places. A single quipu could contain hundreds of cords and thousands of knots. By combining different knots on different cords with different colours, it was possible to record large amounts of mathematical data relating to, for example, tax collection and property ownership.[2]

For hundreds, perhaps thousands of years, quipus were essential to the business of cities, kingdoms and empires.[3] They reached their full potential under the Inca Empire, which ruled 10–12 million people and covered today's Peru, Ecuador and Bolivia, as well as chunks of Chile, Argentina and Colombia. Thanks to quipus, the Incas could save and process large amounts of data, without which they would not have been able to maintain the complex administrative machinery that an empire of that size requires.

In fact, quipus were so effective and accurate that in the early years following the Spanish conquest of South America, the Spaniards themselves employed quipus in the work of administering their new empire. The problem was that the Spaniards did not themselves know how to record and read quipus, making them dependent on local professionals. The continent's new rulers realised that this placed them in a tenuous position – the native quipu experts could easily mislead and cheat their overlords. So once Spain's dominion was more firmly established, quipus were phased out and the new empire's records were kept entirely in Latin script and numerals. Very few quipus survived the Spanish occupation, and most of those remaining are undecipherable, since, unfortunately, the art of reading quipus has been lost.

The Wonders of Bureaucracy

The Mesopotamians eventually started to want to write down things other than monotonous mathematical data. Between 3000 BC and 2500 BC more and more signs were added to the Sumerian system, gradually transforming it into a full script that we today call cuneiform. By 2500 BC, kings were using cuneiform to issue decrees, priests were using it to record oracles, and less exalted citizens were using it to write personal letters. At roughly the same time, Egyptians developed another full script known as hieroglyphics. Other full scripts were developed in China around 1200 BC and in Central America around 1000–500 BC.

From these initial centres, full scripts spread far and wide, taking on various new forms and novel tasks. People began to write poetry, history books, romances, dramas, prophecies and cookbooks. Yet writing's most important task continued to be the storage of reams of mathematical data, and that task remained the prerogative of partial script. The Hebrew Bible, the Greek *Iliad*, the Hindu Mahabharata and the Buddhist Tipitika all began as oral works. For many generations they were transmitted orally and would have lived on even had writing never been invented. But tax registries and complex bureaucracies were born together with partial script, and the two remain inexorably linked to this day like Siamese twins – think of the cryptic entries in computerised databases and spreadsheets.

As more and more things were written, and particularly as administrative archives grew to huge proportions, new problems appeared. Information stored in a person's brain is easy to retrieve. My brain stores billions of bits of data, yet I can quickly, almost instantaneously, recall the name of Italy's capital, immediately afterwards recollect what I did on 11 September 2001, and then reconstruct the route leading from my house to the Hebrew University in Jerusalem. Exactly how the brain does it remains a mystery, but we all know that the brain's retrieval system is amazingly efficient, except when you are trying to remember where you put your car keys.

How, though, do you find and retrieve information stored on quipu cords or clay tablets? If you have just ten tablets or a hundred tablets, it's not a problem. But what if you have accumulated thousands of them, as did one of Hammurabi's contemporaries, King Zimrilim of Mari?

Imagine for a moment that it's 1776 BC. Two Marians are quarrelling over possession of a wheat field. Jacob insists that he bought the field from Esau thirty years ago. Esau retorts that he in fact rented the field to Jacob for a term of thirty years, and that now, the term being up, he intends to reclaim it. They shout and wrangle and start pushing one another before they realise that they can resolve their dispute by going to the royal archive, where are housed the deeds and bills of sale that apply to all the kingdom's real estate. Upon arriving at the archive they are shuttled from one official to another. They wait through several herbal tea breaks, are told to come back tomorrow, and eventually are taken by a grumbling clerk to look for the relevant clay tablet. The clerk opens a door and leads them into a huge room lined, floor to ceiling, with thousands of clay tablets. No wonder the clerk is sour-faced. How is he supposed to locate the deed to the disputed wheat field written thirty years ago? Even if he finds it, how will he be able to cross-check to ensure that the one from thirty years ago is the latest document relating to the field in question? If he can't find it, does that prove that Esau never sold or rented out the field? Or just that the document got lost, or turned to mush when some rain leaked into the archive?

Clearly, just imprinting a document in clay is not enough to guarantee efficient, accurate and convenient data processing. That requires methods of organisation like catalogues, methods of reproduction like photocopy machines, methods of rapid and accurate retrieval like computer algorithms, and pedantic (but hopefully cheerful) librarians who know how to use these tools.

Inventing such methods proved to be far more difficult than inventing writing. Many writing systems developed independently in cultures distant in time and place from each other. Every decade archaeologists discover another few forgotten scripts. Some of them might prove to be even older than the Sumerian scratches in clay. But most of them remain curiosities because those who invented them failed to invent efficient ways of cataloguing and retrieving data. What set apart Sumer, as well as pharaonic Egypt, ancient China and the Inca Empire, is that these cultures developed good techniques of archiving, cataloguing and retrieving written records. They also invested in schools for scribes, clerks, librarians and accountants.

A writing exercise from a school in ancient Mesopotamia discovered by modern archaeologists gives us a glimpse into the lives of these students, some 4,000 years ago:

I went in and sat down, and my teacher read my tablet. He said, 'There's something missing!'

And he caned me.

One of the people in charge said, 'Why did you open your mouth without my permission?'

And he caned me.

The one in charge of rules said, 'Why did you get up without my permission?'

And he caned me.

The gatekeeper said, 'Why are you going out without my permission?'

And he caned me.

The keeper of the beer jug said, 'Why did you get some without my permission?'

And he caned me.

The Sumerian teacher said, 'Why did you speak Akkadian?'*

And he caned me.

My teacher said, 'Your handwriting is no good!'

And he caned me.[4]

Ancient scribes learned not merely to read and write, but also to use catalogues, dictionaries, calendars, forms and tables. They studied and internalised techniques of cataloguing, retrieving and processing information very different from those used by the brain. In the brain, all data is freely associated. When I go with my spouse to sign on a mortgage for our new home, I am reminded of the first place we lived together, which reminds me of our honeymoon in New Orleans, which reminds me of alligators, which remind me of dragons, which remind me of *The Ring of the Nibelungen*, and suddenly, before I know it, there I am humming the Siegfried leitmotif to a puzzled bank clerk. In bureaucracy, things must be kept apart. There is one drawer for home mortgages, another for marriage certificates, a third for tax registers, and a fourth for lawsuits. Otherwise, how can you find anything? Things that belong in more than one drawer, like Wagnerian music dramas (do I file them under 'music', 'theatre', or perhaps invent a new category altogether?), are

* Even after Akkadian became the spoken language, Sumerian remained the language of administration and thus the language recorded with writing. Aspiring scribes thus had to speak Sumerian.

a terrible headache. So one is forever adding, deleting and rearranging drawers.

In order to function, the people who operate such a system of drawers must be reprogrammed to stop thinking as humans and to start thinking as clerks and accountants. As everyone from ancient times till today knows, clerks and accountants think in a non-human fashion. They think like filing cabinets. This is not their fault. If they don't think that way their drawers will all get mixed up and they won't be able to provide the services their government, company or organisation requires. The most important impact of script on human history is precisely this: it has gradually changed the way humans think and view the world. Free association and holistic thought have given way to compartmentalisation and bureaucracy.

The Language of Numbers

As the centuries passed, bureaucratic methods of data processing grew ever more different from the way humans naturally think – and ever more important. A critical step was made sometime before the ninth century AD, when a new partial script was invented, one that could store and process mathematical data with unprecedented efficiency. This partial script was composed of ten signs, representing the numbers from 0 to 9. Confusingly, these signs are known as Arabic numerals even though they were first invented by the Hindus (even more confusingly, modern Arabs use a set of digits that look quite different from Western ones). But the Arabs get the credit because when they invaded India they encountered the system, understood its usefulness, refined it, and spread it through the Middle East and then to Europe. When several other signs were later added to the Arab numerals (such as the signs for addition, subtraction and multiplication), the basis of modern mathematical notation came into being.

$$\ddot{\mathbf{r}}_i = \sum_{j \neq i} \frac{\mu_j(\mathbf{r}_j - \mathbf{r}_i)}{r_{ij}^3} \Bigg\{ 1 - \frac{2(\beta-\gamma)}{c^2} \sum_{l \neq i} \frac{\mu_l}{r_{il}} - \frac{2\beta-1}{c^2} \sum_{k \neq j} \frac{\mu_k}{r_{jk}} + \gamma \left(\frac{s_i}{c}\right)^2$$

$$+ (1-\gamma) \left(\frac{s_j}{c}\right)^2 - \frac{2(1+\gamma)}{c^2} \dot{\mathbf{r}}_i \cdot \dot{\mathbf{r}}_j - \frac{3}{2c^2} \left[\frac{(\mathbf{r}_i - \mathbf{r}_j) \cdot \mathbf{r}_j}{r_{ij}}\right]^2$$

$$+ \frac{1}{2c^2} (\mathbf{r}_j - \mathbf{r}_i) \cdot \ddot{\mathbf{r}}_j \Bigg\}$$

$$+ \frac{1}{c^2} \sum_{j \neq i} \frac{\mu_i}{r_{ij}^3} \{ [\mathbf{r}_i - \mathbf{r}_j] \cdot [(2+2\gamma)\, \dot{r}_i - (1+2\gamma)\, \dot{\mathbf{r}}_j] \} (\dot{\mathbf{r}}_i - \dot{\mathbf{r}}_j)$$

$$+ \frac{3+4\gamma}{2c^2} \sum_{j \neq i} \frac{\mu_j \ddot{\mathbf{r}}_j}{r_{ij}}$$

An equation for calculating the acceleration of mass *i* under the influence of gravity, according to the Theory of Relativity. When most laypeople see such an equation, they usually panic and freeze, like a deer caught in the headlights of a speeding vehicle. The reaction is quite natural, and does not betray a lack of intelligence or curiosity. With rare exceptions, human brains are simply incapable of thinking through concepts like relativity and quantum mechanics. Physicists nevertheless manage to do so, because they set aside the traditional human way of thinking, and learn to think anew with the help of external data-processing systems. Crucial parts of their thought process take place not in the head, but inside computers or on classroom blackboards.

Although this system of writing remains a partial script, it has become the world's dominant language. Almost all states, companies, organisations and institutions – whether they speak Arabic, Hindi, English or Norwegian – use mathematical script to record and process data. Every piece of information that can be translated into mathematical script is stored, spread and processed with mind-boggling speed and efficiency.

A person who wishes to influence the decisions of governments, organisations and companies must therefore learn to speak in numbers. Experts do their best to translate even ideas such as 'poverty', 'happiness' and 'honesty' into numbers ('the poverty line', 'subjective well-being levels', 'credit rating'). Entire fields of knowledge, such as physics and engineering, have already lost almost all touch with the spoken human language, and are maintained solely by mathematical script.

More recently, mathematical script has given rise to an even more revolutionary writing system, a computerised binary script consisting of only two signs: 0 and 1. The words I am now typing on my keyboard are written within my computer by different combinations of 0 and 1.

Writing was born as the maidservant of human consciousness, but is increasingly becoming its master. Our computers have trouble understanding how *Homo sapiens* talks, feels and dreams. So we are teaching *Homo sapiens* to talk, feel and dream in the language of numbers, which can be understood by computers.

Eventually, computers might outperform humans in the very fields that made *Homo sapiens* the ruler of the world: intelligence and communication. The process that began in the Euphrates valley 5,000 years ago, when Sumerian geeks outsourced data-processing from the human brain to a clay tablet, would culminate in Silicon Valley with the victory of the tablet. Humans might still be around, but they could no longer make sense of the world. The new ruler of the world would be a long line of zeros and ones.

8

There Is No Justice in History

UNDERSTANDING HUMAN HISTORY IN THE millennia following the Agricultural Revolution boils down to a single question: how did humans organise themselves in mass-cooperation networks, when they lacked the biological instincts necessary to sustain such networks? The short answer is that humans created imagined orders and devised scripts. These two inventions filled the gaps left by our biological inheritance.

However, the appearance of these networks was, for many, a dubious blessing. The imagined orders sustaining these networks were neither neutral nor fair. They divided people into make-believe groups, arranged in a hierarchy. The upper levels enjoyed privileges and power, while the lower ones suffered from discrimination and oppression. Hammurabi's Code, for example, established a pecking order of superiors, commoners and slaves. Superiors got all the good things in life. Commoners got what was left. Slaves got a beating if they complained.

Despite its proclamation of the equality of all men, the imagined order established by the Americans in 1776 also established a hierarchy. It created a hierarchy between men, who benefited from it, and women, whom it left disempowered. It created a hierarchy between whites, who enjoyed liberty, and blacks and Native Americans, who were considered humans of a lesser type and therefore did not share in the equal rights of men. Many of those who signed the Declaration of Independence were slaveholders. They did not release their slaves upon signing the Declaration, nor did they consider themselves hypocrites. In their view, the rights of *men* had little to do with Negroes.

The American order also consecrated the hierarchy between rich and poor. Most Americans at that time had little problem with the inequality caused by wealthy parents passing their money and businesses on to their children. In their view, equality meant simply that the same laws applied to rich and poor. It had nothing to do with unemployment benefits, integrated education or health insurance. Liberty, too, carried very different connotations than it does today. In 1776, it did not mean that the disempowered (certainly not blacks or Indians or, God forbid, women) could gain and exercise power. It meant simply that the state could not, except in unusual circumstances, confiscate a citizen's private property or tell him what to do with it. The American order thereby upheld the hierarchy of wealth, which some thought was mandated by God and others viewed as representing the immutable laws of nature. Nature, it was claimed, rewarded merit with wealth while penalising indolence.

All the above-mentioned distinctions – between free persons and slaves, between whites and blacks, between rich and poor – are rooted in fictions. (The hierarchy of men and women will be discussed later.) Yet it is an iron rule of history that every imagined hierarchy disavows its fictional origins and claims to be natural and inevitable. For instance, many people who have viewed the hierarchy of free persons and slaves as natural and correct have argued that slavery is not a human invention. Hammurabi saw it as ordained by the gods. Aristotle argued that slaves have a 'slavish nature' whereas free people have a 'free nature'. Their status in society is merely a reflection of their innate nature.

Ask white supremacists about the racial hierarchy, and you are in for a pseudoscientific lecture concerning the biological differences

21. A sign on a South African beach from the period of apartheid, restricting its usage to 'whites' only. People with lighter skin colour are typically more in danger of sunburn than people with darker skin. Yet there was no biological logic behind the division of South African beaches. Beaches reserved for people with lighter skin were not characterised by lower levels of ultraviolet radiation.

between the races. You are likely to be told that there is something in Caucasian blood or genes that makes whites naturally more intelligent, moral and hard-working. Ask a diehard capitalist about the hierarchy of wealth, and you are likely to hear that it is the inevitable outcome of objective differences in abilities. The rich have more money, in this view, because they are more capable and diligent. No one should be bothered, then, if the wealthy get better health care, better education and better nutrition. The rich richly deserve every perk they enjoy.

Hindus who adhere to the caste system believe that cosmic forces have made one caste superior to another. According to a famous Hindu creation myth, the gods fashioned the world out of the body of a primeval being, the Purusa. The sun was created from the Purusa's eye, the moon from the Purusa's brain, the Brahmins (priests) from its mouth, the Kshatriyas (warriors) from its arms, the Vaishyas (peasants and merchants) from its thighs, and the Shudras (servants) from its legs. Accept this explanation and the sociopolitical differences between Brahmins and Shudras are as natural and eternal as the differences between the sun and the moon.[1] The ancient Chinese believed that when the goddess Nü Wa created humans from earth, she kneaded aristocrats from fine yellow soil, whereas commoners were formed from brown mud.[2]

Yet, to the best of our understanding, these hierarchies are all the product of human imagination. Brahmins and Shudras were not really created by the gods from different body parts of a primeval being. Instead, the distinction between the two castes was created by laws and norms invented by humans in northern India about 3,000 years ago. Contrary to Aristotle, there is no known biological difference between slaves and free people. Human laws and norms have turned some people into slaves and others into masters. Between blacks and whites there are some objective biological differences, such as skin colour and hair type, but there is no evidence that the differences extend to intelligence or morality.

Most people claim that their social hierarchy is natural and just, while those of other societies are based on false and ridiculous criteria. Modern Westerners are taught to scoff at the idea of racial hierarchy. They are shocked by laws prohibiting blacks to live in white neighbourhoods, or to study in white schools, or to be treated in white hospitals. But the hierarchy of rich and poor – which mandates that rich people live in separate and more luxurious neighbourhoods, study in separate and more prestigious schools, and receive medical treatment in separate and better-equipped facilities – seems perfectly sensible to many Americans and Europeans. Yet it's a proven fact that most rich people are rich for the simple reason that they were born into a rich family, while most poor people will remain poor throughout their lives simply because they were born into a poor family.

Unfortunately, complex human societies seem to require imagined hierarchies and unjust discrimination. Of course not all hierarchies are morally identical, and some societies suffered from more extreme types of discrimination than others, yet scholars know of no large society that has been able to dispense with discrimination altogether. Time and again people have created order in their societies by classifying the population into imagined categories, such as superiors, commoners and slaves; whites and blacks; patricians and plebeians; Brahmins and Shudras; or rich and poor. These categories have regulated relations between millions of humans by making some people legally, politically or socially superior to others.

Hierarchies serve an important function. They enable complete strangers to know how to treat one another without wasting the time and energy needed to become personally acquainted. A car dealer needs to know immediately how much effort to put into selling vehicles to the dozens of people who enter his dealership each day. He can't make a detailed enquiry into the personality and wallet of each individual. Instead, he uses social cues – the way the person is dressed, his or her age, and perhaps even skin and hair colour. That is how the dealer immediately distinguishes between the rich lawyer who may well buy an expensive luxury car, and a simple office clerk who has come only to look around and dream.

Of course, differences in natural abilities also play a role in the formation of social distinctions. But such diversities of aptitudes and character are usually mediated through imagined hierarchies. This happens in two important ways. First and foremost, most abilities have to be nurtured and developed. Even if somebody is born with a particular talent, that talent will usually remain latent if it is not fostered, honed and exercised. Not all people get the same chance to cultivate and refine their abilities. Whether or not they have such an opportunity will usually depend on their place within their society's imagined hierarchy.

Consider identical twins born in China in 1700, and separated at birth. One brother is raised by a rich merchant family in Beijing, spending his days in school, in the market, or in upper-class social gatherings. The other twin is raised by poor illiterate peasants in a remote village, spending his days in the muddy rice paddies. Despite having exactly the same genes, when they turn twenty they are unlikely to have identical skills in doing business – or in planting rice.

Second, even if people belonging to different classes develop exactly the same abilities, they are unlikely to enjoy equal success because they will have to play the game by different rules. If the peasant brother somehow developed exactly the same business acumen as his rich merchant twin, they still would not have had the same chance of becoming rich. The economic game was rigged by legal restrictions and unofficial glass ceilings. When the peasant brother made his way to the Beijing market, with his torn clothes, rough manners, and incomprehensible dialect, he would quickly have discovered that in the business world manners and connections often speak far louder than genes.

The Vicious Circle

All societies are based on imagined hierarchies, but not necessarily on the same hierarchies. What accounts for the differences? Why did traditional Indian society classify people according to caste, Ottoman society according to religion, and American society according to race? In most cases the hierarchy originated as the result of a set of accidental historical circumstances and was then perpetuated and refined over many generations as different groups developed vested interests in it.

For instance, many scholars surmise that the Hindu caste system took shape when Indo-Aryan people invaded the Indian subcontinent about 3,000 years ago, subjugating the local population. The invaders established a stratified society, in which they – of course – occupied the leading positions (priests and warriors), leaving the natives to live as servants and slaves. The invaders, who were few in number, feared losing their privileged status and unique identity. To forestall this danger, they divided the population into castes, each of which was required to pursue a specific occupation or perform a specific role in society. Each had different legal status, privileges and duties. Mixing of castes – social interaction, marriage, even the sharing of meals – was prohibited. And the distinctions were not just legal – they became an inherent part of religious mythology and practice.

The rulers argued that the caste system reflected an eternal cosmic reality rather than a chance historical development. Concepts of purity and impurity were essential elements in Hindu religion, and they were harnessed to buttress the social pyramid. Pious Hindus were taught that contact with members of a different caste could pollute not only them personally, but society as a whole, and should therefore be abhorred. Such ideas are hardly unique to Hindus. Throughout history, and in almost all societies, concepts of pollution and purity have played a leading role in enforcing social and political divisions and have been exploited by numerous ruling classes to maintain their privileges. The fear of pollution is not a complete fabrication of priests and princes, however. It probably has its roots in biological survival mechanisms that make humans feel an instinctive revulsion towards potential disease carriers, such as sick persons and dead bodies. If you want to keep any human group isolated – women, Jews, Roma, gays, blacks – the best way to do it is convince everyone that these people are a source of pollution.

The Hindu caste system and its attendant laws of purity became deeply embedded in Indian culture. Long after the Indo-Aryan invasion was forgotten, Indians continued to believe in the caste system and to abhor the pollution caused by caste mixing. Castes were not immune to change. In fact, as time went by, large castes were divided into sub-castes. Eventually the original four castes turned into 3,000 different groupings called *jati* (literally 'birth'). But this proliferation of castes did not change the basic principle of the system, according to which every person is born into a particular rank, and any infringement of its rules pollutes the person and society as a whole. A person's *jati* determines her profession, the food she can eat, her place of residence and her eligible marriage partners. Usually a person can marry only within his or her caste, and the resulting children inherit that status.

Whenever a new profession developed or a new group of people appeared on the scene, they had to be recognised as a caste in order to receive a legitimate place within Hindu society. Groups that failed to win recognition as a caste were, literally, outcasts – in this stratified society, they did not even occupy the lowest rung. They became known as Untouchables. They had to live apart from all other people and scrape together a living in humiliating and disgusting ways, such as sifting through garbage dumps for scrap material. Even members of the lowest caste avoided mingling with them, eating with them, touching them and certainly marrying them. In modern India, matters of marriage and work are still heavily influenced by the caste system, despite all attempts by the democratic government of India to break down such distinctions and convince Hindus that there is nothing polluting in caste mixing.[3]

Purity in America

A similar vicious circle perpetuated the racial hierarchy in modern America. From the sixteenth to the eighteenth century, the European conquerors imported millions of African slaves to work the mines and plantations of America. They chose to import slaves from Africa rather than from Europe or East Asia due to three circumstantial factors. Firstly, Africa was closer, so it was cheaper to import slaves from Senegal than from Vietnam.

Secondly, in Africa there already existed a well-developed slave trade (exporting slaves mainly to the Middle East), whereas in Europe slavery was very rare. It was obviously far easier to buy slaves in an existing market than to create a new one from scratch.

Thirdly, and most importantly, American plantations in places such as Virginia, Haiti and Brazil were plagued by malaria and yellow fever, which had originated in Africa. Africans had acquired over the generations a partial genetic immunity to these diseases, whereas Europeans were totally defenceless and died in droves. It was consequently wiser for a plantation owner to invest his money in an African slave than in a European slave or indentured labourer. Paradoxically, genetic superiority (in terms of immunity) translated into social inferiority: precisely because Africans were fitter in tropical climates than Europeans, they ended up as the slaves of European masters! Due to these circumstantial factors, the burgeoning new societies of America were to be divided into a ruling caste of white Europeans and a subjugated caste of black Africans.

But people don't like to say that they keep slaves of a certain race or origin simply because it's economically expedient. Like the Aryan conquerors of India, white Europeans in the Americas wanted to be seen not only as economically successful but also as pious, just and objective. Religious and scientific myths were pressed into service to justify this division. Theologians argued that Africans descend from Ham, son of Noah, saddled by his father with a curse that his offspring would be slaves. Biologists argued that blacks are less intelligent than whites and their moral sense less developed. Doctors alleged that blacks live in filth and spread diseases – in other words, they are a source of pollution.

These myths struck a chord in American culture, and in Western culture generally. They continued to exert their influence long after the conditions that created slavery had disappeared. In the early nineteenth century imperial Britain outlawed slavery and stopped the Atlantic slave trade, and in the decades that followed slavery was gradually outlawed throughout the American continent. Notably, this was the first and only time in history that slaveholding societies voluntarily abolished slavery. But, even though the slaves were freed, the racist myths that justified slavery persisted. Separation of the races was maintained by racist legislation and social custom.

The result was a self-reinforcing cycle of cause and effect, a vicious circle. Consider, for example, the southern United States immediately after the Civil War. In 1865 the Thirteenth Amendment to the US Constitution outlawed slavery and the Fourteenth Amendment mandated that citizenship and the equal protection of the law could not be denied on the basis of race. However, two centuries of slavery meant that most black families were far poorer and far less educated than most white families. A black person born in Alabama in 1865 thus had much less chance of getting a good education and a well-paid job than did his white neighbours. His children, born in the 1880s and 1890s, started life with the same disadvantage – they, too, were born to an uneducated, poor family.

But economic disadvantage was not the whole story. Alabama was also home to many poor whites who lacked the opportunities available to their better-off racial brothers and sisters. In addition, the Industrial Revolution and the waves of immigration made the United States an extremely fluid society, where rags could quickly turn into riches. If money was all that mattered, the sharp divide between the races should soon have blurred, not least through intermarriage.

But that did not happen. By 1865 whites, as well as many blacks, took it to be a simple matter of fact that blacks were less intelligent, more violent and sexually dissolute, lazier and less concerned about personal cleanliness than whites. They were thus the agents of violence, theft, rape and disease – in other words, pollution. If a black Alabaman in 1895 miraculously managed to get a good education and then applied for a respectable job such as a bank teller, his odds of being accepted were far worse than those of an equally qualified white candidate. The stigma that labelled blacks as, by nature, unreliable, lazy and less intelligent conspired against him.

You might think that people would gradually understand that these stigmas were myth rather than fact and that blacks would be able, over time, to prove themselves just as competent, law-abiding and clean as whites. In fact, the opposite happened – these prejudices became more and more entrenched as time went by. Since all the best jobs were held by whites, it became easier to believe that blacks really are inferior. 'Look,' said the average white citizen, 'blacks have been free for generations, yet there are almost no black professors, lawyers, doctors or even bank tellers. Isn't that proof that blacks are simply less intelligent and hard-working?' Trapped in this vicious circle, blacks were not hired for white-collar jobs because they were deemed unintelligent, and the proof of their inferiority was the paucity of blacks in white-collar jobs.

The vicious circle did not stop there. As anti-black stigmas grew stronger, they were translated into a system of 'Jim Crow' laws and norms that were meant to safeguard the racial order. Blacks were forbidden to vote in elections, to study in white schools, to buy in white stores, to eat in white restaurants, to sleep in white hotels. The justification for all of this was that blacks were foul, slothful and vicious, so whites had to be protected from them. Whites did not want to sleep in the same hotel as blacks or to eat in the same restaurant, for fear of diseases. They did not want their children learning in the same school as black children, for fear of brutality and bad influences. They did not want blacks voting in elections, since blacks were ignorant and immoral. These fears were substantiated by scientific studies that 'proved' that blacks were indeed less educated, that various diseases were more common among them, and that their crime rate was far higher (the studies ignored the fact that these 'facts' *resulted* from discrimination against blacks).

By the mid-twentieth century, segregation in the former Confederate states was probably worse than in the late nineteenth century. Clennon King, a black student who applied to the University of Mississippi in 1958, was forcefully committed to a mental asylum. The presiding judge ruled that a black person must surely be insane to think that he could be admitted to the University of Mississippi.

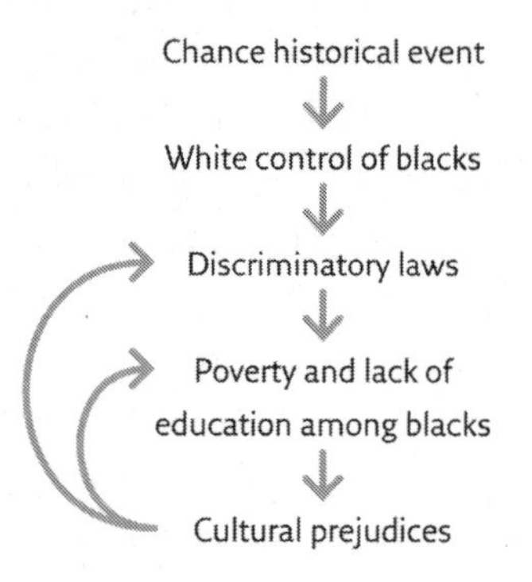

The vicious circle: a chance historical situation is translated into a rigid social system.

Nothing was as revolting to American southerners (and many northerners) as sexual relations and marriage between black men and white women. Sex between the races became the greatest taboo and any violation, or suspected violation, was viewed as deserving immediate and summary punishment in the form of lynching. The Ku Klux Klan, a white supremacist secret society, perpetrated many such killings. They could have taught the Hindu Brahmins a thing or two about purity laws.

With time, the racism spread to more and more cultural arenas. American aesthetic culture was built around white standards of beauty. The physical attributes of the white race – for example light skin, fair and straight hair, a small upturned nose – came to be identified as beautiful. Typical black features – dark skin, dark and bushy hair, a flattened nose – were deemed ugly. These preconceptions ingrained the imagined hierarchy at an even deeper level of human consciousness.

Such vicious circles can go on for centuries and even millennia, perpetuating an imagined hierarchy that sprang from a chance historical occurrence. Unjust discrimination often gets worse, not better, with time. Money comes to money, and poverty to poverty. Education comes to education, and ignorance to ignorance. Those once victimised by history are likely to be victimised yet again. And those whom history has privileged are more likely to be privileged again.

Most sociopolitical hierarchies lack a logical or biological basis – they are nothing but the perpetuation of chance events supported by myths. That is one good reason to study history. If the division into blacks and whites or Brahmins and Shudras was grounded in biological realities – that is, if Brahmins really had better brains than Shudras – biology would be sufficient for understanding human society. Since the biological distinctions between different groups of *Homo sapiens* are, in fact, negligible, biology can't explain the intricacies of Indian society or American racial dynamics. We can only understand those phenomena by studying the events, circumstances, and power relations that transformed figments of imagination into cruel – and very real – social structures.

He and She

Different societies adopt different kinds of imagined hierarchies. Race is very important to modern Americans but was relatively insignificant to medieval Muslims. Caste was a matter of life and death in medieval India, whereas in modern Europe it is practically non-existent. One hierarchy, however, has been of supreme importance in all known human societies: the hierarchy of gender. People everywhere have divided themselves into men and women. And almost everywhere men have got the better deal, at least since the Agricultural Revolution.

Some of the earliest Chinese texts are oracle bones, dating to 1200 BC, used to divine the future. On one was engraved the question: 'Will Lady Hao's childbearing be lucky?' To which was written the reply: 'If the child is born on a *ding* day, lucky; if on a *geng* day, vastly auspicious.' However, Lady Hao was to give birth on a *jiayin* day. The text ends with the morose observation: 'Three weeks and one day later, on *jiayin* day, the child was born. Not lucky. It was a girl.'[4] More than 3,000 years later, when Communist China enacted the 'one child' policy, many Chinese families continued to regard the birth of a girl as a misfortune. Parents would occasionally abandon or murder newborn baby girls in order to have another shot at getting a boy.

In many societies women were simply the property of men, most often their fathers, husbands or brothers. Rape, in many legal systems, falls under property violation – in other words, the victim is not the woman who was raped but the male who owns her. This being the case, the legal remedy was the transfer of ownership – the rapist was required to pay a bride price to the woman's father or brother, upon which she became the rapist's property. The Bible decrees that 'If a man meets a virgin who is not betrothed, and seizes her and lies with her, and they are found, then the man who lay with her shall give to the father of the young woman fifty shekels of silver, and she shall be his wife' (Deuteronomy 22:28–9). The ancient Hebrews considered this a reasonable arrangement.

Raping a woman who did not belong to any man was not considered a crime at all, just as picking up a lost coin on a busy street is not considered theft. And if a husband raped his own wife, he had committed no crime. In fact, the idea that a husband could rape his wife was an oxymoron. To be a husband was to have full control of your wife's sexuality. To say that a husband 'raped' his wife was as illogical as saying that a man stole his own wallet. Such thinking was not confined to the ancient Middle East. As of 2006, there were still fifty-three countries where a husband could not be prosecuted for the rape of his wife. Even in Germany, rape laws were amended only in 1997 to create a legal category of marital rape.[5]

Is the division into men and women a product of the imagination, like the caste system in India and the racial system in America, or is it a natural division with deep biological roots? And if it is indeed a natural division, are there also biological explanations for the preference given to men over women?

Some of the cultural, legal and political disparities between men and women reflect the obvious biological differences between the sexes. Childbearing has always been women's job, because men don't have wombs. Yet around this hard universal kernel, every society accumulated layer upon layer of cultural ideas and norms that have little to do with biology. Societies associate a host of attributes with masculinity and femininity that, for the most part, lack a firm biological basis.

For instance, in democratic Athens of the fifth century BC, an individual possessing a womb had no independent legal status and was forbidden to participate in popular assemblies or to be a judge. With few exceptions, such an individual could not benefit from a good education, nor engage in business or in philosophical discourse. None of Athens' political leaders, none of its great philosophers, orators, artists or merchants had a womb. Does having a womb make a person unfit, biologically, for these professions? The ancient Athenians thought so. Modern Athenians disagree. In present-day Athens, women vote, are elected to public office, make speeches, design everything from jewellery to buildings to software, and go to university. Their wombs do not keep them from doing any of these things as successfully as men do. True, they are still under-represented in politics and business – only about 12 per cent of the members of Greece's parliament are women. But there is no legal barrier to their participation in politics, and most modern Greeks think it is quite normal for a woman to serve in public office.

Many modern Greeks also think that an integral part of being a man is being sexually attracted to women only, and having sexual relations exclusively with the opposite sex. They don't see this as a cultural bias, but rather as a biological reality – relations between two people of the opposite sex are natural, and between two people of the same sex unnatural. In fact, though, Mother Nature does not mind if men are sexually attracted to one another. It's only human mothers steeped in particular cultures who make a scene if their son has a fling with the boy next door. The mother's tantrums are not a biological imperative. A significant number of human cultures have viewed homosexual relations as not only legitimate but even socially constructive, ancient Greece being the most notable example. The *Iliad* does not mention that Thetis had any objection to her son Achilles' relations with Patroclus. Queen Olympias of Macedon was one of the most temperamental and forceful women of the ancient world, and even had her own husband, King Philip, assassinated. Yet she didn't have a fit when her son, Alexander the Great, brought his lover Hephaestion home for dinner.

How can we distinguish what is biologically determined from what people merely try to justify through biological myths? A good rule of thumb is 'Biology enables, culture forbids.' Biology is willing to tolerate a very wide spectrum of possibilities. It's culture that obliges people to realise some possibilities while forbidding others. Biology enables women to have children – some cultures oblige women to realise this possibility. Biology enables men to enjoy sex with one another – some cultures forbid them to realise this possibility.

Culture tends to argue that it forbids only that which is unnatural. But from a biological perspective, nothing is unnatural. Whatever is possible is by definition also natural. A truly unnatural behaviour, one that goes against the laws of nature, simply cannot exist, so it would need no prohibition. No culture has ever bothered to forbid men to photosynthesise, women to run faster than the speed of light, or negatively charged electrons to be attracted to each other.

In truth, our concepts 'natural' and 'unnatural' are taken not from biology, but from Christian theology. The theological meaning of 'natural' is 'in accordance with the intentions of the God who created nature'. Christian theologians argued that God created the human body, intending each limb and organ to serve a particular purpose. If we use our limbs and organs for the purpose envisioned by God, then it is a natural activity. To use them differently than God intends is unnatural. But evolution has no purpose. Organs have not evolved with a purpose, and the way they are used is in constant flux. There is not a single organ in the human body that only does the job its prototype did when it first appeared hundreds of millions of years ago. Organs evolve to perform a particular function, but once they exist, they can be adapted for other usages as well. Mouths, for example, appeared because the earliest multicellular organisms needed a way to take nutrients into their bodies. We still use our mouths for that purpose, but we also use them to kiss, speak and, if we are Rambo, to pull the pins out of hand grenades. Are any of these uses unnatural simply because our worm-like ancestors 600 million years ago didn't do those things with their mouths?

Similarly, wings didn't suddenly appear in all their aerodynamic glory. They developed from organs that served another purpose. According to one theory, insect wings evolved millions of years ago from body protrusions on flightless bugs. Bugs with bumps had a larger surface area than those without bumps, and this enabled them to absorb more sunlight and thus stay warmer. In a slow evolutionary process, these solar heaters grew larger. The same structure that was good for maximum sunlight absorption – lots of surface area, little weight – also, by coincidence, gave the insects a bit of a lift when they skipped and jumped. Those with bigger protrusions could skip and jump further. Some insects started using the things to glide, and from there it was a small step to wings that could actually propel the bug through the air. Next time a mosquito buzzes in your ear, accuse her of unnatural behaviour. If she were well behaved and content with what God gave her, she'd use her wings only as solar panels.

The same sort of multitasking applies to our sexual organs and behavior. Sex first evolved for procreation, and courtship rituals evolved as a way of sizing up the fitness of a potential mate. But many animals now put both to use for a multitude of social purposes that have little to do with creating little copies of themselves. Chimpanzees, for example, use sex to cement political alliances, establish intimacy and defuse tensions. Is that unnatural?

Sex and Gender

There is little sense, then, in arguing that the natural function of women is to give birth, or that homosexuality is unnatural. Most of the laws, norms, rights and obligations that define manhood and womanhood reflect human imagination more than biological reality.

Biologically, humans are divided into males and females. A male *Homo sapiens* has one X chromosome and one Y chromosome; a female *Homo sapiens* has two Xs. But 'man' and 'woman' name social, not biological, categories. While in the great majority of cases in most human societies men are males and women are females, the social terms carry a lot of baggage that has only a tenuous, if any, relationship to the biological terms. A man is not a Sapiens with particular biological qualities such as XY chromosomes, testicles and lots of testosterone. Rather, he fits into a particular slot in his society's imagined human order. His culture's myths assign him particular masculine roles (like engaging in politics), rights (like voting) and duties (like military service). Likewise, a woman is not a Sapiens with two X chromosomes, a womb and plenty of oestrogen.

A female = a biological category		**A woman = a cultural category**	
Ancient Athens	**Modern Athens**	**Ancient Athens**	**Modern Athens**
XX chromosomes	XX chromosomes	Can't vote	Can vote
Womb	Womb	Can't be a judge	Can be a judge
Ovaries	Ovaries	Can't hold government office	Can hold government office
Little testosterone	Little testosterone	Can't decide for herself who to marry	Can decide for herself who to marry
Much oestrogen	Much oestrogen	Typically illiterate	Typically literate
Can produce milk	Can produce milk	Legally owned by father or husband	Legally independent
Exactly the same thing		**Very different things**	

Rather, she is a female member of an imagined human order. The myths of her society assign her unique feminine roles (raising children), rights (protection against violence) and duties (obedience to her husband). Since myths, rather than biology, define the roles, rights and duties of men and women, the meaning of 'manhood' and 'womanhood' have varied immensely from one society to another.

22. Eighteenth-century masculinity: an official portrait of King Louis XIV of France. Note the long wig, stockings, high-heeled shoes, dancer's posture – and huge sword. In contemporary Europe, all these (except for the sword) would be considered marks of effeminacy. But in his time Louis was a European paragon of manhood and virility.

23. Twenty-first-century masculinity: an official portrait of President Barack Obama. What happened to the wig, stockings, high heels – and sword? Dominant men have never looked so dull and dreary as they do today. During most of history, dominant men have been colourful and flamboyant, such as Native American chiefs with their feathered headdresses and Hindu maharajas decked out in silks and diamonds. Throughout the animal kingdom males tend to be more colourful and accessorised than females – think of peacocks' tails and lions' manes.

To make things less confusing, scholars usually distinguish between 'sex', which is a biological category, and 'gender', a cultural category. Sex is divided between males and females, and the qualities of this division are objective and have remained constant throughout history. Gender is divided between men and women (and some cultures recognise other categories). So-called 'masculine' and 'feminine' qualities are inter-subjective and undergo constant changes. For example, there are far-reaching differences in the behaviour, desires, dress and even body posture expected from women in classical Athens and women in modern Athens.[6]

Sex is child's play; but gender is serious business. To get to be a member of the male sex is the simplest thing in the world. You just need to be born with an X and a Y chromosome. To get to be a female is equally simple. A pair of X chromosomes will do it. In contrast, becoming a man or a woman is a very complicated and demanding undertaking. Since most masculine and feminine qualities are cultural rather than biological, no society automatically crowns each male a man, or every female a woman. Nor are these titles laurels that can be rested on once they are acquired. Males must prove their masculinity constantly, throughout their lives, from cradle to grave, in an endless series of rites and performances. And a woman's work is never done – she must continually convince herself and others that she is feminine enough.

Success is not guaranteed. Males in particular live in constant dread of losing their claim to manhood. Throughout history, males have been willing to risk and even sacrifice their lives, just so that people will say, 'He's a real man!'

What's So Good about Men?

At least since the Agricultural Revolution, most human societies have been patriarchal societies that valued men more highly than women. No matter how a society defined 'man' and 'woman', to be a man was always better. Patriarchal societies educate men to think and act in a masculine way and women to think and act in a feminine way, punishing anyone who dares cross those boundaries. Yet they do not equally reward those who conform. Qualities considered masculine are more valued than those considered feminine, and members of a society who personify the feminine ideal get less than those who exemplify the masculine ideal. Fewer resources are invested in the health and education of women; they have fewer economic opportunities, less political power, and less freedom of movement. Gender is a race in which some of the runners compete only for the bronze medal.

True, a handful of women have made it to the alpha position, such as Cleopatra of Egypt, Empress Wu Zetian of China (*c.* AD 700) and Elizabeth I of England. Yet they are the exceptions that prove the rule. Throughout Elizabeth's forty-five-year reign, all Members of Parliament were men, all officers in the Royal Navy and army were men, all judges and lawyers were men, all bishops and archbishops were men, all theologians and priests were men, all doctors and surgeons were men, all students and professors in all universities and colleges were men, all mayors and sheriffs were men, and almost all the writers, architects, poets, philosophers, painters, musicians and scientists were men.

Patriarchy has been the norm in almost all agricultural and industrial societies. It has tenaciously weathered political upheavals, social revolutions and economic transformations. Egypt, for example, was conquered numerous times over the centuries. Assyrians, Persians, Macedonians, Romans, Arabs, Mameluks, Turks and British occupied it – and its society always remained patriarchal. Egypt was governed by pharaonic law, Greek law, Roman law, Muslim law, Ottoman law and British law – and they all discriminated against people who were not 'real men'.

Since patriarchy is so universal, it cannot be the product of some vicious circle that was kick-started by a chance occurrence. It is particularly noteworthy that even before 1492, most societies in both America and Afro-Asia were patriarchal, even though they had been out of contact for thousands of years. If patriarchy in Afro-Asia resulted from some chance occurrence, why were the Aztecs and Incas patriarchal? It is far more likely that even though the precise definition of 'man' and 'woman' varies between cultures, there is some universal biological reason why almost all cultures valued manhood over womanhood. We do not know what this reason is. There are plenty of theories, none of them convincing.

Muscle Power

The most common theory points to the fact that men are stronger than women, and that they have used their greater physical power to force women into submission. A more subtle version of this claim argues that their strength allows men to monopolise tasks that demand hard manual labour, such as ploughing and harvesting. This gives them control of food production, which in turn translates into political clout.

There are two problems with this emphasis on muscle power. First, the statement that 'men are stronger than women' is true only on average, and only with regard to certain types of strength. Women are generally more resistant to hunger, disease and fatigue than men. There are also many women who can run faster and lift heavier weights than many men. Furthermore, and most problematically for this theory, women have, throughout history, been excluded mainly from jobs that require little physical effort (such as the priesthood, law and politics), while engaging in hard manual labour in the fields, in crafts and in the household. If social power were divided in direct relation to physical strength or stamina, women should have got far more of it.

Even more importantly, there simply is no direct relation between physical strength and social power among humans. People in their sixties usually exercise power over people in their twenties, even though twentysomethings are much stronger than their elders. The typical plantation owner in Alabama in the mid-nineteenth century could have been wrestled to the ground in seconds by any of the slaves cultivating his cotton fields. Boxing matches were not used to select Egyptian pharaohs or Catholic popes. In forager societies, political dominance generally resides with the person possessing the best social skills rather than the most developed musculature. In organised crime, the big boss is not necessarily the strongest man. He is often an older man who very rarely uses his own fists; he gets younger and fitter men to do the dirty jobs for him. A guy who thinks that the way to take over the syndicate is to beat up the don is unlikely to live long enough to learn from his mistake. Even among chimpanzees, the alpha male wins his position by building a stable coalition with other males and females, not through mindless violence.

In fact, human history shows that there is often an inverse relation between physical prowess and social power. In most societies, it's the lower classes who do the manual labour. This may reflect *Homo sapiens*' position in the food chain. If all that counted were raw physical abilities, Sapiens would have found themselves on a middle rung of the ladder. But their mental and social skills placed them at the top. It is therefore only natural that the chain of power within the species will also be determined by mental and social abilities more than by brute force. Consequently it sounds improbable that the most influential and most stable social hierarchy in history is founded on men's ability physically to coerce women.

The Scum of Society

Another theory explains that masculine dominance results not from strength but from aggression. Millions of years of evolution have made men far more violent than women. Women can match men as far as hatred, greed and abuse are concerned, but when push comes to shove, the theory goes, men are more willing to engage in raw physical violence. This is why throughout history warfare has been a masculine prerogative.

In times of war, men's control of the armed forces has made them the masters of civilian society, too. They then used their control of civilian society to fight more and more wars, and the greater the number of wars, the greater men's control of society. This feedback loop explains both the ubiquity of war and the ubiquity of patriarchy.

Recent studies of the hormonal and cognitive systems of men and women strengthen the assumption that men indeed have more aggressive and violent tendencies, and are therefore, on average, better suited to serve as common soldiers. Yet granted that the common soldiers are all men, does it follow that the ones managing the war and enjoying its fruits must also be men? That makes no sense. It's like assuming that because all the slaves cultivating cotton fields are black, plantation owners will be black as well. Just as an all-black workforce might be controlled by an all-white management, why couldn't an all-male soldiery be controlled by an all-female or at least partly female government? In fact, in numerous societies throughout history, the top officers did not work their way up from the rank of private. Aristocrats, the wealthy and the educated were automatically assigned officer rank and never served as common soldiers.

When the Duke of Wellington, Napoleon's nemesis, enlisted in the British army at the age of eighteen, he was immediately commissioned as an officer. He didn't think much of the plebeians under his command. 'We have in the service the scum of the earth as common soldiers,' he wrote to a fellow aristocrat during the wars against France. These common soldiers were usually recruited from among the very poorest, or from ethnic minorities (such as the Irish Catholics). Their chances of ascending the military ranks were negligible. The senior ranks were reserved for dukes, princes and kings. But why only for dukes, and not for duchesses?

The French Empire in Africa was established and defended by the sweat and blood of Senegalese, Algerians and working-class Frenchmen. The percentage of well-born Frenchmen within the ranks was negligible. Yet the percentage of well-born Frenchmen within the small elite that led the French army, ruled the empire and enjoyed its fruits was very high. Why just Frenchmen, and not French women?

In China there was a long tradition of subjugating the army to the civilian bureaucracy, so mandarins who had never held a sword often ran the wars. 'You do not waste good iron to make nails,' went a common Chinese saying, meaning that really talented people join the civil bureaucracy, not the army. Why, then, were all of these mandarins men?

One can't reasonably argue that their physical weakness or low testosterone levels prevented women from being successful mandarins, generals and politicians. In order to manage a war, you surely need stamina, but not much physical strength or aggressiveness. Wars are not a pub brawl. They are very complex projects that require an extraordinary degree of organisation, cooperation and appeasement. The ability to maintain peace at home, acquire allies abroad, and understand what goes through the minds of other people (particularly your enemies) is usually the key to victory. Hence an aggressive brute is often the worst choice to run a war. Much better is a cooperative person who knows how to appease, how to manipulate and how to see things from different perspectives. This is the stuff empire-builders are made of. The militarily incompetent Augustus succeeded in establishing a stable imperial regime, achieving something that eluded both Julius Caesar and Alexander the Great, who were much better generals. Both his admiring contemporaries and modern historians often attribute this feat to his virtue of *clementia* – mildness and clemency.

Women are often stereotyped as better manipulators and appeasers than men, and are famed for their superior ability to see things from the perspective of others. If there's any truth in these stereotypes, then women should have made excellent politicians and empire-builders, leaving the dirty work on the battlefields to testosterone-charged but simple-minded machos. Popular myths notwithstanding, this rarely happened in the real world. It is not at all clear why not.

Patriarchal Genes

A third type of biological explanation gives less importance to brute force and violence, and suggests that through millions of years of evolution, men and women evolved different survival and reproduction strategies. As men competed against each other for the opportunity to impregnate fertile women, an individual's chances of reproduction depended above all on his ability to outperform and defeat other men. As time went by, the masculine genes that made it to the next generation were those belonging to the most ambitious, aggressive and competitive men.

A woman, on the other hand, had no problem finding a man willing to impregnate her. However, if she wanted her children to provide her with grandchildren, she needed to carry them in her womb for nine arduous months, and then nurture them for years. During that time she had fewer opportunities to obtain food, and required a lot of help. She needed a man. In order to ensure her own survival and the survival of her children, the woman had little choice but to agree to whatever conditions the man stipulated so that he would stick around and share some of the burden. As time went by, the feminine genes that made it to the next generation belonged to women who were submissive caretakers. Women who spent too much time fighting for power did not leave any of those powerful genes for future generations.

The result of these different survival strategies – so the theory goes – is that men have been programmed to be ambitious and competitive, and to excel in politics and business, whereas women have tended to move out of the way and dedicate their lives to raising children.

But this approach also seems to be belied by the empirical evidence. Particularly problematic is the assumption that women's dependence on external help made them dependent on men, rather than on other women, and that male competitiveness made men socially dominant. There are many species of animals, such as elephants and bonobo chimpanzees, in which the dynamics between dependent females and competitive males results in a *matriarchal* society. Since females need external help, they are obliged to develop their social skills and learn how to cooperate and appease. They construct all-female social networks that help each member raise her children. Males, meanwhile, spend their time fighting and competing. Their social skills and social bonds remain underdeveloped. Bonobo and elephant societies are controlled by strong networks of cooperative females, while the self-centred and uncooperative males are pushed to the sidelines. Though bonobo females are weaker on average than the males, the females often gang up to beat males who overstep their limits.

If this is possible among bonobos and elephants, why not among *Homo sapiens*? Sapiens are relatively weak animals, whose advantage rests in their ability to cooperate in large numbers. If so, we should expect that dependent women, even if they are dependent on men, would use their superior social skills to cooperate among themselves, while outmanoeuvring and manipulating the aggressive, autonomous and self-centred men.

How did it happen that in the one species whose success depends above all on cooperation, individuals who are supposedly less cooperative (men) control individuals who are supposedly more cooperative (women)? At present, we have no good answer. Maybe the common assumptions are just wrong. Maybe males of the species *Homo sapiens* are characterised not by physical strength, aggressiveness and competitiveness, but rather by superior social skills and a greater tendency to cooperate. We just don't know.

What we do know, however, is that during the last century gender roles have undergone a tremendous revolution. More and more societies today not only give men and women equal legal status, political rights and economic opportunities, but also completely rethink their most basic conceptions of gender and sexuality. Though the gender gap is still significant, events have been moving at a breathtaking speed. At the beginning of the twentieth century the idea of giving voting rights to women was generally seen in the USA as outrageous; the prospect of a female cabinet secretary or Supreme Court justice was simply ridiculous; whereas homosexuality was such a taboo subject that it could not even be openly discussed. At the beginning of the twenty-first century women's voting rights are taken for granted; female cabinet secretaries are hardly a cause for comment; and in 2013 five US Supreme Court justices, three of them women, decided in favour of legalising same-sex marriages (overruling the objections of four male justices).

These dramatic changes are precisely what makes the history of gender so bewildering. If, as is being demonstrated today so clearly, the patriarchal system has been based on unfounded myths rather than on biological facts, what accounts for the universality and stability of this system?

Part Three

The Unification of Humankind

24. Pilgrims circling the Ka'aba in Mecca.

9

The Arrow of History

AFTER THE AGRICULTURAL REVOLUTION, human societies grew ever larger and more complex, while the imagined constructs sustaining the social order also became more elaborate. Myths and fictions accustomed people, nearly from the moment of birth, to think in certain ways, to behave in accordance with certain standards, to want certain things, and to observe certain rules. They thereby created artificial instincts that enabled millions of strangers to cooperate effectively. This network of artificial instincts is called 'culture'.

During the first half of the twentieth century, scholars taught that every culture was complete and harmonious, possessing an unchanging essence that defined it for all time. Each human group had its own world view and system of social, legal and political arrangements that ran as smoothly as the planets going around the sun. In this view, cultures left to their own devices did not change. They just kept going at the same pace and in the same direction. Only a force applied from outside could change them. Anthropologists, historians and politicians thus referred to 'Samoan Culture' or 'Tasmanian Culture' as if the same beliefs, norms and values had characterised Samoans and Tasmanians from time immemorial.

Today, most scholars of culture have concluded that the opposite is true. Every culture has its typical beliefs, norms and values, but these are in constant flux. The culture may transform itself in response to changes in its environment or through interaction with neighbouring cultures. But cultures also undergo transitions due to their own internal dynamics. Even a completely isolated culture existing in an ecologically stable environment cannot avoid change. Unlike the laws of physics, which are free of inconsistencies, every man-made order is packed with internal contradictions. Cultures are constantly trying to reconcile these contradictions, and this process fuels change.

For instance, in medieval Europe the nobility believed in both Christianity and chivalry. A typical nobleman went to church in the morning, and listened as the priest held forth on the lives of the saints. 'Vanity of vanities,' said the priest, 'all is vanity. Riches, lust and honour are dangerous temptations. You must rise above them, and follow in Christ's footsteps. Be meek like Him, avoid violence and extravagance, and if attacked – just turn the other cheek.' Returning home in a meek and pensive mood, the nobleman would change into his best silks and go to a banquet in his lord's castle. There the wine flowed like water, the minstrel sang of Lancelot and Guinevere, and the guests exchanged dirty jokes and bloody war tales. 'It is better to die,' declared the barons, 'than to live with shame. If someone questions your honour, only blood can wipe out the insult. And what is better in life than to see your enemies flee before you, and their pretty daughters tremble at your feet?'

The contradiction was never fully resolved. But as the European nobility, clergy and commoners grappled with it, their culture changed. One attempt to figure it out produced the Crusades. On crusade, knights could demonstrate their military prowess and their religious devotion at one stroke. The same contradiction produced military orders such as the Templars and Hospitallers, who tried to mesh Christian and chivalric ideals even more tightly. It was also responsible for a large part of medieval art and literature, such as the tales of King Arthur and the Holy Grail. What was Camelot but an attempt to prove that a good knight can and should be a good Christian, and that good Christians make the best knights?

Another example is the modern political order. Ever since the French Revolution, people throughout the world have gradually come to see both social equality and individual freedom as fundamental values. Yet the two values contradict each other. Equality can be ensured only by curtailing the freedoms of those who are better off. Guaranteeing that every individual will be free to do as he wishes inevitably short-changes equality. The entire political history of the world since 1789 can be seen as a series of attempts to reconcile this contradiction.

Anyone who has read a novel by Charles Dickens knows that the liberal regimes of nineteenth-century Europe gave priority to individual freedom even if it meant throwing insolvent poor families in prison and giving orphans little choice but to join schools for pickpockets. Anyone who has read a novel by Alexander Solzhenitsyn knows how Communism's egalitarian ideal produced brutal tyrannies that tried to control every aspect of daily life.

Contemporary American politics also revolve around this contradiction. Democrats want a more equitable society, even if it means raising taxes to fund programmes to help the poor, elderly and infirm. But that infringes on the freedom of individuals to spend their money as they wish. Why should the government force me to buy health insurance if I prefer using the money to put my kids through college? Republicans, on the other hand, want to maximise individual freedom, even if it means that the income gap between rich and poor will grow wider and that many Americans will not be able to afford health care.

Just as medieval culture did not manage to square chivalry with Christianity, so the modern world fails to square liberty with equality. But this is no defect. Such contradictions are an inseparable part of every human culture. In fact, they are the engines of cultural development, responsible for the creativity and dynamism of our species. Discord in our thoughts, ideas and values compels us to think, re-evaluate and criticise. Consistency is the playground of dull minds. Can you name a single great work of art which is not about conflict?

If tensions, conflicts and irresolvable dilemmas are the spice of every culture, a human being who belongs to any particular culture must hold contradictory beliefs and be riven by incompatible values. It's such an essential feature of any culture that it even has a name: cognitive dissonance. Cognitive dissonance is often considered a failure of the human psyche. In fact, it is a vital asset. Had people been unable to hold contradictory beliefs and values, it would probably have been impossible to establish and maintain any human culture.

If, say, a Christian really wants to understand the Muslims who attend that mosque down the street, he shouldn't look for a pristine set of values that every Muslim holds dear. Rather, he should enquire into the catch-22s of Muslim culture, those places where rules are at war and standards scuffle. It's at the very spot where the Muslims teeter between two imperatives that you'll understand them best.

The Spy Satellite

Human cultures are in constant flux. Is this flux completely random, or does it have some overall pattern? In other words, does history have a direction?

The answer is yes. Over the millennia, small, simple cultures gradually coalesce into bigger and more complex civilisations, so that the world contains fewer and fewer mega-cultures, each of which is bigger and more complex. This is of course a very crude generalisation, true only at the macro level. At the micro level, it seems that for every group of cultures that coalesces into a mega-culture, there's a mega-culture that breaks up into pieces. The Mongol Empire expanded to dominate a huge swathe of Asia and even parts of Europe, only to shatter into fragments. Christianity converted hundreds of millions of people at the same time that it splintered into innumerable sects. The Latin language spread through western and central Europe, then split into local dialects that themselves eventually became national languages. But these break-ups are temporary reversals in an inexorable trend towards unity.

Perceiving the direction of history is really a question of vantage point. When we adopt the proverbial bird's-eye view of history, which examines developments in terms of decades or centuries, it's hard to say whether history moves in the direction of unity or of diversity. However, to understand long-term processes the bird's-eye view is too myopic. We would do better to adopt instead the viewpoint of a cosmic spy satellite, which scans millennia rather than centuries. From such a vantage point it becomes crystal clear that history is moving relentlessly towards unity. The sectioning of Christianity and the collapse of the Mongol Empire are just speed bumps on history's highway.

The best way to appreciate the general direction of history is to count the number of separate human worlds that coexisted at any given moment on planet Earth. Today, we are used to thinking about the whole planet as a single unit, but for most of history, earth was in fact an entire galaxy of isolated human worlds.

Consider Tasmania, a medium-sized island south of Australia. It was cut off from the Australian mainland in about 10,000 BC as the end of the Ice Age caused the sea level to rise. A few thousand hunter-gatherers were left on the island, and had no contact with any other humans until the arrival of the Europeans in the nineteenth century. For 12,000 years, nobody else knew the Tasmanians were there, and they didn't know that there was anyone else in the world. They had their wars, political struggles, social oscillations and cultural developments. Yet as far as the emperors of China or the rulers of Mesopotamia were concerned, Tasmania could just as well have been located on one of Jupiter's moons. The Tasmanians lived in a world of their own.

America and Europe, too, were separate worlds for most of their histories. In AD 378, the Roman emperor Valens was defeated and killed by the Goths at the battle of Adrianople. In the same year, King Chak Tok Ich'aak of Tikal was defeated and killed by the army of Teotihuacan. (Tikal was an important Mayan city state, while Teotihuacan was then the largest city in America, with almost 250,000 inhabitants – of the same order of magnitude as its contemporary, Rome.) There was absolutely no connection between the defeat of Rome and the rise of Teotihuacan. Rome might just as well have been located on Mars, and Teotihuacan on Venus.

How many different human worlds coexisted on earth? Around 10,000 BC our planet contained many thousands of them. By 2000 BC, their numbers had dwindled to the hundreds, or at most a few thousand. By AD 1450, their numbers had declined even more drastically. At that time, just prior to the age of European exploration, earth still contained a significant number of dwarf worlds such as Tasmania. But close to 90 per cent of humans lived in a single mega-world: the world of Afro-Asia. Most of Asia, most of Europe, and most of Africa (including substantial chunks of sub-Saharan Africa) were already connected by significant cultural, political and economic ties.

Most of the remaining tenth of the world's human population was divided between four worlds of considerable size and complexity:

1. The Mesoamerican World, which encompassed most of Central America and parts of North America.

2. The Andean World, which encompassed most of western South America.

3. The Australian World, which encompassed the continent of Australia.

4. The Oceanic World, which encompassed most of the islands of the south-western Pacific Ocean, from Hawaii to New Zealand.

Over the next 300 years, the Afro-Asian giant swallowed up all the other worlds. It consumed the Mesoamerican World in 1521, when the Spanish conquered the Aztec Empire. It took its first bite out of the Oceanic World at the same time, during Ferdinand Magellan's circumnavigation of the globe, and soon after that completed its conquest. The Andean World collapsed in 1532, when Spanish conquistadors crushed the Inca Empire. The first European landed on the Australian continent in 1606, and that pristine world came to an end when British colonisation began in earnest in 1788. Fifteen years later the Britons established their first settlement in Tasmania, thus bringing the last autonomous human world into the Afro-Asian sphere of influence.

It took the Afro-Asian giant several centuries to digest all that it had swallowed, but the process was irreversible. Today almost all humans share the same geopolitical system (the entire planet is divided into internationally recognised states); the same economic system (capitalist market forces shape even the remotest corners of the globe); the same legal system (human rights and international law are valid everywhere, at least theoretically); and the same scientific system (experts in Iran, Israel, Australia and Argentina have exactly the same views about the structure of atoms or the treatment of tuberculosis).

The single global culture is not homogeneous. Just as a single organic body contains many different kinds of organs and cells, so our single global culture contains many different types of lifestyles and people, from New York stockbrokers to Afghan shepherds. Yet they are all closely connected and they influence one another in myriad ways. They still argue and fight, but they argue using the same concepts and fight using the same weapons. A real 'clash of civilisations' is like the proverbial dialogue of the deaf. Nobody can grasp what the other is saying. Today when Iran and the United States rattle swords at one another, they both speak the language of nation states, capitalist economies, international rights and nuclear physics.

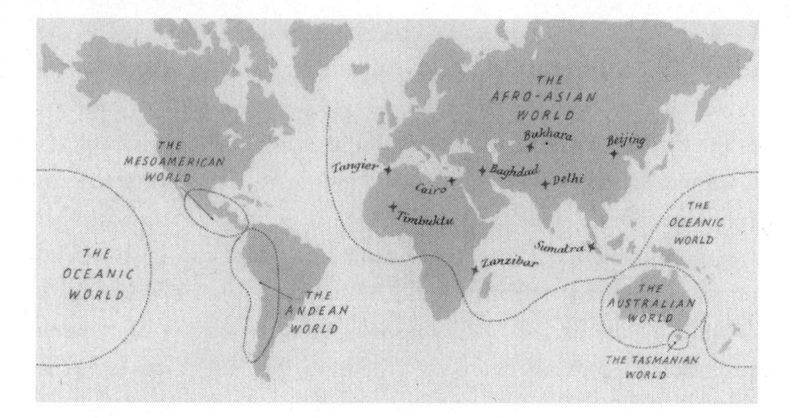

Map 3. Earth in AD 1450. The named locations within the Afro-Asian World were places visited by the fourteenth-century Muslim traveller Ibn Battuta. A native of Tangier, in Morocco, Ibn Battuta visited Timbuktu, Zanzibar, southern Russia, Central Asia, India, China and Indonesia. His travels illustrate the unity of Afro-Asia on the eve of the modern era.

We still talk a lot about 'authentic' cultures, but if by 'authentic' we mean something that developed independently, and that consists of ancient local traditions free of external influences, then there are no authentic cultures left on earth. Over the last few centuries, all cultures were changed almost beyond recognition by a flood of global influences.

One of the most interesting examples of this globalisation is 'ethnic' cuisine. In an Italian restaurant we expect to find spaghetti in tomato sauce; in Polish and Irish restaurants lots of potatoes; in an Argentinian restaurant we can choose between dozens of kinds of beefsteaks; in an Indian restaurant hot chillies are incorporated into just about everything; and the highlight at any Swiss café is thick hot chocolate under an alp of whipped cream. But none of these foods is native to those nations. Tomatoes, chilli peppers and cocoa are all Mexican in origin; they reached Europe and Asia only after the Spaniards conquered Mexico. Julius Caesar and Dante Alighieri never twirled tomato-drenched spaghetti on their forks (even forks hadn't been invented yet), William Tell never tasted chocolate, and Buddha never spiced up his food with chilli. Potatoes reached Poland and Ireland no more than 400 years ago. The only steak you could obtain in Argentina in 1492 was from a llama.

Hollywood films have perpetuated an image of the Plains Indians as brave horsemen, courageously charging the wagons of European pioneers to protect the customs of their ancestors. However, these Native American horsemen were not the defenders of some ancient, authentic culture. Instead, they were the product of a major military and political revolution that swept the plains of western North America in the seventeenth and eighteenth centuries, a consequence of the arrival of European horses. In 1492 there were no horses in America. The culture of the nineteenth-century Sioux and Apache has many appealing features, but it was a modern culture – a result of global forces – much more than 'authentic'.

The Global Vision

From a practical perspective, the most important stage in the process of global unification occurred in the last few centuries, when empires grew and trade intensified. Ever-tightening links were formed between the people of Afro-Asia, America, Australia and Oceania. Thus Mexican chilli peppers made it into Indian food and Spanish cattle began grazing in Argentina. Yet from an ideological perspective, an even more important development occurred during the first millennium BC, when the idea of a universal order took root. For thousands of years previously, history was already moving slowly in the direction of global unity, but the idea of a universal order governing the entire world was still alien to most people.

25. Sioux chiefs (1905). Neither the Sioux nor any other Great Plains tribe had horses prior to 1492.

Homo sapiens evolved to think of people as divided into 'us' and 'them'. 'Us' was the group immediately around you, whoever you were, and 'them' was everyone else. In fact, no social animal is ever guided by the interests of the entire species to which it belongs. No chimpanzee cares about the interests of the chimpanzee species, no snail will lift a tentacle for the global snail community, no lion alpha male makes a bid for becoming the king of all lions, and at the entrance of no beehive can one find the slogan: 'Worker bees of the world – unite!'

But beginning with the Cognitive Revolution, *Homo sapiens* became more and more exceptional in this respect. People began to cooperate on a regular basis with complete strangers, whom they imagined as 'brothers' or 'friends'. Yet this brotherhood was not universal. Somewhere in the next valley, or beyond the mountain range, one could still sense 'them'. When the first pharaoh, Menes, united Egypt around 3000 BC, it was clear to the Egyptians that Egypt had a border, and beyond the border lurked 'barbarians'. The barbarians were alien, threatening, and interesting only to the extent that they had land or natural resources that the Egyptians wanted. All the imagined orders people created tended to ignore a substantial part of humankind.

The first millennium BC witnessed the appearance of three potentially universal orders, whose devotees could for the first time imagine the entire world and the entire human race as a single unit governed by a single set of laws. Everyone was 'us', at least potentially. There was no longer 'them'. The first universal order to appear was economic: the monetary order. The second universal order was political: the imperial order. The third universal order was religious: the order of universal religions such as Buddhism, Christianity and Islam.

Merchants, conquerors and prophets were the first people who managed to transcend the binary evolutionary division, 'us vs them', and to foresee the potential unity of humankind. For the merchants, the entire world was a single market and all humans were potential customers. They tried to establish an economic order that would apply to all, everywhere. For the conquerors, the entire world was a single empire and all humans were potential subjects. And for the prophets, the entire world held a single truth and all humans were potential believers. They too tried to establish an order that would be applicable for everyone everywhere.

During the last three millennia, people made more and more ambitious attempts to realise that global vision. The next three chapters discuss how money, empires and universal religions spread, and how they laid the foundation of the united world of today. We begin with the story of the greatest conqueror in history, a conqueror possessed of extreme tolerance and adaptability, which consequently managed to gain the allegiance of all people. This conqueror is money. People who do not believe in the same god or obey the same king are more than willing to use the same money. Osama bin Laden, for all his hatred of American culture, American religion, and American politics, was very fond of American dollars. How did money succeed where gods and kings failed?

10

The Scent of Money

IN 1519 HERNÁN CORTÉS AND HIS CONquistadors invaded Mexico, hitherto an isolated human world. The Aztecs, as the people who lived there are known to posterity, quickly noticed that the aliens showed an extraordinary interest in a certain yellow metal. In fact, the aliens seemed to never stop talking about it. The natives were not unfamiliar with gold – it was pretty and easy to work, so they used it to make jewellery and statues, and they occasionally used gold dust as a medium of exchange. But when an Aztec wanted to buy something, he generally paid in cocoa beans or bolts of cloth. The Spanish obsession with gold thus seemed inexplicable. What was so important about a metal that could not be eaten, drunk or woven, and was too soft to use for tools or weapons? When the natives questioned Cortés as to why the Spaniards had such a passion for gold, the conquistador answered, 'Because I and my companions suffer from a disease of the heart which can be cured only with gold.'[1]

In the Afro-Asian world from which the Spaniards came, the obsession for gold was indeed an epidemic. Even the bitterest of enemies lusted after the same useless yellow metal. Three centuries before the conquest of Mexico, the ancestors of Cortés and his army waged a bloody war of religion against the Muslim kingdoms in Iberia and North Africa. The followers of Christ and the followers of Allah killed each other by the thousands, devastated fields and orchards, and turned prosperous cities into smouldering ruins – all for the greater glory of Christ or Allah.

As the Christians gradually gained the upper hand, they marked their victories not only by destroying mosques and building churches, but also by issuing new gold and silver coins bearing the sign of the cross and thanking God for His help in combating the infidels. Yet alongside the new currency, the victors minted another type of coin, called the millares, which carried a somewhat different message. These square coins made by the Christian conquerors were emblazoned with flowing Arabic script that declared: 'There is no god except Allah, and Muhammad is Allah's messenger.' Even the Catholic bishops of Melgueil and Agde issued these faithful copies of popular Muslim coins, and God-fearing Christians happily used them.[2]

Tolerance flourished on the other side of the hill too. Muslim merchants in North Africa conducted business using Christian coins such as the Florentine florin, the Venetian ducat and the Neapolitan gigliato. Even Muslim rulers who called for jihad against the infidel Christians were glad to receive taxes in coins that invoked Christ and His Virgin Mother.[3]

How Much Is It?

Hunter-gatherers had no money. Each band hunted, gathered and manufactured almost everything it required, from meat to medicine, from sandals to sorcery. Different band members may have specialised in different tasks, but they shared their goods and services through an economy of favours and obligations. A piece of meat given for free would carry with it the assumption of reciprocity – say, free medical assistance. The band was economically independent; only a few rare items that could not be found locally – seashells, pigments, obsidian and the like – had to be obtained from strangers.

This could usually be done by simple barter: 'We'll give you pretty seashells, and you'll give us high-quality flint.'

Little of this changed with the onset of the Agricultural Revolution. Most people continued to live in small, intimate communities. Much like a hunter-gatherer band, each village was a self-sufficient economic unit, maintained by mutual favours and obligations plus a little barter with outsiders. One villager may have been particularly adept at making shoes, another at dispensing medical care, so villagers knew where to turn when barefoot or sick. But villages were small and their economies limited, so there could be no full-time shoemakers and doctors.

The rise of cities and kingdoms and the improvement in transport infrastructure brought about new opportunities for specialisation. Densely populated cities provided full-time employment not just for professional shoemakers and doctors, but also for carpenters, priests, soldiers and lawyers. Villages that gained a reputation for producing really good wine, olive oil or ceramics discovered that it was worth their while to specialise nearly exclusively in that product and trade it with other settlements for all the other goods they needed. This made a lot of sense. Climates and soils differ, so why drink mediocre wine from your backyard if you can buy a smoother variety from a place whose soil and climate is much better suited to grape vines? If the clay in your backyard makes stronger and prettier pots, then you can make an exchange. Furthermore, full-time specialist vintners and potters, not to mention doctors and lawyers, can hone their expertise to the benefit of all. But specialisation created a problem – how do you manage the exchange of goods between the specialists?

An economy of favours and obligations doesn't work when large numbers of strangers try to cooperate. It's one thing to provide free assistance to a sister or a neighbour, a very different thing to take care of foreigners who might never reciprocate the favour. One can fall back on barter. But barter is effective only when exchanging a limited range of products. It cannot form the basis for a complex economy.[4]

In order to understand the limitations of barter, imagine that you own an apple orchard in the hill country that produces the crispest, sweetest apples in the entire province. You work so hard in your orchard that your shoes wear out. So you harness up your donkey cart and head to the market town down by the river. Your neighbour told you that a shoemaker on the south end of the marketplace made him a really sturdy pair of boots that's lasted him through five seasons. You find the shoemaker's shop and offer to barter some of your apples in exchange for the shoes you need.

The shoemaker hesitates. How many apples should he ask for in payment? Every day he encounters dozens of customers, a few of whom bring along sacks of apples, while others carry wheat, goats or cloth – all of varying quality. Still others offer their expertise in petitioning the king or curing backaches. The last time the shoemaker exchanged shoes for apples was three months ago, and back then he asked for three sacks of apples. Or was it four? But come to think of it, those apples were sour valley apples, rather than prime hill apples. On the other hand, on that previous occasion, the apples were given in exchange for small women's shoes. This fellow is asking for man-size boots. Besides, in recent weeks a disease has decimated the flocks around town, and skins are becoming scarce. The tanners are starting to demand twice as many finished shoes in exchange for the same quantity of leather. Shouldn't that be taken into consideration?

In a barter economy, every day the shoemaker and the apple grower will have to learn anew the relative prices of dozens of commodities. If 100 different commodities are traded in the market, then buyers and sellers will have to know 4,950 different exchange rates. And if 1,000 different commodities are traded, buyers and sellers must juggle 499,500 different exchange rates![5] How do you figure it out?

It gets worse. Even if you manage to calculate how many apples equal one pair of shoes, barter is not always possible. After all, a trade requires that each side want what the other has to offer. What happens if the shoemaker doesn't like apples and if, at the moment in question, what he really wants is a divorce? True, the farmer could look for a lawyer who likes apples and set up a three-way deal. But what if the lawyer is full up on apples but really needs a haircut?

Some societies tried to solve the problem by establishing a central barter system that collected products from specialist growers and manufacturers and distributed them to those who needed them. The largest and most famous such experiment was conducted in the Soviet Union, and it failed miserably. 'Everyone would work according to their abilities, and receive according to their needs' turned in practice into 'everyone would work as little as they can get away with, and receive as much as they could grab'. More moderate and more successful experiments were made on other occasions, for example in the Inca Empire. Yet most societies found a more easy way to connect large numbers of experts – they developed money.

Shells and Cigarettes

Money was created many times in many places. Its development required no technological breakthroughs – it was a purely mental revolution. It involved the creation of a new inter-subjective reality that exists solely in people's shared imagination.

Money is not coins and banknotes. Money is anything that people are willing to use in order to represent systematically the value of other things for the purpose of exchanging goods and services. Money enables people to compare quickly and easily the value of different commodities (such as apples, shoes and divorces), to easily exchange one thing for another, and to store wealth conveniently. There have been many types of money. The most familiar is the coin, which is a standardised piece of imprinted metal. Yet money

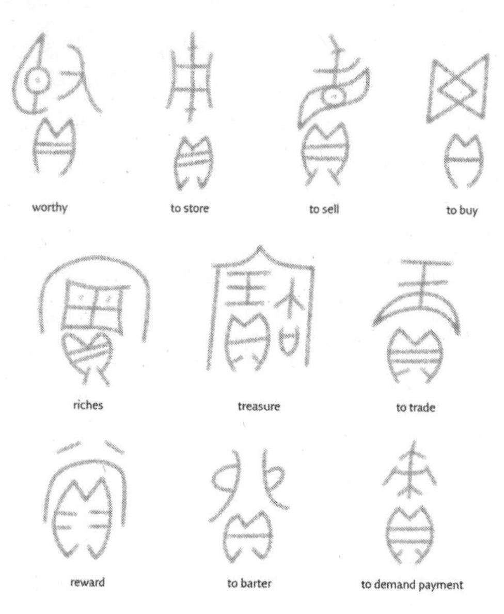

26. In ancient Chinese script the cowry-shell sign represented money, in words such as 'to sell' or 'reward'.

existed long before the invention of coinage, and cultures have prospered using other things as currency, such as shells, cattle, skins, salt, grain, beads, cloth and promissory notes. Cowry shells were used as money for about 4,000 years all over Africa, South Asia, East Asia and Oceania. Taxes could still be paid in cowry shells in British Uganda in the early twentieth century.

In modern prisons and POW camps, cigarettes have often served as money. Even non-smoking prisoners have been willing to accept cigarettes in payment, and to calculate the value of all other goods and services in cigarettes. One Auschwitz survivor described the cigarette currency used in the camp: 'We had our own currency, whose value no one questioned: the cigarette. The price of every article was stated in cigarettes . . . In "normal" times, that is, when the candidates to the gas chambers were coming in at a regular pace, a loaf of bread cost twelve cigarettes; a 300-gram package of margarine, thirty; a watch, eighty to 200; a litre of alcohol, 400 cigarettes!'[6]

In fact, even today coins and banknotes are a rare form of money. The sum total of money in the world is about $60 trillion, yet the sum total of coins and banknotes is less than $6 trillion.[7] More than 90 per cent of all money – more than $50 trillion appearing in our accounts – exists only on computer servers. Accordingly, most business transactions are executed by moving electronic data from one computer file to another, without any exchange of physical cash. Only a criminal buys a house, for example, by handing over a suitcase full of banknotes. As long as people are willing to trade goods and services in exchange for electronic data, it's even better than shiny coins and crisp banknotes – lighter, less bulky, and easier to keep track of.

For complex commercial systems to function, some kind of money is indispensable. A shoemaker in a money economy needs to know only the prices charged for various kinds of shoes – there is no need to memorise the exchange rates between shoes and apples or goats. Money also frees apple experts from the need to search out apple-craving shoemakers, because everyone always wants money. This is perhaps its most basic quality. Everyone always wants money because everyone else also always wants money, which means you can exchange money for whatever you want or need. The shoemaker will always be happy to take your money, because no matter what he really wants – apples, goats or a divorce – he can get it in exchange for money.

Money is thus a universal medium of exchange that enables people to convert almost everything into almost anything else. Brawn gets converted to brain when a discharged soldier finances his college tuition with his military benefits. Land gets converted into loyalty when a baron sells property to support his retainers. Health is converted to justice when a physician uses her fees to hire a lawyer – or bribe a judge. It is even possible to convert sex into salvation, as fifteenth-century prostitutes did when they slept with men for money, which they in turn used to buy indulgences from the Catholic Church.

Ideal types of money enable people not merely to turn one thing into another, but to store wealth as well. Many valuables cannot be stored – such as time or beauty. Some things can be stored only for a short time, such as strawberries. Other things are more durable, but take up a lot of space and require expensive facilities and care. Grain, for example, can be stored for years, but to do so you need to build huge storehouses and guard against rats, mould, water, fire and thieves. Money, whether paper, computer bits or cowry shells, solves these problems. Cowry shells don't rot, are unpalatable to rats, can survive fires and are compact enough to be locked up in a safe.

In order to use wealth it is not enough just to store it. It often needs to be transported from place to place. Some forms of wealth, such as real estate, cannot be transported at all. Commodities such as wheat and rice can be transported only with difficulty. Imagine a wealthy farmer living in a moneyless land who emigrates to a distant province. His wealth consists mainly of his house and rice paddies. The farmer cannot take with him the house or the paddies. He might exchange them for tons of rice, but it would be very burdensome and expensive to transport all that rice. Money solves these problems. The farmer can sell his property in exchange for a sack of cowry shells, which he can easily carry wherever he goes.

Because money can convert, store and transport wealth easily and cheaply, it made a vital contribution to the appearance of complex commercial networks and dynamic markets. Without money, commercial networks and markets would have been doomed to remain very limited in their size, complexity and dynamism.

How Does Money Work?

Cowry shells and dollars have value only in our common imagination. Their worth is not inherent in the chemical structure of the shells and paper, or their colour, or their shape. In other words, money isn't a material reality – it is a psychological construct. It works by converting matter into mind. But why does it succeed? Why should anyone be willing to exchange a fertile rice paddy for a handful of useless cowry shells? Why are you willing to flip hamburgers, sell health insurance or babysit three obnoxious brats when all you get for your exertions is a few pieces of coloured paper?

People are willing to do such things when they trust the figments of their collective imagination. Trust is the raw material from which all types of money are minted. When a wealthy farmer sold his possessions for a sack of cowry shells and travelled with them to another province, he trusted that upon reaching his destination other people would be willing to sell him rice, houses and fields in exchange for the shells. Money is accordingly a system of mutual trust, and not just any system of mutual trust: *money is the most universal and most efficient system of mutual trust ever devised.*

What created this trust was a very complex and long-term network of political, social and economic relations. Why do I believe in the cowry shell or gold coin or dollar bill? Because my neighbours believe in them. And my neighbours believe in them because I believe in them. And we all believe in them because our king believes in them and demands them in taxes, and because our priest believes in them and demands them in tithes. Take a dollar bill and look at it carefully. You will see that it is simply a colourful piece of paper with the signature of the US secretary of the treasury on one side, and the slogan 'In God we trust' on the other. We accept the dollar in payment, because we trust in God and the US secretary of the treasury. The crucial role of trust explains why our financial systems are so tightly bound up with our political, social and ideological systems, why financial crises are often triggered by political developments, and why the stock market can rise or fall depending on the way traders feel on a particular morning.

Initially, when the first versions of money were created, people didn't have this sort of trust, so it was necessary to define as 'money' things that had real intrinsic value. History's first known money – Sumerian barley money – is a good example. It appeared in Sumer around 3000 BC, at the same time and place, and under the same circumstances, in which writing appeared. Just as writing developed to answer the needs of intensifying administrative activities, so barley money developed to answer the needs of intensifying economic activities.

Barley money was simply barley – fixed amounts of barley grains used as a universal measure for evaluating and exchanging all other goods and services. The most common measurement was the sila, equivalent to roughly one litre. Standardised bowls, each capable of containing one sila, were mass-produced so that whenever people needed to buy or sell anything, it was easy to measure the necessary amounts of barley. Salaries, too, were set and paid in silas of barley. A male labourer earned sixty silas a month, a female labourer thirty silas. A foreman could earn between 1,200 and 5,000 silas. Not even the most ravenous foreman could eat 5,000 litres of barley a month, but he could use the silas he didn't eat to buy all sorts of other commodities – oil, goats, slaves, and something else to eat besides barley.[8]

Even though barley has intrinsic value, it was not easy to convince people to use it as *money* rather than as just another commodity. In order to understand why, just think what would happen if you took a sack full of barley to your local shopping centre, and tried to buy a shirt or a pizza. The vendors would probably call security. Still, it was somewhat easier to build trust in barley as the first type of money, because barley has an inherent biological value. Humans can eat it. On the other hand, it was difficult to store and transport barley. The real breakthrough in monetary history occurred when people gained trust in money that lacked inherent value, but was easier to store and transport. Such money appeared in ancient Mesopotamia in the middle of the third millennium BC. This was the silver shekel.

The silver shekel was not a coin, but rather 8.33 grams of silver.

When Hammurabi's Code declared that a superior man who killed a slave woman must pay her owner twenty silver shekels, it meant that he had to pay 166 grams of silver, not twenty coins. Most monetary terms in the Old Testament are given in terms of silver rather than coins. Joseph's brothers sold him to the Ishmaelites for twenty silver shekels, or rather 166 grams of silver (the same price as a slave woman – he was a youth, after all).

Unlike the barley sila, the silver shekel had no inherent value. You cannot eat, drink or clothe yourself in silver, and it's too soft for making useful tools – ploughshares or swords of silver would crumple almost as fast as ones made out of aluminium foil. When they are used for anything, silver and gold are made into jewellery, crowns and other status symbols – luxury goods that members of a particular culture identify with high social status. Their value is purely cultural.

Set weights of precious metals eventually gave birth to coins. The first coins in history were struck around 640 BC by King Alyattes of Lydia, in western Anatolia. These coins had a standardised weight of gold or silver, and were imprinted with an identification mark. The mark testified to two things. First, it indicated how much precious metal the coin contained. Second, it identified the authority that issued the coin and that guaranteed its contents. Almost all coins in use today are descendants of the Lydian coins.

Coins had two important advantages over unmarked metal ingots. First, the latter had to be weighed for every transaction. Second, weighing the ingot is not enough. How does the shoemaker know that the silver ingot I put down for my boots is really made of pure silver, and not of lead covered on the outside by a thin silver coating? Coins help solve these problems. The mark imprinted on them testifies to their exact value, so the shoemaker doesn't have to keep a scale on his cash register. More importantly, the mark on the coin is the signature of some political authority that guarantees the coin's value.

The shape and size of the mark varied tremendously throughout

27. One of the earliest coins in history, from Lydia of the seventh century BC.

history, but the message was always the same: 'I, the Great King So-And-So, give you my personal word that this metal disc contains exactly five grams of gold. If anyone dares counterfeit this coin, it means he is fabricating my own signature, which would be a blot on my reputation. I will punish such a crime with the utmost severity.' That's why counterfeiting money has always been considered a much more serious crime than other acts of deception. Counterfeiting is not just cheating – it's a breach of sovereignty, an act of subversion against the power, privileges and person of the king. The legal term is lese-majesty (violating majesty), and was typically punished by torture and death. As long as people trusted the power and integrity of the king, they trusted his coins. Total strangers could easily agree on the worth of a Roman denarius coin, because they trusted the power and integrity of the Roman emperor, whose name and picture adorned it.

In turn, the power of the emperor rested on the denarius. Just think how difficult it would have been to maintain the Roman Empire without coins – if the emperor had to raise taxes and pay salaries in barley and wheat. It would have been impossible to collect barley taxes in Syria, transport the funds to the central treasury in

Rome, and transport them again to Britain in order to pay the legions there. It would have been equally difficult to maintain the empire if the inhabitants of the city of Rome believed in gold coins, but the subject populations rejected this belief, putting their trust instead in cowry shells, ivory beads or rolls of cloth.

The Gospel of Gold

The trust in Rome's coins was so strong that even outside the empire's borders, people were happy to receive payment in denarii. In the first century AD, Roman coins were an accepted medium of exchange in the markets of India, even though the closest Roman legion was thousands of kilometres away. The Indians had such a strong confidence in the denarius and the image of the emperor that when local rulers struck coins of their own they closely imitated the denarius, down to the portrait of the Roman emperor! The name 'denarius' became a generic name for coins. Muslim caliphs Arabicised this name and issued 'dinars'. The dinar is still the official name of the currency in Jordan, Iraq, Serbia, Macedonia, Tunisia and several other countries.

As Lydian-style coinage was spreading from the Mediterranean to the Indian Ocean, China developed a slightly different monetary system, based on bronze coins and unmarked silver and gold ingots. Yet the two monetary systems had enough in common (especially the reliance on gold and silver) that close monetary and commercial relations were established between the Chinese zone and the Lydian zone. Muslim and European merchants and conquerors gradually spread the Lydian system and the gospel of gold to the far corners of the earth. By the late modern era the entire world was a single monetary zone, relying first on gold and silver, and later on a few trusted currencies such as the British pound and the American dollar.

The appearance of a single transnational and transcultural monetary zone laid the foundation for the unification of Afro-Asia, and eventually of the entire globe, into a single economic and political sphere. People continued to speak mutually incomprehensible languages, obey different rulers and worship distinct gods, but all believed in gold and silver and in gold and silver coins. Without this shared belief, global trading networks would have been virtually impossible. The gold and silver that sixteenth-century conquistadors found in America enabled European merchants to buy silk, porcelain and spices in East Asia, thereby moving the wheels of economic growth in both Europe and East Asia. Most of the gold and silver mined in Mexico and the Andes slipped through European fingers to find a welcome home in the purses of Chinese silk and porcelain manufacturers. What would have happened to the global economy if the Chinese hadn't suffered from the same 'disease of the heart' that afflicted Cortés and his companions – and had refused to accept payment in gold and silver?

Yet why should Chinese, Indians, Muslims and Spaniards – who belonged to very different cultures that failed to agree about much of anything – nevertheless share the belief in gold? Why didn't it happen that Spaniards believed in gold, while Muslims believed in barley, Indians in cowry shells, and Chinese in rolls of silk? Economists have a ready answer. Once trade connects two areas, the forces of supply and demand tend to equalise the prices of transportable goods. In order to understand why, consider a hypothetical case. Assume that when regular trade opened between India and the Mediterranean, Indians were uninterested in gold, so it was almost worthless. But in the Mediterranean, gold was a coveted status symbol, hence its value was high. What would happen next?

Merchants travelling between India and the Mediterranean would notice the difference in the value of gold. In order to make a profit, they would buy gold cheaply in India and sell it dearly in the Mediterranean. Consequently, the demand for gold in India would skyrocket, as would its value. At the same time the Mediterranean would experience an influx of gold, whose value would consequently drop. Within a short time the value of gold in India and the Mediterranean would be quite similar. The mere fact that Mediterranean people believed in gold would cause Indians to start believing in it as well. Even if Indians still had no real use for gold, the fact that Mediterranean people wanted it would be enough to make the Indians value it.

Similarly, the fact that another person believes in cowry shells, or dollars, or electronic data, is enough to strengthen our own belief in them, even if that person is otherwise hated, despised or ridiculed by us. Christians and Muslims who could not agree on religious beliefs could nevertheless agree on a monetary belief, because whereas religion asks us to believe in something, money asks us to believe that *other people believe in something.*

For thousands of years, philosophers, thinkers and prophets have besmirched money and called it the root of all evil. Be that as it may, money is also the apogee of human tolerance. Money is more open-minded than language, state laws, cultural codes, religious beliefs and social habits. Money is the only trust system created by humans that can bridge almost any cultural gap, and that does not discriminate on the basis of religion, gender, race, age or sexual orientation. Thanks to money, even people who don't know each other and don't trust each other can nevertheless cooperate effectively.

The Price of Money

Money is based on two universal principles:

a. Universal convertibility: with money as an alchemist, you can turn land into loyalty, justice into health, and violence into knowledge.

b. Universal trust: with money as a go-between, any two people can cooperate on any project.

These principles have enabled millions of strangers to cooperate effectively in trade and industry. But these seemingly benign principles have a dark side. When everything is convertible, and when trust depends on anonymous coins and cowry shells, it corrodes local traditions, intimate relations and human values, replacing them with the cold laws of supply and demand.

Human communities and families have always been based on belief in 'priceless' things, such as honour, loyalty, morality and love. These things lie outside the domain of the market, and they shouldn't be bought or sold for money. Even if the market offers a good price, certain things just aren't done. Parents mustn't sell their children into slavery; a devout Christian must not commit a mortal sin; a loyal knight must never betray his lord; and ancestral tribal lands should never be sold to foreigners.

Money has always tried to break through these barriers, like water seeping through cracks in a dam. Parents have been reduced to selling some of their children into slavery in order to buy food for the others. Devout Christians have murdered, stolen and cheated – and later used their spoils to buy forgiveness from the Church. Ambitious knights auctioned their allegiance to the highest bidder, while securing the loyalty of their own followers by cash payments. Tribal lands were sold to foreigners from the other side of the world in order to purchase an entry ticket into the global economy.

Money has an even darker side. For although money builds universal trust between strangers, this trust is invested not in humans, communities or sacred values, but in money itself and in the impersonal systems that back it. We do not trust the stranger, or the next-door neighbour – we trust the coin they hold. If they run out of coins, we run out of trust. As money brings down the dams of community, religion and state, the world is in danger of becoming one big and rather heartless marketplace.

Hence the economic history of humankind is a delicate dance. People rely on money to facilitate cooperation with strangers, but they're afraid it will corrupt human values and intimate relations. With one hand people willingly destroy the communal dams that held at bay the movement of money and commerce for so long. Yet with the other hand they build new dams to protect society, religion and the environment from enslavement to market forces.

It is common nowadays to believe that the market always prevails, and that the dams erected by kings, priests and communities cannot long hold back the tides of money. This is naive. Brutal warriors, religious fanatics and concerned citizens have repeatedly managed to trounce calculating merchants, and even to reshape the economy. It is therefore impossible to understand the unification of humankind as a purely economic process. In order to understand how thousands of isolated cultures coalesced over time to form the global village of today, we must take into account the role of gold and silver, but we cannot disregard the equally crucial role of steel.

11

Imperial Visions

THE ANCIENT ROMANS WERE USED TO being defeated. Like the rulers of most of history's great empires, they could lose battle after battle but still win the war. An empire that cannot sustain a blow and remain standing is not really an empire. Yet even the Romans found it hard to stomach the news arriving from northern Iberia in the middle of the second century BC. A small, insignificant mountain town called Numantia, inhabited by the peninsula's native Celts, had dared to throw off the Roman yoke. Rome at the time was the unquestioned master of the entire Mediterranean basin, having vanquished the Macedonian and Seleucid empires, subjugated the proud city states of Greece, and turned Carthage into a smouldering ruin. The Numantians had nothing on their side but their fierce love of freedom and their inhospitable terrain. Yet they forced legion after legion to surrender or retreat in shame.

Eventually, in 134 BC, Roman patience snapped. The Senate decided to send Scipio Aemilianus, Rome's foremost general and the man who had levelled Carthage, to take care of the Numantians. He was given a massive army of more than 30,000 soldiers. Scipio, who respected the fighting spirit and martial skill of the Numantians, preferred not to waste his soldiers in unnecessary combat. Instead, he encircled Numantia with a line of fortifications, blocking the town's contact with the outside world. Hunger did his work for him. After more than a year, the food supply ran out. When the

Numantians realised that all hope was lost, they burned down their town; according to Roman accounts, most of them killed themselves so as not to become Roman slaves.

Numantia later became a symbol of Spanish independence and courage. Miguel de Cervantes, the author of *Don Quixote*, wrote a tragedy called *The Siege of Numantia* which ends with the town's destruction, but also with a vision of Spain's future greatness. Poets composed paeans to its fierce defenders and painters committed majestic depictions of the siege to canvas. In 1882, its ruins were declared a 'national monument' and became a pilgrimage site for Spanish patriots. In the 1950s and 1960s, the most popular comic books in Spain weren't about Superman and Spider-Man – they told of the adventures of El Jabato, an imaginary ancient Iberian hero who fought against the Roman oppressors. The ancient Numantians are to this day Spain's paragons of heroism and patriotism, cast as role models for the country's young people.

Yet Spanish patriots extol the Numantians in *Spanish* – a romance language that is a progeny of Scipio's Latin. The Numantians spoke a now-dead lost Celtic language. Cervantes wrote *The Siege of Numantia* in Latin script, and the play follows Graeco-Roman artistic models. Numantia had no theatres. Spanish patriots who admire Numantian heroism tend also to be loyal followers of the Roman Catholic Church – don't miss that first word – a church whose leader still sits in Rome and whose God prefers to be addressed in Latin. Similarly, modern Spanish law derives from Roman law; Spanish politics is built on Roman foundations; and Spanish cuisine and architecture owe a far greater debt to Roman legacies than to those of the Celts of Iberia. Nothing is really left of Numantia save ruins. Even its story has reached us thanks only to the writings of Roman historians. It was tailored to the tastes of Roman audiences which relished tales of freedom-loving barbarians. The victory of Rome over Numantia was so complete that the victors co-opted the very memory of the vanquished.

It's not our kind of story. We like to see underdogs win. But there is no justice in history. Most past cultures have sooner or later fallen prey to the armies of some ruthless empire, which have consigned them to oblivion. Empires, too, ultimately fall, but they tend to leave behind rich and enduring legacies. Almost all people in the twenty-first century are the offspring of one empire or another.

What Is an Empire?

An empire is a political order with two important characteristics. First, to qualify for that designation you have to rule over a significant number of distinct peoples, each possessing a different cultural identity and a separate territory. How many peoples exactly? Two or three is not sufficient. Twenty or thirty is plenty. The imperial threshold passes somewhere in between.

Second, empires are characterised by flexible borders and a potentially unlimited appetite. They can swallow and digest more and more nations and territories without altering their basic structure or identity. The British state of today has fairly clear borders that cannot be exceeded without altering the fundamental structure and identity of the state. A century ago almost any place on earth could have become part of the British Empire.

Cultural diversity and territorial flexibility give empires not only their unique character, but also their central role in history. It's thanks to these two characteristics that empires have managed to unite diverse ethnic groups and ecological zones under a single political umbrella, thereby fusing together larger and larger segments of the human species and of planet Earth.

It should be stressed that an empire is defined solely by its cultural diversity and flexible borders, rather than by its origins, its form of government, its territorial extent, or the size of its population. An empire need not emerge from military conquest. The Athenian Empire began its life as a voluntary league, and the Habsburg Empire was born in wedlock, cobbled together by a string of shrewd marriage alliances. Nor must an empire be ruled by an autocratic emperor. The British Empire, the largest empire in history, was ruled by a democracy. Other democratic (or at least republican) empires have included the modern Dutch, French, Belgian and American empires, as well as the premodern empires of Novgorod, Rome, Carthage and Athens.

Size, too, does not really matter. Empires can be puny. The Athenian Empire at its zenith was much smaller in size and population than today's Greece. The Aztec Empire was smaller than today's Mexico. Both were nevertheless empires, whereas modern Greece and modern Mexico are not, because the former gradually subdued dozens and even hundreds of different polities while the latter have not. Athens lorded it over more than a hundred formerly independent city states, whereas the Aztec Empire, if we can trust its taxation records, ruled 371 different tribes and peoples.[1]

How was it possible to squeeze such a human potpourri into the territory of a modest modern state? It was possible because in the past there were many more distinct peoples in the world, each of which had a smaller population and occupied less territory than today's typical people. The land between the Mediterranean and the Jordan River, which today struggles to satisfy the ambitions of just two peoples, easily accommodated in biblical times dozens of nations, tribes, petty kingdoms and city states.

Empires were one of the main reasons for the drastic reduction in human diversity. The imperial steamroller gradually obliterated the unique characteristics of numerous peoples (such as the Numantians), forging out of them new and much larger groups.

Evil Empires?

In our time, 'imperialist' ranks second only to 'fascist' in the lexicon of political swear words. The contemporary critique of empires commonly takes two forms:

1. Empires do not work. In the long run, it is not possible to rule effectively over a large number of conquered peoples.

2. Even if it can be done, it should not be done, because empires are evil engines of destruction and exploitation. Every people has a right to self-determination, and should never be subject to the rule of another.

From a historical perspective, the first statement is plain nonsense, and the second is deeply problematic.

The truth is that empire has been the world's most common form of political organisation for the last 2,500 years. Most humans during these two and a half millennia have lived in empires. Empire is also a very stable form of government. Most empires have found it alarmingly easy to put down rebellions. In general, they have been toppled only by external invasion or by a split within the ruling elite. Conversely, conquered peoples don't have a very good record of freeing themselves from their imperial overlords. Most have remained subjugated for hundreds of years. Typically, they have been slowly digested by the conquering empire, until their distinct cultures fizzled out.

For example, when the Western Roman Empire finally fell to invading Germanic tribes in AD 476, the Numantians, Arverni, Helvetians, Samnites, Lusitanians, Umbrians, Etruscans and hundreds of other forgotten peoples whom the Romans conquered centuries earlier did not emerge from the empire's eviscerated carcass like Jonah from the belly of the great fish. None of them were left. The biological descendants of the people who had identified themselves as members of those nations, who had spoken their languages, worshipped their gods and told their myths and legends, now thought, spoke and worshipped as Romans.

In many cases, the destruction of one empire hardly meant independence for subject peoples. Instead, a new empire stepped into the vacuum created when the old one collapsed or retreated. Nowhere has this been more obvious than in the Middle East. The current political constellation in that region – a balance of power between many independent political entities with more or less stable borders – is almost without parallel at any time in the last several millennia. The last time the Middle East experienced such a situation was in the eighth century BC – almost 3,000 years ago! From the rise of the Neo-Assyrian Empire in the eighth century BC until the collapse of the British and French empires in the mid-twentieth century AD, the Middle East passed from the hands of one empire into the hands of another, like a baton in a relay race. And by the time the British and French finally dropped the baton, the Aramaeans, the Ammonites, the Phoenicians, the Philistines, the Moabites, the Edomites and the other peoples conquered by the Assyrians had long disappeared.

True, today's Jews, Armenians and Georgians claim with some measure of justice that they are the offspring of ancient Middle Eastern peoples. Yet these are only exceptions that prove the rule, and even these claims are somewhat exaggerated. It goes without saying that the political, economic and social practices of modern Jews, for example, owe far more to the empires under which they lived during the past two millennia than to the traditions of the ancient kingdom of Judaea. If King David were to show up in an ultra-Orthodox synagogue in present-day Jerusalem, he would be utterly bewildered to find people dressed in east European clothes, speaking in a German dialect (Yiddish) and having endless arguments about the meaning of a Babylonian text (the Talmud). There were neither synagogues, volumes of Talmud, nor even Torah scrolls in ancient Judaea.

*

Building and maintaining an empire usually required the vicious slaughter of large populations and the brutal oppression of everyone who was left. The standard imperial toolkit included wars, enslavement, deportation and genocide. When the Romans invaded Scotland in AD 83, they were met by fierce resistance from local Caledonian tribes, and reacted by laying waste to the country. In reply to Roman peace offers, the chieftain Calgacus called the Romans 'the ruffians of the world', and said that 'to plunder, slaughter and robbery they give the lying name of empire; they make a desert and call it peace'.[2]

This does not mean, however, that empires leave nothing of value in their wake. To colour all empires black and to disavow all imperial legacies is to reject most of human culture. Imperial elites used the profits of conquest to finance not only armies and forts but also philosophy, art, justice and charity. A significant proportion of humanity's cultural achievements owe their existence to the exploitation of conquered populations. The profits and prosperity brought by Roman imperialism provided Cicero, Seneca and St Augustine with the leisure and wherewithal to think and write; the Taj Mahal could not have been built without the wealth accumulated by Mughal exploitation of their Indian subjects; and the Habsburg Empire's profits from its rule over its Slavic-, Hungarian- and Romanian-speaking provinces paid Haydn's salaries and Mozart's commissions. No Caledonian writer preserved Calgacus' speech for posterity. We know of it thanks to the Roman historian Tacitus. In fact, Tacitus probably made it up. Most scholars today agree that Tacitus not only fabricated the speech but invented the character of Calgacus, the Caledonian chieftain, to serve as a mouthpiece for what he and other upper-class Romans thought about their own country.

Even if we look beyond elite culture and high art, and focus instead on the world of common people, we find imperial legacies in the majority of modern cultures. Today most of us speak, think and dream in imperial languages that were forced upon our ancestors by the sword. Most East Asians speak and dream in the language of the Han Empire. No matter what their origins, nearly all the inhabitants of the two American continents, from Alaska's Barrow Peninsula to the Straits of Magellan, communicate in one of four imperial languages: Spanish, Portuguese, French or English. Present-day Egyptians speak Arabic, think of themselves as Arabs, and identify wholeheartedly with the Arab Empire that conquered Egypt in the seventh century and crushed with an iron fist the repeated revolts that broke out against its rule. About 10 million Zulus in South Africa hark back to the Zulu age of glory in the nineteenth century, even though most of them descend from tribes who fought *against* the Zulu Empire, and were incorporated into it only through bloody military campaigns.

It's for Your Own Good

The first empire about which we have definitive information was the Akkadian Empire of Sargon the Great (*c.*2250 BC). Sargon began his career as the king of Kish, a small city state in Mesopotamia. Within a few decades he managed to conquer not only all other Mesopotamian city states, but also large territories outside the Mesopotamian heartland. Sargon boasted that he had conquered the entire world. In reality, his dominion stretched from the Persian Gulf to the Mediterranean, and included most of today's Iraq and Syria, along with a few slices of modern Iran and Turkey.

The Akkadian Empire did not last long after its founder's death, but Sargon left behind an imperial mantle that seldom remained unclaimed. For the next 1,700 years, Assyrian, Babylonian and Hittite kings adopted Sargon as a role model, boasting that they, too, had conquered the entire world. Then, around 550 BC, Cyrus the Great of Persia came along with an even more impressive boast.

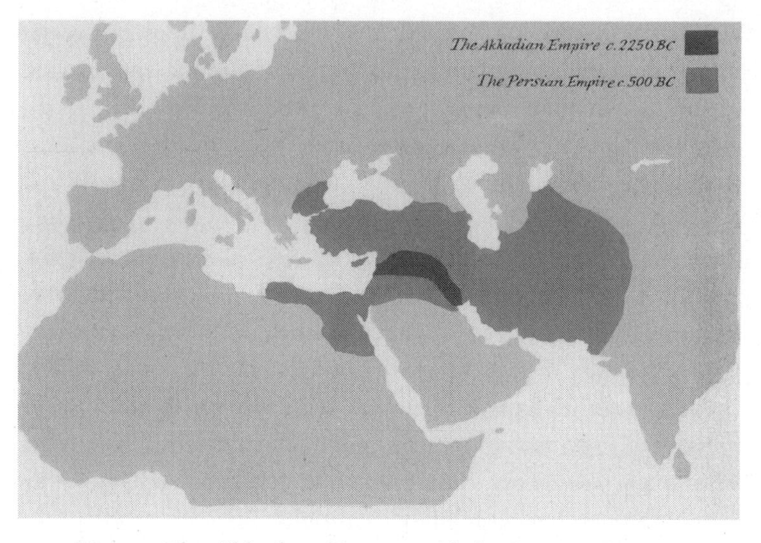

Map 4. The Akkadian Empire and the Persian Empire.

The kings of Assyria always remained the kings of Assyria. Even when they claimed to rule the entire world, it was obvious that they were doing it for the greater glory of Assyria, and they were not apologetic about it. Cyrus, on the other hand, claimed not merely to rule the whole world, but to do so for the sake of all people. 'We are conquering you for your own benefit,' said the Persians. Cyrus wanted the peoples he subjected to love him and to count themselves lucky to be Persian vassals. The most famous example of Cyrus' innovative efforts to gain the approbation of a nation living under the thumb of his empire was his command that the Jewish exiles in Babylonia be allowed to return to their Judaean homeland and rebuild their temple. He even offered them financial assistance. Cyrus did not see himself as a Persian king ruling over Jews – he was also the king of the Jews, and thus responsible for their welfare.

The presumption to rule the entire world for the benefit of all its inhabitants was startling. Evolution has made *Homo sapiens*, like other social mammals, a xenophobic creature. Sapiens instinctively divide humanity into two parts, 'us' and 'them'. 'Us' is people like you and me, who share our language, religion and customs. We are all responsible for each other, but not responsible for 'them'. We were always distinct from them, and owe them nothing. We don't want to see any of them in our territory, and we don't care an iota what happens in their territory. They are barely even human. In the language of the Dinka people of the Sudan, 'Dinka' simply means 'people'. People who are not Dinka are not people. The Dinka's bitter enemies are the Nuer. What does the word Nuer mean in Nuer language? It means 'original people'. Thousands of kilometres from the Sudan deserts, in the frozen ice-lands of Alaska and north-eastern Siberia, live the Yupiks. What does Yupik mean in Yupik language? It means 'real people'.[3]

In contrast with this ethnic exclusiveness, imperial ideology from Cyrus onward has tended to be inclusive and all-encompassing. Even though it has often emphasised racial and cultural differences between rulers and ruled, it has still recognised the basic unity of the entire world, the existence of a single set of principles governing all places and times, and the mutual responsibilities of all human beings. Humankind is seen as a large family: the privileges of the parents go hand in hand with responsibility for the welfare of the children.

This new imperial vision passed from Cyrus and the Persians to Alexander the Great, and from him to Hellenistic kings, Roman emperors, Muslim caliphs, Indian dynasts, and eventually even to Soviet premiers and American presidents. This benevolent imperial vision has justified the existence of empires, and negated not only attempts by subject peoples to rebel, but also attempts by independent peoples to resist imperial expansion.

Similar imperial visions were developed independently of the Persian model in other parts of the world, most notably in Central America, in the Andean region, and in China. According to traditional Chinese political theory, Heaven (*Tian*) is the source of all legitimate authority on earth. Heaven chooses the most worthy person or family and gives them the Mandate of Heaven. This person or family then rules over All Under Heaven (*Tianxia*) for the benefit of all its inhabitants. Thus, a legitimate authority is – by definition – universal. If a ruler lacks the Mandate of Heaven, then he lacks legitimacy to rule even a single city. If a ruler enjoys the mandate, he is obliged to spread justice and harmony to the entire world. The Mandate of Heaven could not be given to several candidates simultaneously, and consequently one could not legitimise the existence of more than one independent state.

The first emperor of the united Chinese empire, Qín Shǐ Huángdì, boasted that 'throughout the six directions [of the universe] everything belongs to the emperor . . . wherever there is a human footprint, there is not one who did not become a subject [of the emperor] . . . his kindness reaches even oxen and horses. There is not one who did not benefit. Every man is safe under his own roof.'[4] In Chinese political thinking as well as Chinese historical memory, imperial periods were henceforth seen as golden ages of order and justice. In contradiction to the modern Western view that a just world is composed of separate nation states, in China periods of political fragmentation were seen as dark ages of chaos and injustice. This perception has had far-reaching implications for Chinese history. Every time an empire collapsed, the dominant political theory goaded the powers that be not to settle for paltry independent principalities, but to attempt reunification. Sooner or later these attempts always succeeded.

When They Become Us

Empires have played a decisive part in amalgamating many small cultures into fewer big cultures. Ideas, people, goods and technology spread more easily within the borders of an empire than in a politically fragmented region. Often enough, it was the empires themselves which deliberately spread ideas, institutions, customs and norms. One reason was to make life easier for themselves. It is difficult to rule an empire in which every little district has its own set of laws, its own form of writing, its own language and its own money. Standardisation was a boon to emperors.

A second and equally important reason why empires actively spread a common culture was to gain legitimacy. At least since the days of Cyrus and Qín Shǐ Huángdì, empires have justified their actions – whether road-building or bloodshed – as necessary to spread a superior culture from which the conquered benefit even more than the conquerors.

The benefits were sometimes salient – law enforcement, urban planning, standardisation of weights and measures – and sometimes questionable – taxes, conscription, emperor worship. But most imperial elites earnestly believed that they were working for the general welfare of all the empire's inhabitants. China's ruling class treated their country's neighbours and its foreign subjects as miserable barbarians to whom the empire must bring the benefits of culture. The Mandate of Heaven was bestowed upon the emperor not in order to exploit the world, but in order to educate humanity. The Romans, too, justified their dominion by arguing that they were endowing the barbarians with peace, justice and refinement. The wild Germans and painted Gauls had lived in squalor and ignorance until the Romans tamed them with law, cleaned them up in public bathhouses, and improved them with philosophy. The Mauryan Empire in the third century BC took as its mission the dissemination of Buddha's teachings to an ignorant world. The Muslim caliphs received a divine mandate to spread the Prophet's revelation, peacefully if possible but by the sword if necessary. The Spanish and Portuguese empires proclaimed that it was not riches they sought in the Indies and America, but converts to the true faith. The sun never set on the British mission to spread the twin gospels of liberalism and free trade. The Soviets felt duty-bound to facilitate the inexorable historical march from capitalism towards the utopian dictatorship of the proletariat. Many Americans nowadays maintain that their government has a moral imperative to bring Third

World countries the benefits of democracy and human rights, even if these goods are delivered by cruise missiles and F-16s.

The cultural ideas spread by empire were seldom the exclusive creation of the ruling elite. Since the imperial vision tends to be universal and inclusive, it was relatively easy for imperial elites to adopt ideas, norms and traditions from wherever they found them, rather than to stick fanatically to a single hidebound tradition. While some emperors sought to purify their cultures and return to what they viewed as their roots, for the most part empires have begot hybrid civilisations that absorbed much from their subject peoples. The imperial culture of Rome was Greek almost as much as Roman. The imperial Abbasid culture was part Persian, part Greek, part Arab. Imperial Mongol culture was a Chinese copycat. In the imperial United States, an American president of Kenyan blood can munch on Italian pizza while watching his favourite film, *Lawrence of Arabia*, a British epic about the Arab rebellion against the Turks.

Not that this cultural melting pot made the process of cultural assimilation any easier for the vanquished. The imperial civilisation may well have absorbed numerous contributions from various conquered peoples, but the hybrid result was still alien to the vast majority. The process of assimilation was often painful and traumatic. It is not easy to give up a familiar and loved local tradition, just as it is difficult and stressful to understand and adopt a new culture. Worse still, even when subject peoples were successful in adopting the imperial culture, it could take decades, if not centuries, until the imperial elite accepted them as part of 'us'. The generations between conquest and acceptance were left out in the cold. They had already lost their beloved local culture, but they were not allowed to take an equal part in the imperial world. On the contrary, their adopted culture continued to view them as barbarians.

Imagine an Iberian of good stock living a century after the fall of Numantia. He speaks his native Celtic dialect with his parents, but has acquired impeccable Latin, with only a slight accent, because he needs it to conduct his business and deal with the authorities. He indulges his wife's penchant for elaborately ornate baubles, but is a bit embarrassed that she, like other local women, retains this relic of Celtic taste – he'd rather have her adopt the clean simplicity of the jewellery worn by the Roman governor's wife. He himself wears Roman tunics and, thanks to his success as a cattle merchant, due in no small part to his expertise in the intricacies of Roman commercial law, he has been able to build a Roman-style villa. Yet, even though he can recite Book III of Virgil's *Georgics* by heart, the Romans still treat him as though he's semi-barbarian. He realises with frustration that he'll never get a government appointment, or one of the really good seats in the amphitheatre.

In the late nineteenth century, many educated Indians were taught the same lesson by their British masters. One famous anecdote tells of an ambitious Indian who mastered the intricacies of the English language, took lessons in Western-style dance, and even became accustomed to eating with a knife and fork. Equipped with his new manners, he travelled to England, studied law at University College London, and became a qualified barrister. Yet this young man of law, bedecked in suit and tie, was thrown off a train in the British colony of South Africa for insisting on travelling first class instead of settling for third class, where 'coloured' men like him were supposed to ride. His name was Mohandas Karamchand Gandhi.

In some cases the processes of acculturation and assimilation eventually broke down the barriers between the newcomers and the old elite. The conquered no longer saw the empire as an alien system of occupation, and the conquerors came to view their subjects as equal to themselves. Rulers and ruled alike came to see 'them' as 'us'. All the subjects of Rome eventually, after centuries of imperial rule, were granted Roman citizenship. Non-Romans rose to occupy the top ranks in the officer corps of the Roman legions and were appointed to the Senate. In AD 48 the emperor Claudius admitted to the Senate several Gallic notables, who, he noted in a speech, through 'customs, culture, and the ties of marriage have blended with ourselves'. Snobbish senators protested introducing these former enemies into the heart of the Roman political system. Claudius reminded them of an inconvenient truth. Most of their own senatorial families descended from Italian tribes who once fought against Rome, and were later granted Roman citizenship. Indeed, the emperor reminded them, his own family was of Sabine ancestry.[5]

During the second century AD, Rome was ruled by a line of emperors born in Iberia, in whose veins probably flowed at least a few drops of local Iberian blood. The reigns of the Iberian emperors – from Trajan to Marcus Aurelius – are often seen as the empire's golden age. After that, all the ethnic dams were let down. Emperor Septimius Severus (193–211) was the scion of a Punic family from Libya. Elagabalus (218–22) was a Syrian. Emperor Philip (244–9) was known colloquially as 'Philip the Arab'. The empire's new citizens adopted Roman imperial culture with such zest that, for centuries and even millennia after the empire collapsed, they continued to speak the empire's language, to live by the empire's laws, and to believe in the Christian God that the empire had adopted from one of its Levantine provinces.

A similar process occurred in the Arab Empire. When it was established in the mid-seventh century AD, it was based on a sharp division between the ruling Arab–Muslim elite and the subjugated Egyptians, Syrians, Iranians and Berbers, who were neither Arabs nor Muslim. Many of the empire's subjects gradually adopted the Muslim faith, the Arabic language and a hybrid imperial culture. The old Arab elite looked upon these parvenus with deep hostility, fearing to lose its unique status and identity. The frustrated converts clamoured for an equal share within the empire and in the world of Islam. Eventually they got their way. Egyptians, Syrians and Mesopotamians were increasingly seen as 'Arabs'. Arabs, in their turn – whether 'authentic' Arabs from Arabia or newly minted Arabs from Egypt and Syria – came to be increasingly dominated by non-Arab Muslims, in particular by Iranians, Turks and Berbers. The great success of the Arab imperial project was that the imperial culture it created was wholeheartedly adopted by numerous non-Arab people, who continued to uphold it, develop it and spread it – even after the original empire collapsed and the Arabs as an ethnic group lost their dominion.

In China the success of the imperial project was even more thorough. For more than 2,000 years, a welter of ethnic and cultural groups first termed barbarians were successfully integrated into imperial Chinese culture and became Han Chinese (so named after the Han Empire that ruled China from 206 BC to AD 220). The ultimate achievement of the Chinese Empire is that it is still alive and kicking, yet it is hard to see it as an empire except in outlying areas such as Tibet and Xinjiang. More than 90 per cent of the population of China are seen by themselves and by others as Han.

We can understand the decolonisation process of the last few decades in a similar way. During the modern era Europeans conquered much of the globe under the guise of spreading a superior Western culture. They were so successful that billions of people gradually adopted significant parts of that culture. Indians, Africans, Arabs, Chinese and Maori learned French, English and Spanish. They began to believe in human rights and the principle of self-determination, and they adopted Western ideologies such as liberalism, capitalism, Communism, feminism and nationalism.

During the twentieth century, local groups that had adopted Western values claimed equality with their European conquerors in the name of these very values. Many anti-colonial struggles were waged under the banners of self-determination, socialism and human rights, all of which are Western legacies. Just as Egyptians, Iranians and Turks adopted and adapted the imperial culture that they inherited from the original Arab conquerors, so today's Indians, Africans and Chinese have accepted much of the imperial culture of their former Western overlords, while seeking to mould it in accordance with their needs and traditions.

The Imperial Cycle

Stage	Rome	Islam	European imperialism
A small group establishes a big empire	The Romans establish the Roman Empire	The Arabs establish the Arab caliphate	The Europeans establish the European empires
An imperial culture is forged	Graeco-Roman culture	Arab–Muslim culture	Western culture
The imperial culture is adopted by the subject peoples	The subject peoples adopt Latin, Roman law, Roman political ideas, etc.	The subject peoples adopt Arabic, Islam, etc.	The subject peoples adopt English and French, socialism, nationalism, human rights, etc.
The subject peoples demand equal status in the name of common imperial values	Illyrians, Gauls and Punics demand equal status with the Romans in the name of common Roman values	Egyptians, Iranians and Berbers demand equal status with the Arabs in the name of common Muslim values	Indians, Chinese and Africans demand equal status with Europeans in the name of common Western values such as nationalism, socialism and human rights

Stage	Rome	Islam	European imperialism
The empire's founders lose their dominance	Romans cease to exist as a unique ethnic group. Control of the empire passes to a new multi-ethnic elite	Arabs lose control of the Muslim world, in favour of a multi-ethnic Muslim elite	Europeans lose control of the global world, in favour of a multi-ethnic elite largely committed to Western values and ways of thinking
The imperial culture continues to flourish and develop	The Illyrians, Gauls and Punics continue to develop their adopted Roman culture	The Egyptians, Iranians and Berbers continue to develop their adopted Muslim culture	The Indians, Chinese, and Africans continue to develop their adopted Western culture

Good Guys and Bad Guys in History

It is tempting to divide history neatly into good guys and bad guys, with all empires among the bad guys. For the vast majority of empires were founded on blood, and maintained their power through oppression and war. Yet most of today's cultures are based on imperial legacies. If empires are by definition bad, what does that say about us?

There are schools of thought and political movements that seek to purge human culture of imperialism, leaving behind what they claim is a pure, authentic civilisation, untainted by sin. These ideologies are at best naive; at worst they serve as disingenuous window-dressing for crude nationalism and bigotry. Perhaps you could make a case that some of the myriad cultures that emerged at the dawn of recorded history were pure, untouched by sin and unadulterated by other societies. But no culture since that dawn can reasonably make that claim, certainly no culture that exists now on earth. All human cultures are at least in part the legacy of empires and imperial civilisations, and no academic or political surgery can cut out the imperial legacies without killing the patient.

Think, for example, about the love–hate relationship between the independent Indian republic of today and the British Raj. The British conquest and occupation of India cost the lives of millions of Indians, and was responsible for the continuous humiliation and exploitation of hundreds of millions more. Yet many Indians adopted, with the zest of converts, Western ideas such as self-determination and human rights, and were dismayed when the British refused to live up to their own declared values by granting native Indians either equal rights as British subjects or independence.

Nevertheless, the modern Indian state is a child of the British Empire. The British killed, injured and persecuted the inhabitants of the subcontinent, but they also united a bewildering mosaic of warring kingdoms, principalities and tribes, creating a shared national consciousness and a country that functioned more or less as a single political unit. They laid the foundations of the Indian judicial system, created its administrative structure, and built the railroad network that was critical for economic integration. Independent India adopted Western democracy, in its British incarnation, as its form of government. English is still the subcontinent's lingua franca, a neutral tongue that native speakers of Hindi, Tamil and Malayalam can use to communicate. Indians are passionate cricket players and chai (tea) drinkers, and both game and beverage are British legacies. Commercial tea farming did not exist in India until the mid-nineteenth century, when it was

28. The Chhatrapati Shivaji train station in Mumbai. It began its life as Victoria Station, Bombay. The British built it in the Neo-Gothic style that was popular in late nineteenth-century Britain. A Hindu nationalist government changed the names of both city and station, but showed no appetite for razing such a magnificent building, even if it was built by foreign oppressors.

introduced by the British East India Company. It was the snobbish British sahibs who spread the custom of tea drinking throughout the subcontinent.

How many Indians today would want to call a vote to divest themselves of democracy, English, the railway network, the legal system, cricket and tea on the grounds that they are imperial legacies? And if they did, wouldn't the very act of calling a vote to decide the issue demonstrate their debt to their former overlords?

Even if we were to completely disavow the legacy of a brutal empire in the hope of reconstructing and safeguarding the 'authentic' cultures that preceded it, in all probability what we will be defending

29. The Taj Mahal. An example of 'authentic' Indian culture, or the alien creation of Muslim imperialism?

is nothing but the legacy of an older and no less brutal empire. Those who resent the mutilation of Indian culture by the British Raj inadvertently sanctify the legacies of the Mughal Empire and the conquering sultanate of Delhi. And whoever attempts to rescue 'authentic Indian culture' from the alien influences of these Muslim empires sanctifies the legacies of the Gupta Empire, the Kushan Empire and the Maurya Empire. If an extreme Hindu nationalist were to destroy all the buildings left by the British conquerors, such as Mumbai's main train station, what about the structures left by India's Muslim conquerors, such as the Taj Mahal?

Nobody really knows how to solve this thorny question of cultural inheritance. Whatever path we take, the first step is to acknowledge the complexity of the dilemma and to accept that simplistically dividing the past into good guys and bad guys leads nowhere. Unless, of course, we are willing to admit that we usually follow the lead of the bad guys.

The New Global Empire

Since around 200 BC, most humans have lived in empires. It seems likely that in the future, too, most humans will live in one. This empire will not necessarily be ruled by a single country or ethnic group. Like the Late Roman Empire or the Chinese empires, it might be ruled by a multi-ethnic elite, held together by common interests and a common culture.

In the early twenty-first century, the world is still divided into about 200 states. But none of these states is truly independent. They all depend on one another. Their economies form a single global network of trade and finance, shaped by immensely powerful currents of capital, labour and information. An economic crisis in China or a new technology coming out of the USA can instantaneously disrupt economies on the other side of the planet.

Cultural trends also spread with lightning speed. Almost everywhere you go you can eat Indian curry, watch Hollywood movies, play English-style association football, or listen to the latest K-pop hit. A multi-ethnic global society is forming over and above the individual states. Entrepreneurs, engineers, bankers and scholars throughout the world speak the same language and share similar views and interests.

Most importantly, the 200 states increasingly share the same global problems. Intercontinental ballistic missiles and atom bombs recognise no borders, and no nation can prevent nuclear war by itself. Climate change too threatens the prosperity and survival of all humans, and no government can single-handedly stop global warming.

An even greater challenge is posed by new technologies such as bioengineering and artificial intelligence. As we shall see in the last chapter, these technologies could be used to re-engineer not just our weapons and vehicles, but even our bodies and minds. Indeed, they could be used to create completely new types of life forms, and change the future course of evolution. Who will decide what to do with such divine powers of creation?

It is unlikely that humankind can deal with these challenges without global cooperation. It remains to be seen how such cooperation could be secured. Perhaps global cooperation can only be secured through violent clashes and the imposition of a new conquering empire. Perhaps humans could find a more peaceful way to unite. For 2,500 years since Cyrus the Great numerous empires promised to build a universal political order for the benefit of all humans. They all lied, and they all failed. No empire was truly universal, and no empire really served the benefit of all humans. Will a future empire do better?

12

The Law of Religion

IN THE MEDIEVAL MARKET IN SAMARKAND, a city built on a Central Asian oasis, Syrian merchants ran their hands over fine Chinese silks, fierce tribesmen from the steppes displayed the latest batch of straw-haired slaves from the far west, and shopkeepers pocketed shiny gold coins imprinted with exotic scripts and the profiles of unfamiliar kings. Here, at one of that era's major crossroads between east and west, north and south, the unification of humankind was an everyday fact. The same process could be observed at work when Kublai Khan's army mustered to invade Japan in 1281. Mongol cavalrymen in skins and furs rubbed shoulders with Chinese foot soldiers in bamboo hats, drunken Korean auxiliaries picked fights with tattooed sailors from the South China Sea, engineers from Central Asia listened with dropping jaws to the tall tales of European adventurers, and all obeyed the command of a single emperor.

Meanwhile, around the holy Ka'aba in Mecca, human unification was proceeding by other means. Had you been a pilgrim to Mecca, circling Islam's holiest shrine in the year 1300, you might have found yourself in the company of a party from Mesopotamia, their robes floating in the wind, their eyes blazing with ecstasy, and their mouths repeating one after the other the ninety-nine names of God. Just ahead you might have seen a weather-beaten Turkish patriarch from the Asian steppes, hobbling on a stick and stroking his beard thoughtfully. To one side, gold jewellery shining against jet-black skin, might have been a group of Muslims from the African kingdom of Mali. The aroma of clove, turmeric, cardamom and sea salt would have signalled the presence of brothers from India, or perhaps from the mysterious spice islands further east.

Today religion is often considered a source of discrimination, disagreement and disunion. Yet, in fact, religion has been the third great unifier of humankind, alongside money and empires. Since all social orders and hierarchies are imagined, they are all fragile, and the larger the society, the more fragile it is. The crucial historical role of religion has been to give superhuman legitimacy to these fragile structures. Religions assert that our laws are not the result of human caprice, but are ordained by an absolute and indisputable authority. This helps place at least some fundamental laws beyond challenge, thereby ensuring social stability.

Religion can thus be defined as *a system of human laws that is founded on a belief in superhuman laws.* This involves two distinct criteria:

1. Religion is an entire system of laws, rather than an isolated custom or belief. Knocking on wood for good luck isn't a religion. Even a belief in reincarnation does not constitute a religion, as long as it does not validate some concrete laws and norms.

2. To be considered a religion, the system of laws must claim to be based on superhuman laws rather than on human decisions. Professional football is not a religion, because despite its many rules, rites and often bizarre rituals, everyone knows that human beings invented football themselves, and FIFA may at any moment enlarge the size of the goal or cancel the offside rule.

Despite their ability to legitimise widespread social and political orders, not all religions have actuated this potential. In order to unite under its aegis a large expanse of territory inhabited by disparate groups of human beings, a religion must possess two further qualities. First, it must espouse a *universal* superhuman order that is true always and everywhere. Second, it must insist on spreading this belief to everyone. In other words, it must be universal and missionary.

The best-known religions of history, such as Islam and Buddhism, are universal and missionary. Consequently people tend to believe that all religions are like them. In fact, the majority of ancient religions were local and exclusive. Their followers believed in local deities and spirits, and had no interest in converting the entire human race. As far as we know, universal and missionary religions began to appear only in the first millennium BC. Their emergence was one of the most important revolutions in history, and made a vital contribution to the unification of humankind, much like the emergence of universal empires and universal money.

Silencing the Lambs

When animism was the dominant belief system, human norms and values had to take into consideration the outlook and interests of a multitude of other beings, such as animals, plants, fairies and ghosts. For example, a forager band in the Ganges Valley may have established a rule forbidding people to cut down a particularly large fig tree, lest the fig-tree spirit become angry and take revenge. Another forager band living in the Indus Valley may have forbidden people from hunting white-tailed foxes, because a white-tailed fox once revealed to a wise old woman where the band might find precious obsidian.

Such religions tended to be very local in outlook, and to emphasise the unique features of specific locations, climates and phenomena. Most foragers spent their entire lives within an area of no more than a thousand square kilometres. In order to survive, the inhabitants of a particular valley needed to understand the superhuman order that regulated their valley, and to adjust their behaviour accordingly. It was pointless to try to convince the inhabitants of some distant valley to follow the same rules. The people of the Indus did not bother to send missionaries to the Ganges to convince locals not to hunt white-tailed foxes.

The Agricultural Revolution seems to have been accompanied by a religious revolution. Hunter-gatherers picked and pursued wild plants and animals, which could be seen as equal in status to *Homo sapiens*. The fact that man hunted sheep did not make sheep inferior to man, just as the fact that tigers hunted man did not make man inferior to tigers. Beings communicated with one another directly and negotiated the rules governing their shared habitat. In contrast, farmers owned and manipulated plants and animals, and could hardly degrade themselves by negotiating with their possessions. Hence the first religious effect of the Agricultural Revolution was to turn plants and animals from equal members of a spiritual round table into property.

This, however, created a big problem. Farmers may have desired absolute control of their sheep, but they knew perfectly well that their control was limited. They could lock the sheep in pens, castrate rams and selectively breed ewes, yet they could not ensure that the ewes conceived and gave birth to healthy lambs, nor could they prevent the eruption of deadly epidemics. How then to safeguard the fecundity of the flocks?

A leading theory about the origin of the gods argues that gods gained importance because they offered a solution to this problem. Gods such as the fertility goddess, the sky god and the god of medicine took centre stage when plants and animals lost their ability to speak, and the gods' main role was to mediate between humans and the mute plants and animals. Much of ancient mythology is in fact a legal contract in which humans promise everlasting devotion to the gods in exchange for mastery over plants and animals – the first chapters of the book of Genesis are a prime example. For thousands of years after the Agricultural Revolution, religious liturgy consisted mainly of humans sacrificing lambs, wine and cakes to divine powers, who in exchange promised abundant harvests and fecund flocks.

The Agricultural Revolution initially had a far smaller impact on the status of other members of the animist system, such as rocks, springs, ghosts and demons. However, these too gradually lost status in favour of the new gods. As long as people lived their entire lives within limited territories of a few hundred square kilometres, most of their needs could be met by local spirits. But once kingdoms and trade networks expanded, people needed to contact entities whose power and authority encompassed a whole kingdom or an entire trade basin.

The attempt to answer these needs led to the appearance of polytheistic religions (from the Greek: *poly* = many, *theos* = god). These religions understood the world to be controlled by a group of powerful gods, such as the fertility goddess, the rain god and the war god. Humans could appeal to these gods and the gods might, if they received devotions and sacrifices, deign to bring rain, victory and health.

Animism did not entirely disappear at the advent of polytheism. Demons, fairies, ghosts, holy rocks, holy springs and holy trees remained an integral part of almost all polytheist religions. These spirits were far less important than the great gods, but for the mundane needs of many ordinary people, they were good enough. While the king in his capital city sacrificed dozens of fat rams to the great war god, praying for victory over the barbarians, the peasant in his hut lit a candle to the fig-tree fairy, praying that she help cure his sick son.

Yet the greatest impact of the rise of great gods was not on sheep or demons, but upon the status of *Homo sapiens*. Animists thought that humans were just one of many creatures inhabiting the world. Polytheists, on the other hand, increasingly saw the world as a reflection of the relationship between gods and humans. Our prayers, our sacrifices, our sins and our good deeds determined the fate of the entire ecosystem. A terrible flood might wipe out billions of ants, grasshoppers, turtles, antelopes, giraffes and elephants, just because a few stupid Sapiens made the gods angry. Polytheism thereby exalted not only the status of the gods, but also that of humankind. Less fortunate members of the old animist system lost their stature and became either extras or silent decor in the great drama of man's relationship with the gods.

The Benefits of Idolatry

Two thousand years of monotheistic brainwashing have caused most Westerners to see polytheism as ignorant and childish idolatry. This is an unjust stereotype. In order to understand the inner logic of polytheism, it is necessary to grasp the central idea buttressing the belief in many gods.

Polytheism does not necessarily dispute the existence of a single power or law governing the entire universe. In fact, most polytheist and even animist religions recognised such a supreme power that stands behind all the different gods, demons and holy rocks. In classical Greek polytheism, Zeus, Hera, Apollo and their colleagues were subject to an omnipotent and all-encompassing power – Fate (Moira, Ananke). Nordic gods, too, were in thrall to Fate, which doomed them to perish in the cataclysm of Ragnarök (the Twilight of the Gods). In the polytheistic religion of the Yoruba of West Africa, all gods were born of the supreme god Olodumare, and remained subject to him. In Hindu polytheism, a single principle, Atman, controls the myriad gods and spirits, humankind, and the biological and physical world. Atman is the eternal essence or soul of the entire universe, as well as of every individual and every phenomenon.

The fundamental insight of polytheism, which distinguishes it from monotheism, is that the supreme power governing the world is devoid of interests and biases, and therefore it is unconcerned with the mundane desires, cares and worries of humans. It's pointless to ask this power for victory in war, for health or for rain, because from its all-encompassing vantage point, it makes no difference whether a particular kingdom wins or loses, whether a particular city prospers or withers, whether a particular person recuperates or dies. The Greeks did not waste any sacrifices on Fate, and Hindus built no temples to Atman.

The only reason to approach the supreme power of the universe would be to renounce all desires and embrace the bad along with the good – to embrace even defeat, poverty, sickness and death. Thus some Hindus, known as Sadhus or Sannyasis, devote their lives to uniting with Atman, thereby achieving enlightenment. They strive to see the world from the viewpoint of this fundamental principle, to realise that from its eternal perspective all mundane desires and fears are meaningless and ephemeral phenomena.

Most Hindus, however, are not Sadhus. They are sunk deep in the morass of mundane concerns, where Atman is not much help. For assistance in such matters, Hindus approach the gods with their partial powers. Precisely because their powers are partial rather than all-encompassing, gods such as Ganesha, Lakshmi and Saraswati have interests and biases. Humans can therefore make deals with these partial powers and rely on their help in order to win wars and recuperate from illness. There are necessarily many of these smaller powers, since once you start dividing up the all-encompassing power of a supreme principle, you'll inevitably end up with more than one deity. Hence the plurality of gods.

The insight of polytheism is conducive to far-reaching religious tolerance. Since polytheists believe, on the one hand, in one supreme and completely disinterested power, and on the other hand in many partial and biased powers, there is no difficulty for the devotees of one god to accept the existence and efficacy of other gods. Polytheism is inherently open-minded, and rarely persecutes 'heretics' and 'infidels'.

Even when polytheists conquered huge empires, they did not try to convert their subjects. The Egyptians, the Romans and the Aztecs did not send missionaries to foreign lands to spread the worship of Osiris, Jupiter or Huitzilopochtli (the chief Aztec god), and they certainly didn't dispatch armies for that purpose. Subject peoples throughout the empire were expected to respect the empire's gods and rituals, since these gods and rituals protected and legitimised the empire. Yet they were not required to give up their local gods and rituals. In the Aztec Empire, subject peoples were obliged to build temples for Huitzilopochtli, but these temples were built alongside those of local gods, rather than in their stead. In many cases the imperial elite itself adopted the gods and rituals of subject people. The Romans happily added the Asian goddess Cybele and the Egyptian goddess Isis to their pantheon.

The only god that the Romans long refused to tolerate was the monotheistic and evangelising god of the Christians. The Roman Empire did not require the Christians to give up their beliefs and rituals, but it did expect them to pay respect to the empire's protector gods and to the divinity of the emperor. This was seen as a declaration of political loyalty. When the Christians vehemently refused to do so, and went on to reject all attempts at compromise, the Romans reacted by persecuting what they understood to be a politically subversive faction. And even this was done half-heartedly. In the 300 years from the crucifixion of Christ to the conversion of Emperor Constantine, polytheistic Roman emperors initiated no more than four general persecutions of Christians. Local administrators and governors incited some anti-Christian violence of their own. Still, if we combine all the victims of all these persecutions, it turns out that in these three centuries, the polytheistic Romans killed no more than a few thousand Christians.[1] In contrast, over the course of the next 1,500 years, Christians slaughtered Christians by the millions to defend slightly different interpretations of the religion of love and compassion.

The religious wars between Catholics and Protestants that swept Europe in the sixteenth and seventeenth centuries are particularly notorious. All those involved accepted Christ's divinity and His gospel of compassion and love. However, they disagreed about the nature of this love. Protestants believed that the divine love is so great that God was incarnated in flesh and allowed Himself to be tortured and crucified, thereby redeeming the original sin and opening the gates of heaven to all those who professed faith in Him. Catholics maintained that faith, while essential, was not enough. To enter heaven, believers had to participate in church rituals and do good deeds. Protestants refused to accept this, arguing that this quid pro quo belittles God's greatness and love. Whoever thinks that entry to heaven depends upon his or her own good deeds magnifies his own importance, and implies that Christ's suffering on the cross and God's love for humankind are not enough.

These theological disputes turned so violent that during the sixteenth and seventeenth centuries, Catholics and Protestants killed each other by the hundreds of thousands. On 24 August 1572, French Catholics who stressed the importance of good deeds attacked communities of French Protestants who highlighted God's love for humankind. In this attack, the St Bartholomew's Day Massacre, between 5,000 and 10,000 Protestants were slaughtered in less than twenty-four hours. When the pope in Rome heard the news from France, he was so overcome by joy that he organised festive prayers to celebrate the occasion and commissioned Giorgio Vasari to decorate one of the Vatican's rooms with a fresco of the massacre (the room is currently off-limits to visitors).[2] More Christians were killed by fellow Christians in those twenty-four hours than by the polytheistic Roman Empire throughout its entire existence.

God Is One

With time some followers of polytheist gods became so fond of their particular patron that they drifted away from the basic polytheist insight. They began to believe that their god was the only god, and that He was in fact the supreme power of the universe. Yet at the same time they continued to view Him as possessing interests and biases, and believed that they could strike deals with Him. Thus were born monotheist religions, whose followers beseech the supreme power of the universe to help them recover from illness, win the lottery and gain victory in war.

The first monotheist religion known to us appeared in Egypt, *c.*1350 BC, when Pharaoh Akhenaten declared that one of the minor deities of the Egyptian pantheon, the god Aten, was, in fact, the supreme power ruling the universe. Akhenaten institutionalised the worship of Aten as the state religion and tried to check the worship of all other gods. His religious revolution, however, was unsuccessful. After his death, the worship of Aten was abandoned in favour of the old pantheon.

Polytheism continued to give birth here and there to other monotheist religions, but they remained marginal, not least because they failed to digest their own universal message. Judaism, for example, argued that the supreme power of the universe has interests and biases, yet His chief interest is in the tiny Jewish nation and in the obscure land of Israel. Judaism had little to offer other nations, and throughout most of its existence it has not been a missionary religion. This stage can be called the stage of 'local monotheism'.

The big breakthrough came with Christianity. This faith began as an esoteric Jewish sect that sought to convince Jews that Jesus of Nazareth was their long-awaited messiah. However, one of the sect's first leaders, Paul of Tarsus, reasoned that if the supreme power of the universe has interests and biases, and if He had bothered to incarnate Himself in the flesh and to die on the cross for the salvation of humankind, then this is something everyone should hear about, not just Jews. It was thus necessary to spread the good word – the gospel – about Jesus throughout the world.

Paul's arguments fell on fertile ground. Christians began organising widespread missionary activities aimed at all humans. In one of history's strangest twists, this esoteric Jewish sect took over the mighty Roman Empire.

Christian success served as a model for another monotheist religion that appeared in the Arabian peninsula in the seventh century – Islam. Like Christianity, Islam, too, began as a small sect in a remote corner of the world, but in an even stranger and swifter historical surprise it managed to break out of the deserts of Arabia and conquer an immense empire stretching from the Atlantic Ocean to India. Henceforth, the monotheist idea played a central role in world history.

Monotheists have tended to be far more fanatical and missionary than polytheists. A religion that recognises the legitimacy of other faiths implies either that its god is not the supreme power of the universe, or that it received from God just part of the universal truth. Since monotheists have usually believed that they are in possession of the entire message of the one and only God, they have been compelled to discredit all other religions. Over the last two millennia, monotheists repeatedly tried to strengthen their hand by violently exterminating all competition.

It worked. At the beginning of the first century AD, there were hardly any monotheists in the world. Around AD 500, one of the world's largest empires – the Roman Empire – was a Christian polity, and missionaries were busy spreading Christianity to other parts of Europe, Asia and Africa. By the end of the first millennium AD, most people in Europe, West Asia and North Africa were monotheists, and empires from the Atlantic Ocean to the Himalayas claimed to be ordained by the single great God. By the early sixteenth century, monotheism dominated most of Afro-Asia, with the exception of East Asia and the southern parts of Africa, and it began extending long tentacles towards South Africa, America and Oceania. Today most people outside East Asia adhere to one monotheist religion or another, and the global political order is built on monotheistic foundations.

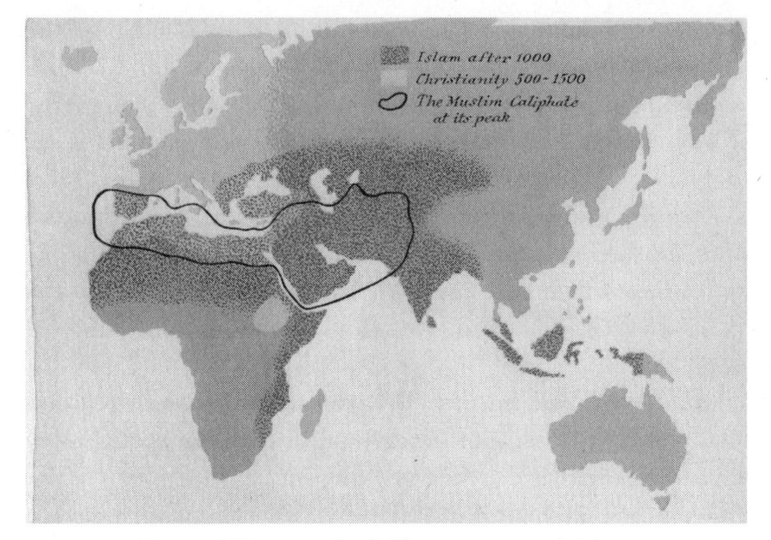

Map 5. The spread of Christianity and Islam.

Yet just as animism continued to survive within polytheism, so polytheism continued to survive within monotheism. In theory, once a person believes that the supreme power of the universe has interests and biases, what's the point in worshipping partial powers? Who would want to approach a lowly bureaucrat when the president's office is open to you? Indeed, monotheist theology tends to deny the existence of all gods except the supreme God, and to pour hellfire and brimstone over anyone who dares worship them.

Yet there has always been a chasm between theological theories and historical realities. Most people have found it difficult to digest the monotheist idea fully. They have continued to divide the world into 'us' and 'them', and to see the supreme power of the universe as too distant and alien for their mundane needs. The monotheist religions expelled the gods through the front door with a lot of fanfare, only to take them back in through the side window. Christianity, for example, developed its own pantheon of saints, whose cults differed little from those of the polytheistic gods.

Just as the god Jupiter defended Rome and Huitzilopochtli protected the Aztec Empire, so every Christian kingdom had its own patron saint who helped it overcome difficulties and win wars. England was protected by St George, Scotland by St Andrew, Hungary by St Stephen, and France had St Martin. Cities and towns, professions, and even diseases – each had their own saint. The city of Milan had St Ambrose, while St Mark watched over Venice. St Florian protected chimney cleaners, whereas St Matthew lent a hand to tax collectors in distress. If you suffered from headaches you had to pray to St Agathius, but if from toothaches then St Apollonia was a much better audience.

The Christian saints did not merely resemble the old polytheistic gods. Often they were these very same gods in disguise. For example, the chief goddess of Celtic Ireland prior to the coming of Christianity was Brigid. When Ireland was Christianised, Brigid too was baptised. She became St Brigit, who to this day is the most revered saint in Catholic Ireland.

The Battle of Good and Evil

Polytheism gave birth not merely to monotheist religions, but also to dualistic ones. Dualistic religions espouse the existence of two opposing powers: good and evil. Unlike monotheism, dualism believes that evil is an independent power, neither created by the good God, nor subordinate to it. Dualism explains that the entire universe is a battleground between these two forces, and that everything that happens in the world is part of the struggle.

Dualism is a very attractive world view because it has a short and simple answer to the famous Problem of Evil, one of the fundamental concerns of human thought. 'Why is there evil in the world? Why is there suffering? Why do bad things happen to good people?' Monotheists have to practise intellectual gymnastics to explain how an all-knowing, all-powerful and perfectly good God allows so much suffering in the world. One well-known explanation is that this is God's way of allowing for human free will. Were there no evil, humans could not choose between good and evil, and hence there would be no free will. This, however, is a non-intuitive answer that immediately raises a host of new questions. Freedom of will allows humans to choose evil. Many indeed choose evil and, according to the standard monotheist account, this choice must bring divine punishment in its wake. If God knew in advance that a particular person would use her free will to choose evil, and that as a result she would be punished for this by eternal tortures in hell, why did God create her? Theologians have written countless books to answer such questions. Some find the answers convincing. Some don't. What's undeniable is that monotheists have a hard time dealing with the Problem of Evil.

For dualists, it's easy to explain evil. Bad things happen even to good people because the world is not governed single-handedly by a good God. There is an independent evil power loose in the world. The evil power does bad things.

Dualism has its own drawbacks. While solving the Problem of Evil, it is unnerved by the Problem of Order. If the world was created by a single God, it's clear why it is such an orderly place, where everything obeys the same laws. But if Good and Evil battle for control of the world, who enforces the laws governing this cosmic war? Two rival states can fight one another because both obey the same laws of physics. A missile launched from Pakistan can hit targets in India because gravity works the same way in both countries. When Good and Evil fight, what common laws do they obey, and who decreed these laws?

So, monotheism explains order, but is mystified by evil. Dualism explains evil, but is puzzled by order. There is one logical way of solving the riddle: to argue that there is a single omnipotent God who created the entire universe – and He's evil. But nobody in history has had the stomach for such a belief.

*

Dualistic religions flourished for more than a thousand years. Sometime between 1500 BC and 1000 BC a prophet named Zoroaster (Zarathustra) was active somewhere in Central Asia. His creed passed from generation to generation until it became the most important of dualistic religions – Zoroastrianism. Zoroastrians saw the world as a cosmic battle between the good god Ahura Mazda and the evil god Angra Mainyu. Humans had to help the good god in this battle. Zoroastrianism was an important religion during the Achaemenid Persian Empire (550–330 BC) and later became the official religion of the Sassanid Persian Empire (AD 224–651). It exerted a major influence on almost all subsequent Middle Eastern and Central Asian religions, and it inspired a number of other dualist religions, such as Gnosticism and Manichaeism.

During the third and fourth centuries AD, the Manichaean creed spread from China to North Africa, and for a moment it appeared that it would beat Christianity to achieve dominance in the Roman Empire. Yet the Manichaeans lost the soul of Rome to the Christians, the Zoroastrian Sassanid Empire was overrun by the monotheistic Muslims, and the dualist wave subsided. Today only a handful of dualist communities survive in India and the Middle East.

Nevertheless, the rising tide of monotheism did not really wipe out dualism. Jewish, Christian and Muslim monotheism absorbed numerous dualist beliefs and practices, and some of the most basic ideas of what we call 'monotheism' are, in fact, dualist in origin and spirit. Countless Christians, Muslims and Jews believe in a powerful evil force – like the one Christians call the Devil or Satan – who can act independently, fight against the good God, and wreak havoc without God's permission.

How can a monotheist adhere to such a dualistic belief (which, by the way, is nowhere to be found in the Old Testament)? Logically, it is impossible. Either you believe in a single omnipotent God or you believe in two opposing powers, neither of which is omnipotent. Still, humans have a wonderful capacity to believe in contradictions.

So it should not come as a surprise that millions of pious Christians, Muslims and Jews manage to believe at one and the same time in an omnipotent God and an independent Devil. Countless Christians, Muslims and Jews have gone so far as to imagine that the good God even needs our help in its struggle against the Devil, which inspired among other things the call for jihads and crusades.

Another key dualistic concept, particularly in Gnosticism and Manichaeism, was the sharp distinction between body and soul, between matter and spirit. Gnostics and Manichaeans argued that the good god created the spirit and the soul, whereas matter and bodies are the creation of the evil god. Man, according to this view, serves as a battleground between the good soul and the evil body. From a monotheistic perspective, this is nonsense – why distinguish so sharply between body and soul, or matter and spirit? And why argue that body and matter are evil? After all, everything was created by the same good God. But monotheists could not help but be captivated by dualist dichotomies, precisely because they helped them address the problem of evil. So such oppositions eventually became cornerstones of Christian and Muslim thought. Belief in heaven (the realm of the good god) and hell (the realm of the evil god) was also dualist in origin. There is no trace of this belief in the Old Testament, which also never claims that the souls of people continue to live after the death of the body.

In fact, monotheism, as it has played out in history, is a kaleidoscope of monotheist, dualist, polytheist and animist legacies, jumbling together under a single divine umbrella. The average Christian believes in the monotheist God, but also in the dualist Devil, in polytheist saints, and in animist ghosts. Scholars of religion have a name for this simultaneous avowal of different and even contradictory ideas and the combination of rituals and practices taken from different sources. It's called syncretism. Syncretism might, in fact, be the single great world religion.

The Law of Nature

All the religions we have discussed so far share one important characteristic: they all focus on a belief in gods and other supernatural entities. This seems obvious to Westerners, who are familiar mainly with monotheistic and polytheist creeds. In fact, however, the religious history of the world does not boil down to the history of gods. During the first millennium BC, religions of an altogether new kind began to spread through Afro-Asia. The newcomers, such as Jainism and Buddhism in India, Daoism and Confucianism in China, and Stoicism, Cynicism and Epicureanism in the Mediterranean basin, were characterised by their disregard of gods.

These creeds maintained that the superhuman order governing the world is the product of natural laws rather than of divine wills and whims. Some of these natural-law religions continued to espouse the existence of gods, but their gods were subject to the laws of nature no less than humans, animals and plants were. Gods had their niche in the ecosystem, just as elephants and porcupines had theirs, but could no more change the laws of nature than elephants can. A prime example is Buddhism, the most important of the ancient natural-law religions, which remains one of the major faiths.

The central figure of Buddhism is not a god but a human being, Siddhartha Gautama. According to Buddhist tradition, Gautama was heir to a small Himalayan kingdom, sometime around 500 BC. The young prince was deeply affected by the suffering evident all around him. He saw that men and women, children and old people, all suffer not just from occasional calamities such as war and plague, but also from anxiety, frustration and discontent, all of which seem to be an inseparable part of the human condition. People pursue wealth and power, acquire knowledge and possessions, beget sons and daughters, and build houses and palaces. Yet no matter what they achieve, they are never content. Those who live in poverty dream of riches. Those who have a million want 2 million. Those who have 2 million want 10 million. Even the rich and famous are rarely satisfied. They too are haunted by ceaseless cares and worries, until sickness, old age and death put a bitter end to them. Everything that one has accumulated vanishes like smoke. Life is a pointless rat race. But how to escape it?

At the age of twenty-nine Gautama slipped away from his palace in the middle of the night, leaving behind his family and possessions. He travelled as a homeless vagabond throughout northern India, searching for a way out of suffering. He visited ashrams and sat at the feet of gurus but nothing liberated him entirely – some dissatisfaction always remained. He did not despair. He resolved to investigate suffering on his own until he found a method for complete liberation. He spent six years meditating on the essence, causes and cures for human anguish. In the end he came to the realisation that suffering is not caused by ill fortune, by social injustice or by divine whims. Rather, suffering is caused by the behaviour patterns of one's own mind.

Gautama's insight was that no matter what the mind experiences, it usually reacts with craving, and craving always involves dissatisfaction. When the mind experiences something distasteful it craves to be rid of the irritation. When the mind experiences something pleasant, it craves that the pleasure will remain and will intensify. Therefore, the mind is always dissatisfied and restless. This is very clear when we experience unpleasant things, such as pain. As long as the pain continues, we are dissatisfied and do all we can to avoid it. Yet even when we experience pleasant things we are never content. We either fear that the pleasure might disappear, or we hope that it will intensify. People dream for years about finding love but are rarely satisfied when they find it. Some become anxious that their partner will leave; others feel that they have settled cheaply, and could have found someone better. And we all know people who manage to do both.

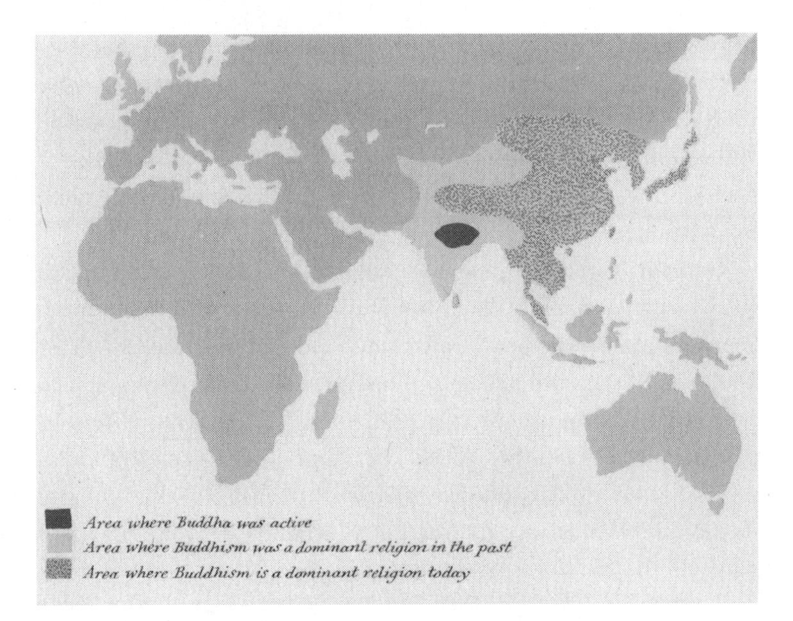

Map 6. The spread of Buddhism.

Great gods can send us rain, social institutions can provide justice and good health care, and lucky coincidences can turn us into millionaires, but none of them can change our basic mental patterns. Hence even the greatest kings are doomed to live in angst, constantly fleeing grief and anguish, forever chasing after greater pleasures.

Gautama found that there was a way to exit this vicious circle. If, when the mind experiences something pleasant or unpleasant, it simply understands things as they are, then there is no suffering. If you experience sadness without craving that the sadness go away, you continue to feel sadness but you do not suffer from it. There can actually be richness in the sadness. If you experience joy without craving that the joy linger and intensify, you continue to feel joy without losing your peace of mind.

But how do you get the mind to accept things as they are, without craving? To accept sadness as sadness, joy as joy, pain as pain? Gautama developed a set of meditation techniques that train the mind to experience reality as it is, without craving. These practices train the mind to focus all its attention on the question 'What am I experiencing now?' rather than on 'What would I rather be experiencing?' It is difficult to achieve this state of mind, but not impossible.

Gautama grounded these meditation techniques in a set of ethical rules meant to make it easier for people to focus on actual experience and to avoid falling into cravings and fantasies. He instructed his followers to avoid killing, promiscuous sex and theft, since such acts necessarily stoke the fire of craving (for power, for sensual pleasure or for wealth). When the flames are completely extinguished, craving is replaced by a state of perfect contentment and serenity, known as nirvana (the literal meaning of which is 'extinguishing the fire'). Those who have attained nirvana are fully liberated from all suffering. They experience reality with the utmost clarity, free of fantasies and delusions. While they will most likely still encounter unpleasantness and pain, such experiences cause them no misery. A person who does not crave cannot suffer.

According to Buddhist tradition, Gautama himself attained nirvana and was fully liberated from suffering. Henceforth he was known as 'Buddha', which means 'the Enlightened One'. Buddha spent the rest of his life explaining his discoveries to others so that everyone could be freed from suffering. He encapsulated his teachings in a single law: suffering arises from craving; the only way to be fully liberated from suffering is to be fully liberated from craving; and the only way to be liberated from craving is to train the mind to experience reality as it is.

This law, known as *dharma* or *dhamma*, is seen by Buddhists as a universal law of nature. That 'suffering arises from craving' is always and everywhere true, just as in modern physics E always equals mc^2. Buddhists are people who believe in this law and make it the fulcrum of all their activities. Belief in gods, on the other hand, is of minor importance to them. The first principle of monotheist religions is 'God exists. What does He want from me?' The first principle of Buddhism is 'Suffering exists. How do I escape it?'

Buddhism does not deny the existence of gods – they are described as powerful beings who can bring rains and victories – but they have no influence on the law that suffering arises from craving. If the mind of a person is free of all craving, no god can make him miserable. Conversely, once craving arises in a person's mind, all the gods in the universe cannot save him from suffering.

Yet much like the monotheist religions, premodern natural-law religions such as Buddhism never really rid themselves of the worship of gods. Buddhism told people that they should aim for the ultimate goal of complete liberation from suffering, rather than for stops along the way such as economic prosperity and political power. However, 99 per cent of Buddhists did not attain nirvana, and even if they hoped to do so in some future lifetime, they devoted most of their present lives to the pursuit of mundane achievements. So they continued to worship various gods, such as the Hindu gods in India, the Bon gods in Tibet, and the Shinto gods in Japan.

Moreover, as time went by several Buddhist sects developed pantheons of Buddhas and bodhisattvas. These are human and non-human beings with the capacity to achieve full liberation from suffering but who forego this liberation out of compassion, in order to help the countless beings still trapped in the cycle of misery. Instead of worshipping gods, many Buddhists began worshipping these enlightened beings, asking them for help not only in attaining nirvana, but also in dealing with mundane problems. Thus we find many Buddhas and bodhisattvas throughout East Asia who spend their time bringing rain, stopping plagues, and even winning bloody wars – in exchange for prayers, colourful flowers, fragrant incense and gifts of rice and candy.

The Worship of Man

The last 300 years are often depicted as an age of growing secularism, in which religions have increasingly lost their importance. If we are talking about theist religions, this is largely correct. But if we take into consideration natural-law religions, then modernity turns out to be an age of intense religious fervour, unparalleled missionary efforts, and the bloodiest wars of religion in history. The modern age has witnessed the rise of a number of new natural-law religions, such as liberalism, Communism, capitalism, nationalism and Nazism. These creeds do not like to be called religions, and refer to themselves as ideologies. But this is just a semantic exercise. If a religion is a system of human norms and values that is founded on belief in a superhuman order, then Soviet Communism was no less a religion than Islam.

Islam is of course different from Communism, because Islam sees the superhuman order governing the world as the edict of an omnipotent creator god, whereas Soviet Communism did not believe in gods. But Buddhism too gives short shrift to gods, and yet we commonly classify it as a religion. Like Buddhists, Communists believed in a superhuman order of natural and immutable laws that should guide human actions. Whereas Buddhists believe that the law of nature was discovered by Siddhartha Gautama, Communists believed that the law of nature was discovered by Karl Marx, Friedrich Engels and Vladimir Ilyich Lenin. The similarity does not end there. Like other religions, Communism too had its holy scripts and prophetic books, such as Marx's *Das Kapital*, which foretold that history would soon end with the inevitable victory of the proletariat. Communism had its holidays and festivals, such as the First of May and the anniversary of the October Revolution. It had theologians adept at Marxist dialectics, and every unit in the Soviet army had a chaplain, called a commissar, who monitored the piety of soldiers and officers. Communism had martyrs, holy wars and heresies, such

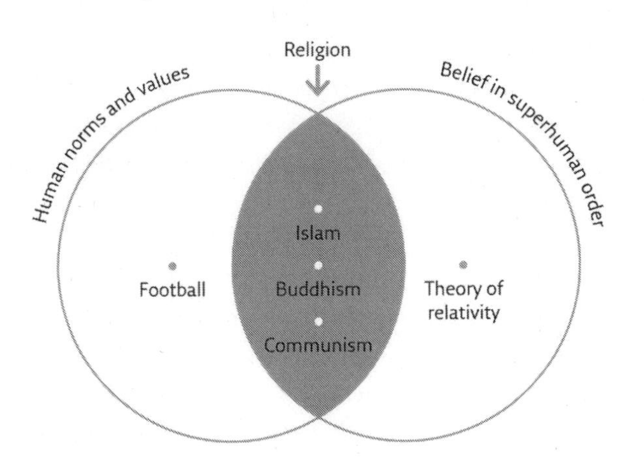

Religion is a system of human norms and values that is founded on belief in a superhuman order. The theory of relativity is not a religion, because (at least so far) there are no human norms and values that are founded on it. Football is not a religion because nobody argues that its rules reflect superhuman edicts. Islam, Buddhism and Communism are all religions, because all are systems of human norms and values that are founded on belief in a superhuman order. (Note the difference between 'superhuman' and 'supernatural'. The Buddhist law of nature and the Marxist laws of history are superhuman, since they were not legislated by humans. Yet they are not supernatural.)

as Trotskyism. Soviet Communism was a fanatical and missionary religion. A devout Communist could not be a Christian or a Buddhist, and was expected to spread the gospel of Marx and Lenin even at the price of his or her life.

Some readers may feel very uncomfortable with this line of reasoning. If it makes you feel better, you are free to go on calling Communism an ideology rather than a religion. It makes no difference. We can divide creeds into god-centred religions and godless ideologies that claim to be based on natural laws. But then, to be consistent, we would need to catalogue at least some Buddhist, Daoist and Stoic sects as ideologies rather than religions. Conversely, we should note that belief in gods persists within many modern ideologies, and that some of them, most notably liberalism, make little sense without this belief.

It would be impossible to survey here the history of all the new modern creeds, especially because there are no clear boundaries between them. They are no less syncretic than monotheism and popular Buddhism. Just as a Buddhist could worship Hindu deities, and just as a monotheist could believe in the existence of Satan, so the typical American nowadays is simultaneously a nationalist (she believes in the existence of an American nation with a special role to play in history), a free-market capitalist (she believes that open competition and the pursuit of self-interest are the best ways to create a prosperous society), and a liberal humanist (she believes that humans have been endowed by their creator with certain inalienable rights). Nationalism will be discussed in Chapter 18. Capitalism – the most successful of the modern religions – gets a whole chapter, Chapter 16, which expounds its principal beliefs and rituals. In the remaining pages of this chapter I will address the humanist religions.

Theist religions sanctify the gods. Humanist religions sanctify humanity, or more correctly, *Homo sapiens*. Humanism is a belief that *Homo sapiens* has a unique and sacred nature, which is fundamentally different from the nature of all other animals and of all other phenomena. Humanists believe that the unique nature of *Homo sapiens* is the most important thing in the world, and it determines the meaning of everything that happens in the universe. The supreme good is the good of *Homo sapiens*. The rest of the world and all other beings exist solely for the benefit of this species.

All humanists sanctify humanity, but they do not agree on its definition. Humanism has split into three rival sects that fight over the exact definition of 'humanity', just as rival Christian sects fought over the exact definition of God. Today, the most important humanist sect is liberal humanism, which believes that 'humanity' is a quality of individual humans, and that the liberty of individuals is therefore sacrosanct. According to liberals, the sacred nature of humanity resides within each and every individual *Homo sapiens*. The inner core of individual humans gives meaning to the world, and is the source for all ethical and political authority. If we encounter an ethical or political dilemma, we should look inside and listen to our inner voice – the voice of humanity. The chief commandments of liberal humanism are meant to protect the liberty of this inner voice against intrusion or harm. These commandments are collectively known as 'human rights'.

This, for example, is why liberals object to torture and the death penalty. In early modern Europe, murderers were thought to violate and destabilise the cosmic order. To bring the cosmos back to balance, it was necessary to torture and publicly execute the criminal, so that everyone could see the order re-established. Attending gruesome executions was a favourite pastime for Londoners and Parisians in the era of Shakespeare and Molière. In today's Europe, murder is seen as a violation of the sacred nature of humanity. In order to restore order, present-day Europeans do not torture and execute criminals. Instead, they punish a murderer in what they see as the most 'humane' way possible, thus safeguarding and even rebuilding his human sanctity. By honouring the human nature of the murderer, everyone is reminded of the sanctity of humanity, and order is restored. By defending the murderer, we right what the murderer has wronged.

Even though liberal humanism sanctifies humans, it does not deny the existence of God, and is, in fact, founded on monotheist beliefs. The liberal belief in the free and sacred nature of each individual is a direct legacy of the traditional Christian belief in free and eternal individual souls. Without recourse to eternal souls and a Creator God, it becomes embarrassingly difficult for liberals to explain what is so special about individual Sapiens.

Another important sect is socialist humanism. Socialists believe that 'humanity' is collective rather than individualistic. They hold as sacred not the inner voice of each individual, but the species *Homo sapiens* as a whole. Whereas liberal humanism seeks as much freedom as possible for individual humans, socialist humanism seeks equality between all humans. According to socialists, inequality is the worst blasphemy against the sanctity of humanity, because it privileges peripheral qualities of humans over their universal essence. For example, when the rich are privileged over the poor, it means that we value money more than the universal essence of all humans, which is the same for rich and poor alike.

Like liberal humanism, socialist humanism is built on monotheist foundations. The idea that all humans are equal is a revamped version of the monotheist conviction that all souls are equal before God. The only humanist sect that has actually broken loose from traditional monotheism is evolutionary humanism, whose most famous representatives were the Nazis. What distinguished the Nazis from other humanist sects was a different definition of 'humanity', one deeply influenced by the theory of evolution. In contrast to other humanists, the Nazis believed that humankind is not something universal and eternal, but rather a mutable species that can evolve or degenerate. Man can evolve into superman, or degenerate into a subhuman.

The main ambition of the Nazis was to protect humankind from degeneration and encourage its progressive evolution. This is why the Nazis said that the Aryan race, the most advanced form of humanity, had to be protected and fostered, while degenerate kinds of *Homo sapiens* like Jews, Roma, homosexuals and the mentally ill had to be quarantined and even exterminated. The Nazis explained that *Homo sapiens* itself appeared when one 'superior' population of ancient humans evolved, whereas 'inferior' populations such as the

Neanderthals became extinct. These different populations were at first no more than different races, but developed independently along their own evolutionary paths. This might well happen again. According to the Nazis, *Homo sapiens* had already divided into several distinct races, each with its own unique qualities. One of these races, the Aryan race, had the finest qualities – rationalism, beauty, integrity, diligence. The Aryan race therefore had the potential to turn man into superman. Other races, such as Jews and blacks, were today's Neanderthals, possessing inferior qualities. If allowed to breed, and in particular to intermarry with Aryans, they would adulterate all human populations and doom *Homo sapiens* to extinction.

Biologists have since debunked Nazi racial theory. In particular, genetic research conducted after 1945 has demonstrated that the differences between the various human lineages are far smaller than the Nazis postulated. But these conclusions are relatively new. Given the state of scientific knowledge in 1933, Nazi beliefs were hardly outside the pale. The existence of different human races, the superiority of the white race, and the need to protect and cultivate this superior race were widely held beliefs among most Western elites. Scholars in the most prestigious Western universities, using the orthodox scientific methods of the day, published studies that allegedly proved that members of the white race were more intelligent, more ethical and more skilled than Africans or Indians. Politicians in Washington, London and Canberra took it for granted that it was their job to prevent the adulteration and degeneration of the white race, by, for example, restricting immigration from China or even Italy to 'Aryan' countries such as the USA and Australia.

These positions did not change simply because new scientific research was published. Sociological and political developments were far more powerful engines of change. In this sense, Hitler dug not just his own grave but that of racism in general. When he launched the Second World War, he compelled his enemies to make clear distinctions

Humanist Religions – Religions that Sanctify Humanity

Liberal humanism	**Socialist humanism**	**Evolutionary humanism**
Homo sapiens has a unique and sacred nature that is fundamentally different from the nature of all other beings and phenomena. The supreme good is the good of humanity.		
'Humanity' is individualistic and resides within each individual *Homo sapiens.*	'Humanity' is collective and resides within the species *Homo sapiens* as a whole.	'Humanity' is a mutable species. Humans might degenerate into subhumans or evolve into superhumans.
The supreme commandment is to protect the inner core and freedom of each individual *Homo sapiens.*	The supreme commandment is to protect equality within the species *Homo sapiens.*	The supreme commandment is to protect humankind from degenerating into subhumans, and to encourage its evolution into superhumans.

between 'us' and 'them'. Afterwards, precisely because Nazi ideology was so racist, racism became discredited in the West. But the change took time. White supremacy remained a mainstream ideology in American politics at least until the 1960s. The White Australia policy which restricted immigration of non-white people to Australia remained in force until 1966. Aboriginal Australians did not receive equal political rights until the 1960s, and most were prevented from voting in elections because they were deemed unfit to function as citizens.

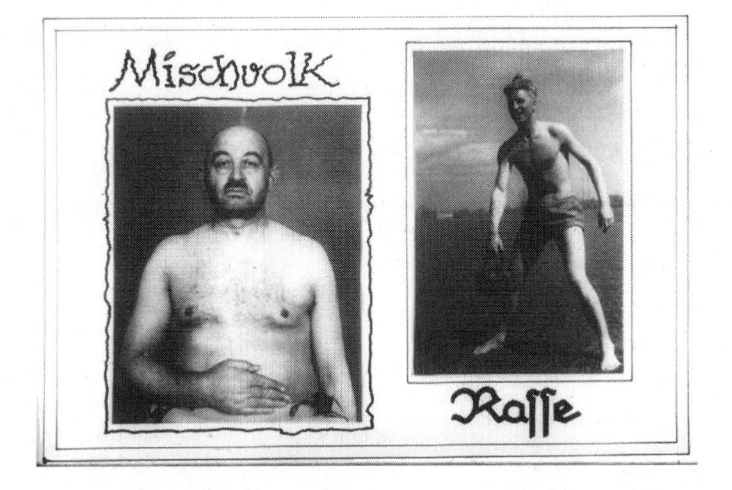

30. A Nazi propaganda poster showing on the right a 'racially pure Aryan' and on the left a 'cross-breed'. Nazi admiration for the human body is evident, as is their fear that the lower races might pollute humanity and cause its degeneration.

The Nazis did not loathe humanity. They fought liberal humanism, human rights and Communism precisely because they admired humanity and believed in the great potential of the human species. But following the logic of Darwinian evolution, they argued that natural selection must be allowed to weed out unfit individuals and leave only the fittest to survive and reproduce. By succouring the weak, liberalism and Communism not only allowed unfit individuals to survive, they actually gave them the opportunity to reproduce, thereby undermining natural selection. In such a world, the fittest humans would inevitably drown in a sea of unfit degenerates. Humankind would become less and less fit with each passing generation – which could lead to its extinction.

A 1942 German biology textbook explains in the chapter 'The Laws of Nature and Mankind' that the supreme law of nature is that all beings are locked in a remorseless struggle for survival. After

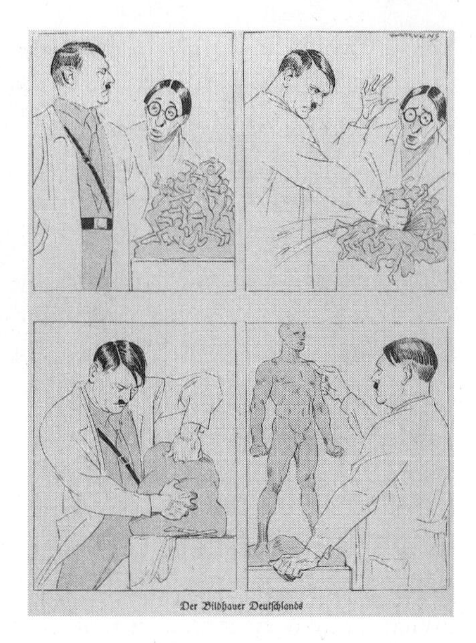

31. A Nazi cartoon of 1933. Hitler is presented as a sculptor who creates the superman. A bespectacled liberal intellectual is appalled by the violence needed to create the superman. (Note also the erotic glorification of the human body.)

describing how plants struggle for territory, how beetles struggle to find mates and so forth, the textbook concludes that:

The battle for existence is hard and unforgiving, but is the only way to maintain life. This struggle eliminates everything that is unfit for life, and selects everything that is able to survive . . . These natural laws are incontrovertible; living creatures demonstrate them by their very survival. They are unforgiving. Those who resist them will be wiped out. Biology not only tells us about animals and plants, but also shows us the laws we must follow in our lives, and steels our wills to live and fight according to these laws. The meaning of life is struggle. Woe to him who sins against these laws.

Then follows a quotation from *Mein Kampf*: 'The person who attempts to fight the iron logic of nature thereby fights the principles he must thank for his life as a human being. To fight against nature is to bring about one's own destruction.'[3]

*

At the dawn of the third millennium, the future of evolutionary humanism is unclear. For sixty years after the end of the war against Hitler it was taboo to link humanism with evolution and to advocate using biological methods to 'upgrade' *Homo sapiens*. But today such projects are back in vogue. No one speaks about exterminating lower races or inferior people, but many contemplate using our increasing knowledge of human biology to create superhumans.

At the same time, a huge gulf is opening between the tenets of liberal humanism and the latest findings of the life sciences, a gulf we cannot ignore much longer. Our liberal political and judicial systems are founded on the belief that every individual has a sacred inner nature, indivisible and immutable, which gives meaning to the world, and which is the source of all ethical and political authority. This is a reincarnation of the traditional Christian belief in a free and eternal soul that resides within each individual. Yet over the last 200 years, the life sciences have thoroughly undermined this belief. Scientists studying the inner workings of the human organism have found no soul there. They increasingly argue that human behaviour is determined by hormones, genes and synapses, rather than by free will – the same forces that determine the behaviour of chimpanzees, wolves and ants. Our judicial and political systems largely try to sweep such inconvenient discoveries under the carpet. But in all frankness, how long can we maintain the wall separating the department of biology from the departments of law and political science?

13

The Secret of Success

COMMERCE, EMPIRES AND UNIVERSAL religions eventually brought virtually every Sapiens on every continent into the global world we live in today. Not that this process of expansion and unification was linear or without interruptions. Looking at the bigger picture, though, the transition from many small cultures to a few large cultures and finally to a single global society was probably an inevitable result of the dynamics of human history.

But saying that a global society is inevitable is not the same as saying that the end result had to be the particular kind of global society we now have. We can certainly imagine other outcomes. Why is English so widespread today, and not Danish? Why are there about 2 billion Christians and 1.25 billion Muslims, but only 150,000 Zoroastrians and no Manichaeans? If we could go back in time to 10,000 years ago and set the process going again, time after time, would we always see the rise of monotheism and the decline of dualism?

We can't do such an experiment, so we don't really know. But an examination of two crucial characteristics of history can provide us with some clues.

1. The Hindsight Fallacy

Every point in history is a crossroads. A single travelled road leads from the past to the present, but myriad paths fork off into the future. Some of those paths are wider, smoother and better marked, and are thus more likely to be taken, but sometimes history – or the people who make history – takes unexpected turns.

At the beginning of the fourth century AD, the Roman Empire faced a wide horizon of religious possibilities. It could have stuck to its traditional and variegated polytheism. But its emperor, Constantine, looking back on a fractious century of civil war, seems to have thought that a single religion with a clear doctrine could help unify his ethnically diverse realm. He could have chosen any of a number of contemporary cults to be his national faith – Manichaeism, Mithraism, the cults of Isis or Cybele, Zoroastrianism, Judaism and even Buddhism were all available options. Why did he opt for Jesus? Was there something in Christian theology that attracted him personally, or perhaps an aspect of the faith that made him think it would be easier to use for his purposes? Did he have a religious experience, or did some of his advisers suggest that the Christians were quickly gaining adherents and that it would be best to jump on that wagon? Historians can speculate, but not provide any definitive answer. They can describe *how* Christianity took over the Roman Empire, but they cannot explain *why* this particular possibility was realised.

What is the difference between describing 'how' and explaining 'why'? To describe 'how' means to reconstruct the series of specific events that led from one point to another. To explain 'why' means to find causal connections that account for the occurrence of this particular series of events to the exclusion of all others.

Some scholars do indeed provide deterministic explanations of events such as the rise of Christianity. They attempt to reduce human history to the workings of biological, ecological or economic forces. They argue that there was something about the geography, genetics or economy of the Roman Mediterranean that made the rise of a monotheist religion inevitable. Yet most historians tend to be sceptical of such deterministic theories. This is one of the distinguishing marks of history as an academic discipline – the better you know a particular historical period, the *harder* it becomes to explain why things happened one way and not another. Those who have only a superficial knowledge of a certain period tend to focus only on the possibility that was eventually realised. They offer a just-so story to explain with hindsight why that outcome was inevitable. Those more deeply informed about the period are much more cognisant of the roads not taken.

In fact, the people who knew the period best – those alive at the time – were the most clueless of all. For the average Roman in Constantine's time, the future was a fog. It is an iron rule of history that what looks inevitable in hindsight was far from obvious at the time. Today is no different. Are we out of the global economic crisis, or is the worst still to come? Will China continue growing until it becomes the leading superpower? Will the United States lose its hegemony? Is the upsurge of monotheistic fundamentalism the wave of the future or a local whirlpool of little long-term significance? Are we heading towards ecological disaster or technological paradise? There are good arguments to be made for all of these outcomes, but no way of knowing for sure. In a few decades, people will look back and think that the answers to all of these questions were obvious.

It is particularly important to stress that possibilities which seem very unlikely to contemporaries often get realised. When Constantine assumed the throne in 306, Christianity was little more than an esoteric Eastern sect. If you were to suggest then that it was about to become the Roman state religion, you'd have been laughed out of the room just as you would be today if you were to suggest that by the year 2050 Hare Krishna would be the state religion of the USA. In

October 1913, the Bolsheviks were a small radical Russian faction. No reasonable person would have predicted that within a mere four years they would take over the country. In AD 600, the notion that a band of desert-dwelling Arabs would soon conquer an expanse stretching from the Atlantic Ocean to India was even more preposterous. Indeed, had the Byzantine army been able to repel the initial onslaught, Islam would probably have remained an obscure cult of which only a handful of cognoscenti were aware. Scholars would then have a very easy job explaining why a faith based on a revelation to a middle-aged Meccan merchant could never have caught on.

Not that everything is possible. Geographical, biological and economic forces create constraints. Yet these constraints leave ample room for surprising developments, which do not seem bound by any deterministic laws.

This conclusion disappoints many people, who prefer history to be deterministic. Determinism is appealing because it implies that our world and our beliefs are a natural and inevitable product of history. It is natural and inevitable that we live in nation states, organise our economy along capitalist principles, and fervently believe in human rights. To acknowledge that history is not deterministic is to acknowledge that it is just a coincidence that most people today believe in nationalism, capitalism and human rights.

History cannot be explained deterministically and it cannot be predicted because it is chaotic. So many forces are at work and their interactions are so complex that extremely small variations in the strength of the forces and the way they interact produce huge differences in outcomes. Not only that, but history is what is called a 'level two' chaotic system. Chaotic systems come in two shapes. Level one chaos is chaos that does not react to predictions about it. The weather, for example, is a level one chaotic system. Though it is influenced by myriad factors, we can build computer models that take more and more of them into consideration, and produce better and better weather forecasts.

Level two chaos is chaos that reacts to predictions about it, and therefore can never be predicted accurately. Markets, for example, are a level two chaotic system. What will happen if we develop a computer program that forecasts with 100 per cent accuracy the price of oil tomorrow? The price of oil will immediately react to the forecast, which would consequently fail to materialise. If the current price of oil is $90 a barrel, and the infallible computer program predicts that tomorrow it will be $100, traders will rush to buy oil so that they can profit from the predicted price rise. As a result, the price will shoot up to $100 a barrel today rather than tomorrow. Then what will happen tomorrow? Nobody knows.

Politics, too, is a second-order chaotic system. Many people criticise Sovietologists for failing to predict the 1989 revolutions and castigate Middle East experts for not anticipating the Arab Spring revolutions of 2011. This is unfair. Revolutions are, by definition, unpredictable. A predictable revolution never erupts.

Why not? Imagine that it's 2010 and some genius political scientists in cahoots with a computer wizard have developed an infallible algorithm that, incorporated into an attractive interface, can be marketed as a revolution predictor. They offer their services to President Hosni Mubarak of Egypt and, in return for a generous down payment, tell Mubarak that according to their forecasts a revolution would certainly break out in Egypt during the course of the following year. How would Mubarak react? Most likely, he would immediately lower taxes, distribute billions of dollars in handouts to the citizenry – and also beef up his secret police force, just in case. The pre-emptive measures work. The year comes and goes and, surprise, there is no revolution. Mubarak demands his money back. 'Your algorithm is worthless!' he shouts at the scientists. 'In the end I could have built another palace instead of giving all that money away!' 'But the reason the revolution didn't happen is because we predicted it,' the scientists say in their defence. 'Prophets who predict things that don't happen?' Mubarak remarks as he motions his guards to grab them. 'I could have picked up a dozen of those for next to nothing in the Cairo marketplace.'

So why study history? Unlike physics or economics, history is not a means for making accurate predictions. We study history not to know the future but to widen our horizons, to understand that our present situation is neither natural nor inevitable, and that we consequently have many more possibilities before us than we imagine. For example, studying how Europeans came to dominate Africans enables us to realise that there is nothing natural or inevitable about the racial hierarchy, and that the world might well be arranged differently.

2. Blind Clio

We cannot explain the choices that history makes, but we can say something very important about them: history's choices are not made for the benefit of humans. There is absolutely no proof that human well-being inevitably improves as history rolls along. There is no proof that cultures that are beneficial to humans must inexorably succeed and spread, while less beneficial cultures disappear. There is no proof that Christianity was a better choice than Manichaeism, or that the Arab Empire was more beneficial than that of the Sassanid Persians.

There is no proof that history is working for the benefit of humans because we lack an objective scale on which to measure such benefit. Different cultures define the good differently, and we have no objective yardstick by which to judge between them. The victors, of course, always believe that their definition is correct. But why should we believe the victors? Christians believe that the victory of Christianity over Manichaeism was beneficial to humankind, but if we do not accept the Christian world view then there is no reason to agree with them. Muslims believe that the fall of the Sassanid Empire into Muslim hands was beneficial to humankind. But these benefits are evident only if we accept the Muslim world view. It may well be that we'd all be better off if Christianity and Islam had been forgotten or defeated.

Ever more scholars see cultures as a kind of mental infection or parasite, with humans as its unwitting host. Organic parasites, such as viruses, live inside the body of their hosts. They multiply and spread from one host to the other, feeding off their hosts, weakening them, and sometimes even killing them. As long as the hosts live long enough to pass along the parasite, it cares little about the condition of its host. In just this fashion, cultural ideas live inside the minds of humans. They multiply and spread from one host to another, occasionally weakening the hosts and sometimes even killing them. A cultural idea – such as belief in Christian heaven above the clouds or Communist paradise here on earth – can compel a human to dedicate his or her life to spreading that idea, even at the price of death. The human dies, but the idea spreads. According to this approach, cultures are not conspiracies concocted by some people in order to take advantage of others (as Marxists tend to think). Rather, cultures are mental parasites that emerge accidentally, and thereafter take advantage of all people infected by them.

This approach is sometimes called memetics. It assumes that, just as organic evolution is based on the replication of organic information units called 'genes', so cultural evolution is based on the replication of cultural information units called 'memes'.[1] Successful cultures are those that excel in reproducing their memes, irrespective of the costs and benefits to their human hosts.

Most scholars in the humanities disdain memetics, seeing it as an amateurish attempt to explain cultural processes with crude biological analogies. But many of these same scholars adhere to memetics' twin sister – postmodernism. Postmodernist thinkers speak about discourses rather than memes as the building blocks of culture. Yet they too see cultures as propagating themselves with little regard for the benefit of humankind. For example, postmodernist thinkers describe nationalism as a deadly plague that spread throughout the world in the nineteenth and twentieth centuries, causing wars, oppression, hate and genocide. The moment people in one country were infected with it, those in neighbouring countries were also likely to catch the virus. The nationalist virus presented itself as being beneficial for humans, yet it has been beneficial mainly to itself.

Similar arguments are common in the social sciences, under the aegis of game theory. Game theory explains how in multi-player systems, views and behaviour patterns that harm *all* players nevertheless manage to take root and spread. Arms races are a famous example. Many arms races bankrupt all those who take part in them, without really changing the military balance of power. When Pakistan buys advanced aeroplanes, India responds in kind. When India develops nuclear bombs, Pakistan follows suit. When Pakistan enlarges its navy, India counters. At the end of the process, the balance of power may remain much as it was, but meanwhile billions of dollars that could have been invested in education or health are spent on weapons. Yet the arms race dynamic is hard to resist. 'Arms racing' is a pattern of behaviour that spreads itself like a virus from one country to another, harming everyone, but benefiting itself, under the evolutionary criteria of survival and reproduction. (Keep in mind that an arms race, like a gene, has no awareness – it does not consciously seek to survive and reproduce. Its spread is the unintended result of a powerful dynamic.)

No matter what you call it – game theory, postmodernism or memetics – the dynamics of history are not directed towards enhancing human well-being. There is no basis for thinking that the most successful cultures in history are necessarily the best ones for *Homo sapiens*. Like evolution, history disregards the happiness of individual organisms. And individual humans, for their part, are usually far too ignorant and weak to influence the course of history to their own advantage.

History proceeds from one junction to the next, choosing for some mysterious reason to follow first this path, then another. Around

AD 1500, history made its most momentous choice, changing not only the fate of humankind, but arguably the fate of all life on earth. We call it the Scientific Revolution. It began in western Europe, a large peninsula on the western tip of Afro-Asia, which up till then played no important role in history. Why did the Scientific Revolution begin there of all places, and not in China or India? Why did it begin at the midpoint of the second millennium AD rather than two centuries before or three centuries later? We don't know. Scholars have proposed dozens of theories, but none of them is particularly convincing.

History has a very wide horizon of possibilities, and many possibilities are never realised. It is conceivable to imagine history going on for generations upon generations while bypassing the Scientific Revolution, just as it is conceivable to imagine history without Christianity, without a Roman Empire, and without gold coins.

Part Four
The Scientific Revolution

32. Alamogordo, 16 July 1945, 05:29:53. Eight seconds after the first atomic bomb was detonated. The nuclear physicist Robert Oppenheimer, upon seeing the explosion, quoted from the Bhagavad Gita: 'Now I am become Death, the destroyer of worlds.'

14

The Discovery of Ignorance

WERE, SAY, A SPANISH PEASANT TO HAVE fallen asleep in AD 1000 and woken up 500 years later, to the din of Columbus' sailors boarding the *Niña*, *Pinta* and *Santa Maria*, the world would have seemed to him quite familiar. Despite many changes in technology, manners and political boundaries, this medieval Rip Van Winkle would have felt at home. But had one of Columbus' sailors fallen into a similar slumber and woken up to the ringtone of a twenty-first-century iPhone, he would have found himself in a world strange beyond comprehension. 'Is this heaven?' he might well have asked himself. 'Or perhaps – hell?'

The last 500 years have witnessed a phenomenal and unprecedented growth in human power. In the year 1500, there were about 500 million *Homo sapiens* in the entire world. Today, there are 7 billion.[1] The total value of goods and services produced by humankind in the year 1500 is estimated at $250 billion, in today's dollars.[2] Nowadays the value of a year of human production is close to $60 trillion.[3] In 1500, humanity consumed about 13 trillion calories of energy per day. Today, we consume 1,500 trillion calories a day.[4] (Take a second look at those figures – human population has increased fourteenfold, production 240-fold, and energy consumption 115-fold.)

Suppose a single modern battleship got transported back to Columbus' time. In a matter of seconds it could make driftwood out of the *Niña*, *Pinta* and *Santa Maria* and then sink the navies of every great world power of the time without sustaining a scratch. Five modern freighters could have taken on board all the cargo borne by the whole world's merchant fleets.[5] A modern computer could easily store every word and number in all the codex books and scrolls in every single medieval library, with room to spare. Any large bank today holds more money than all the world's premodern kingdoms put together.[6]

In 1500, few cities had more than 100,000 inhabitants. Most buildings were constructed of mud, wood and straw; a three-storey building was a skyscraper. The streets were rutted dirt tracks, dusty in summer and muddy in winter, plied by pedestrians, horses, goats, chickens and a few carts. The most common urban noises were human and animal voices, along with the occasional hammer and saw. At sunset, the cityscape went black, with only an occasional candle or torch flickering in the gloom. If an inhabitant of such a city could see modern Tokyo, New York or Mumbai, what would she think?

Prior to the sixteenth century, no human had circumnavigated the earth. This changed in 1522, when Magellan's expedition returned to Spain after a journey of 72,000 kilometres. It took three years and cost the lives of almost all the crew members, Magellan included. In 1873, Jules Verne could imagine that Phileas Fogg, a wealthy British adventurer, might just be able to make it around the world in eighty days. Today anyone with a middle-class income can safely and easily circumnavigate the globe in just forty-eight hours.

In 1500, humans were confined to the earth's surface. They could build towers and climb mountains, but the sky was reserved for birds, angels and deities. On 20 July 1969 humans landed on the moon. This was not merely a historical achievement, but an evolutionary and even cosmic feat. During the previous 4 billion years of evolution, no organism managed even to leave the earth's atmosphere, and certainly none left a foot or tentacle print on the moon.

For most of history, humans knew nothing about 99.99 per cent of the organisms on the planet – namely, the microorganisms. This was not because they were of no concern to us. Each of us bears billions of one-celled creatures within us, and not just as free-riders. They are our best friends, and deadliest enemies. Some of them digest our food and clean our guts, while others cause illnesses and epidemics. Yet it was only in 1674 that a human eye first saw a microorganism, when Anton van Leeuwenhoek took a peek through his home-made microscope and was startled to see an entire world of tiny creatures milling about in a drop of water. During the subsequent 300 years, humans have made the acquaintance of a huge number of microscopic species. We've managed to defeat most of the deadliest contagious diseases they cause, and have harnessed microorganisms in the service of medicine and industry. Today we engineer bacteria to produce medications, manufacture biofuel and kill parasites.

But the single most remarkable and defining moment of the past 500 years came at 05:29:45 on 16 July 1945. At that precise second, American scientists detonated the first atomic bomb at Alamogordo, New Mexico. From that point onward, humankind had the capability not only to change the course of history, but to end it.

The historical process that led to Alamogordo and to the moon is known as the Scientific Revolution. During this revolution humankind has obtained enormous new powers by investing resources in scientific research. It is a revolution because, until about AD 1500, humans the world over doubted their ability to obtain new medical, military and economic powers. While government and wealthy patrons allocated funds to education and scholarship, the aim was, in general, to preserve existing capabilities rather than acquire new ones. The typical premodern ruler gave money to priests, philosophers and poets in the hope that they would legitimise his rule and maintain the social order. He did not expect them to discover new medications, invent new weapons or stimulate economic growth.

During the last five centuries, humans increasingly came to believe that they could increase their capabilities by investing in scientific research. This wasn't just blind faith – it was repeatedly proven empirically. The more proofs there were, the more resources wealthy people and governments were willing to put into science. We would never have been able to walk on the moon, engineer microorganisms and split the atom without such investments. The US government, for example, has in recent decades allocated billions of dollars to the study of nuclear physics. The knowledge produced by this research has made possible the construction of nuclear power stations, which provide cheap electricity for American industries, which pay taxes to the US government, which uses some of these taxes to finance further research in nuclear physics.

Why did modern humans develop a growing belief in their ability to obtain new powers through research? What forged the bond between science, politics and economics? This chapter looks at the

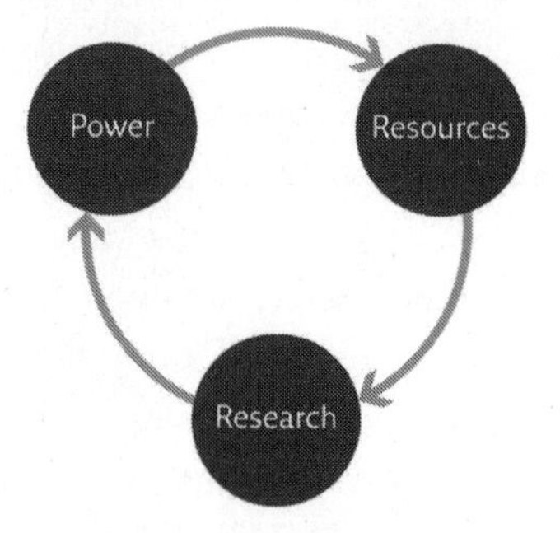

The Scientific Revolution's feedback loop. Science needs more than just research to make progress. It depends on the mutual reinforcement of science, politics and economics. Political and economic institutions provide the resources without which scientific research is almost impossible. In return, scientific research provides new powers that are used, among other things, to obtain new resources, some of which are reinvested in research.

unique nature of modern science in order to provide part of the answer. The next two chapters examine the formation of the alliance between science, the European empires and the economics of capitalism.

Ignoramus

Humans have sought to understand the universe at least since the Cognitive Revolution. Our ancestors put a great deal of time and effort into trying to discover the rules that govern the natural world. But modern science differs from all previous traditions of knowledge in three critical ways:

a. **The willingness to admit ignorance.** Modern science is based on the Latin injunction *ignoramus* – 'we do not know'. It assumes that we don't know everything. Even more critically, it accepts that the things that we think we know could be proven wrong as we gain more knowledge. No concept, idea or theory is sacred and beyond challenge.

b. **The centrality of observation and mathematics.** Having admitted ignorance, modern science aims to obtain new knowledge. It does so by gathering observations and then using mathematical tools to connect these observations into comprehensive theories.

c. **The acquisition of new powers.** Modern science is not content with creating theories. It uses these theories in order to acquire new powers, and in particular to develop new technologies.

The Scientific Revolution has not been a revolution of knowledge. It has been above all a revolution of ignorance. The great discovery that launched the Scientific Revolution was the discovery that humans do not know the answers to their most important questions.

Premodern traditions of knowledge such as Islam, Christianity, Buddhism and Confucianism asserted that everything that is important to know about the world was already known. The great gods, or the one almighty God, or the wise people of the past possessed all-encompassing wisdom, which they revealed to us in scriptures and oral traditions. Ordinary mortals gained knowledge by delving into these ancient texts and traditions and understanding them properly. It was inconceivable that the Bible, the Qur'an or the Vedas were missing out on a crucial secret of the universe – a secret that might yet be discovered by flesh-and-blood creatures.

Ancient traditions of knowledge admitted only two kinds of ignorance. First, an *individual* might be ignorant of something important. To obtain the necessary knowledge, all he needed to do was ask somebody wiser. There was no need to discover something that nobody yet knew. For example, if a peasant in some thirteenth-century Yorkshire village wanted to know how the human race originated, he assumed that Christian tradition held the definitive answer. All he had to do was ask the local priest.

Second, an *entire tradition* might be ignorant of *unimportant* things. By definition, whatever the great gods or the wise people of the past did not bother to tell us was unimportant. For example, if our Yorkshire peasant wanted to know how spiders weave their webs, it was pointless to ask the priest, because there was no answer to this question in any of the Christian Scriptures. That did not mean, however, that Christianity was deficient. Rather, it meant that understanding how spiders weave their webs was unimportant. After all, God knew perfectly well how spiders do it. If this were a vital piece of information, necessary for human prosperity and salvation, God would have included a comprehensive explanation in the Bible.

Christianity did not forbid people to study spiders. But spider scholars – if there were any in medieval Europe – had to accept their peripheral role in society and the irrelevance of their findings to the eternal truths of Christianity. No matter what a scholar might discover about spiders or butterflies or Galapagos finches, that knowledge was little more than trivia, with no bearing on the fundamental truths of society, politics and economics.

In fact, things were never quite that simple. In every age, even the most pious and conservative, there were people who argued that there were *important* things of which their *entire tradition* was ignorant. Yet such people were usually marginalised or persecuted – or else they founded a new tradition and began arguing that *they* knew everything there is to know. For example, the prophet Muhammad began his religious career by condemning his fellow Arabs for living in ignorance of the divine truth. Yet Muhammad himself very quickly began to argue that *he* knew the full truth, and his followers began calling him 'the Seal of the Prophets'. Henceforth, there was no need of revelations beyond those given to Muhammad.

Modern-day science is a unique tradition of knowledge, inasmuch as it openly admits *collective* ignorance regarding *the most important questions*. Darwin never argued that he was 'the Seal of the Biologists', and that he had solved the riddle of life once and for all. After centuries of extensive scientific research, biologists admit that they still don't have any good explanation for how brains produce consciousness. Physicists admit that they don't know what caused the Big Bang, or how to reconcile quantum mechanics with the theory of general relativity.

In other cases, competing scientific theories are vociferously debated on the basis of constantly emerging new evidence. A prime example is the debates about how best to run the economy. Though individual economists may claim that their method is the best, orthodoxy changes with every financial crisis and stock-exchange bubble, and it is generally accepted that the final word on economics is yet to be said.

In still other cases, particular theories are supported so consistently by the available evidence, that all alternatives have long since fallen by the wayside. Such theories are accepted as true – yet everyone agrees that were new evidence to emerge that contradicts the theory, it would have to be revised or discarded. Good examples of these are the plate tectonics theory and the theory of evolution.

The willingness to admit ignorance has made modern science more dynamic, supple and inquisitive than any previous tradition of knowledge. This has hugely expanded our capacity to understand how the world works and our ability to invent new technologies. But it presents us with a serious problem that most of our ancestors did not have to cope with. Our current assumption that we do not know everything, and that even the knowledge we possess is tentative, extends to the shared myths that enable millions of strangers to cooperate effectively. If the evidence shows that many of those myths are doubtful, how can we hold society together? How can our communities, countries and international system function?

All modern attempts to stabilise the sociopolitical order have had no choice but to rely on either of two unscientific methods:

a. Take a scientific theory, and in opposition to common scientific practices, *declare that it is a final and absolute truth.* This was the method used by Nazis (who claimed that their racial policies were the corollaries of biological facts) and Communists (who claimed that Marx and Lenin had divined absolute economic truths that could never be refuted).

b. Leave science out of it and live in accordance with *a non-scientific absolute truth.* This has been the strategy of liberal humanism, which is built on a dogmatic belief in the unique worth and rights of human beings – a doctrine which has embarrassingly little in common with the scientific study of *Homo sapiens.*

But that shouldn't surprise us. Even science itself has to rely on religious and ideological beliefs to justify and finance its research.

Modern culture has nevertheless been willing to embrace ignorance to a much greater degree than has any previous culture. One of the things that has made it possible for modern social orders to hold together is the spread of an almost religious belief in technology and in the methods of scientific research, which have replaced to some extent the belief in absolute truths.

The Scientific Dogma

Modern science has no dogma. Yet it has a common core of research methods, which are all based on collecting empirical observations – those we can observe with at least one of our senses – and putting them together with the help of mathematical tools.

People throughout history collected empirical observations, but the importance of these observations was usually limited. Why waste precious resources obtaining new observations when we already have all the answers we need? But as modern people came to admit that they did not know the answers to some very important questions, they found it necessary to look for *completely new* knowledge. Consequently, the dominant modern research method takes for granted the insufficiency of old knowledge. Instead of studying old traditions, emphasis is now placed on new observations and experiments. When present observation collides with past tradition, we give precedence to the observation. Of course, physicists analysing the spectra of distant galaxies, archaeologists analysing the finds from a Bronze Age city, and political scientists studying the emergence of capitalism do not disregard tradition. They start by studying what the wise people of the past have said and written. But from their first year in college, aspiring physicists, archaeologists and political scientists are taught that it is their mission to go beyond what Einstein, Heinrich Schliemann and Max Weber ever knew.

Mere observations, however, are not knowledge. In order to understand the universe, we need to connect observations into comprehensive theories. Earlier traditions usually formulated their theories in terms of stories. Modern science uses mathematics.

There are very few equations, graphs and calculations in the Bible, the Qur'an, the Vedas or the Confucian classics. When traditional mythologies and scriptures laid down general laws, these were presented in narrative rather than mathematical form. Thus a fundamental principle of Manichaean religion asserted that the world is a battleground between good and evil. An evil force created matter, while a good force created spirit. Humans are caught between these two forces, and should choose good over evil. Yet the prophet Mani made no attempt to offer a mathematical formula that could be used to predict human choices by quantifying the respective strength of these two forces. He never calculated that 'the force acting on a man is equal to the acceleration of his spirit divided by the mass of his body'.

This is exactly what scientists seek to accomplish. In 1687, Isaac Newton published *The Mathematical Principles of Natural Philosophy*, arguably the most important book in modern history. Newton presented a general theory of movement and change. The greatness of Newton's theory was its ability to explain and predict the movements of all bodies in the universe, from falling apples to shooting stars, using three very simple mathematical laws:

1. $\sum \vec{F} = 0$

2. $\sum \vec{F} = m\vec{a}$

3. $\vec{F}_{1,2} = \vec{F}_{2,1}$

Henceforth, anyone who wished to understand and predict the movement of a cannonball or a planet simply had to make measurements of the object's mass, direction and acceleration, and the forces acting on it. By inserting these numbers into Newton's equations, the future position of the object could be predicted. It worked like magic. Only around the end of the nineteenth century did scientists come across a few observations that did not fit well with Newton's laws, and these led to the next revolutions in physics – the theory of relativity and quantum mechanics.

Newton showed that the book of nature is written in the language of mathematics. Some chapters (physics, for example) boil down to clear-cut equations; but scholars who attempted to reduce biology, economics and psychology to neat Newtonian equations have discovered that these fields have a level of complexity that makes such an aspiration futile. This did not mean, however, that they gave up on mathematics. A new branch of mathematics was developed over the last 200 years to deal with the more complex aspects of reality: statistics.

In 1744, two Presbyterian clergymen in Scotland, Alexander Webster and Robert Wallace, decided to set up a life-insurance fund that would provide pensions for the widows and orphans of dead clergymen. They proposed that each of their church's ministers would pay a small portion of his income into the fund, which would invest the money. If a minister died, his widow would receive dividends on the fund's profits. This would allow her to live comfortably for the rest of her life. But to determine how much the ministers had to pay in so that the fund would have enough money to live up to its obligations, Webster and Wallace had to be able to predict how many ministers would die each year, how many widows and orphans they would leave behind, and by how many years the widows would outlive their husbands.

Take note of what the two churchmen did not do. They did not pray to God to reveal the answer. Nor did they search for an answer in the Holy Scriptures or among the works of ancient theologians. Nor did they enter into an abstract philosophical disputation. Being Scots, they were practical types. So they contacted a professor of mathematics from the University of Edinburgh, Colin Maclaurin. The three of them collected data on the ages at which people died and used these to calculate how many ministers were likely to pass away in any given year.

Their work was founded on several recent breakthroughs in the fields of statistics and probability. One of these was Jacob Bernoulli's Law of Large Numbers. Bernoulli had codified the principle that while it might be difficult to predict with certainty a single event, such as the death of a particular person, it was possible to predict with great accuracy the average outcome of many similar events. That is, while Maclaurin could not use maths to predict whether Webster and Wallace would die next year, he could, given enough data, tell Webster and Wallace how many Presbyterian ministers in Scotland would almost certainly die next year. Fortunately, they had ready-made data that they could use. Actuary tables published fifty years previously by Edmond Halley proved particularly useful. Halley had analysed records of 1,238 births and 1,174 deaths that he obtained from the city of Breslau, Germany. Halley's tables made it possible to see that, for example, a twenty-year-old person has a 1:100 chance of dying in a given year, but a fifty-year-old person has a 1:39 chance.

Processing these numbers, Webster and Wallace concluded that, on average, there would be 930 living Scottish Presbyterian ministers at any given moment, and an average of twenty-seven ministers would die each year, eighteen of whom would be survived by widows. Five of those who did not leave widows would leave orphaned children, and two of those survived by widows would also be outlived by children from previous marriages who had not yet reached the age of sixteen. They further computed how much time was likely to go by before the widows' death or remarriage (in both these eventualities, payment of the pension would cease). These figures enabled Webster and Wallace to determine how much money the ministers who joined their fund had to pay in order to provide for their loved ones. By contributing £2 12*s.* 2*d.* a year, a minister could guarantee that his widowed wife would receive at least £10 a year – a hefty sum in those days. If he thought that was not enough he could choose to pay in more, up to a level of £6 11*s.* 3*d.* a year – which would guarantee his widow the even more handsome sum of £25 a year.

According to their calculations, by the year 1765 the Fund for a Provision for the Widows and Children of the Ministers of the Church of Scotland would have capital totalling £58,348. Their calculations proved amazingly accurate. When that year arrived, the fund's capital stood at £58,347 – just £1 less than the prediction! This was even better than the prophecies of Habakkuk, Jeremiah or St John. Today, Webster and Wallace's fund, known simply as Scottish Widows, is one of the largest pension and insurance companies in the world. With assets worth £100 billion, it insures not only Scottish widows, but anyone willing to buy its policies.[7]

Probability calculations such as those used by the two Scottish ministers became the foundation not merely of actuarial science, which is central to the pension and insurance business, but also of the science of demography (founded by another clergyman, the Anglican Robert Malthus). Demography in its turn was the cornerstone on which Charles Darwin (who almost became an Anglican pastor) built his theory of evolution. While there are no equations that predict what kind of organism will evolve under a specific set of conditions, geneticists use probability calculations to compute the likelihood that a particular mutation will spread in a given population. Similar probabilistic models have become central to economics, sociology, psychology, political science and the other social and natural sciences. Even physics eventually supplemented Newton's classical equations with the probability clouds of quantum mechanics.

We need merely look at the history of education to realise how far this process has taken us. Throughout most of history, mathematics was an esoteric field that even educated people rarely studied seriously. In medieval Europe, logic, grammar and rhetoric formed the educational core, while the teaching of mathematics seldom went beyond simple arithmetic and geometry. Nobody studied statistics. The undisputed monarch of all sciences was theology.

Today few students study rhetoric; logic is restricted to philosophy departments, and theology to seminaries. But more and more students are motivated – or forced – to study mathematics. There is an irresistible drift towards the exact sciences – defined as 'exact' by their use of mathematical tools. Even fields of study that were traditionally part of the humanities, such as the study of human language (linguistics) and the human psyche (psychology), rely increasingly on mathematics and seek to present themselves as exact sciences. Statistics courses are now part of the basic requirements not just in physics and biology, but also in psychology, sociology, economics and political science.

In the course catalogue of the psychology department at my own university, the first required course in the curriculum is 'Introduction to Statistics and Methodology in Psychological Research'. Second-year psychology students must take 'Statistical Methods in Psychological Research'. Confucius, Buddha, Jesus and Muhammad would have been bewildered if you'd told them that in order to understand the human mind and cure its illnesses you must first study statistics.

Knowledge is Power

Most people have a hard time digesting modern science because its mathematical language is difficult for our minds to grasp, and its findings often contradict common sense. Out of the 7 billion people in the world, how many really understand quantum mechanics, cell biology or macroeconomics? Science nevertheless enjoys immense prestige because of the new powers it gives us. Presidents and generals may not understand nuclear physics, but they have a good grasp of what nuclear bombs can do.

In 1620 Francis Bacon published a scientific manifesto titled *The*

New Instrument. In it he argued that 'knowledge is power'. The real test of 'knowledge' is not whether it is true, but whether it empowers us. Scientists usually assume that no theory is 100 per cent correct. Consequently, truth is a poor test for knowledge. The real test is utility. A theory that enables us to do new things constitutes knowledge.

Over the centuries, science has offered us many new tools. Some are mental tools, such as those used to predict death rates and economic growth. Even more important are technological tools. The connection forged between science and technology is so strong that today people tend to confuse the two. We often think that it is impossible to develop new technologies without scientific research, and that there is little point in research if it does not result in new technologies.

In fact, the relationship between science and technology is a very recent phenomenon. Prior to 1500, science and technology were totally separate fields. When Bacon connected the two in the early seventeenth century, it was a revolutionary idea. During the seventeenth and eighteenth centuries this relationship tightened, but the knot was tied only in the nineteenth century. Even in 1800, most rulers who wanted a strong army, and most business magnates who wanted a successful business, did not bother to finance research in physics, biology or economics.

I don't mean to claim that there is no exception to this rule. A good historian can find precedent for everything. But an even better historian knows when these precedents are but curiosities that cloud the big picture. Generally speaking, most premodern rulers and business people did not finance research about the nature of the universe in order to develop new technologies, and most thinkers did not try to translate their findings into technological gadgets. Rulers financed educational institutions whose mandate was to spread traditional knowledge for the purpose of buttressing the existing order.

Here and there people did develop new technologies, but these were usually created by uneducated craftsmen using trial and error, not by scholars pursuing systematic scientific research. Cart manufacturers built the same carts from the same materials year in, year out. They did not set aside a percentage of their annual profits in order to research and develop new cart models. Cart design occasionally improved, but it was usually thanks to the ingenuity of some local carpenter who never set foot in a university and did not even know how to read.

This was true of the public as well as the private sector. Whereas modern states call in their scientists to provide solutions in almost every area of national policy, from energy to health to waste disposal, ancient kingdoms seldom did so. The contrast between then and now is most pronounced in weaponry. When outgoing President Dwight Eisenhower warned in 1961 of the growing power of the military–industrial complex, he left out a part of the equation. He should have alerted his country to the military–industrial–scientific complex, because today's wars are scientific productions. The world's military forces initiate, fund and steer a large part of humanity's scientific research and technological development.

When the First World War bogged down into interminable trench warfare, both sides called in the scientists to break the deadlock and save the nation. The men in white answered the call, and out of the laboratories rolled a constant stream of new wonder-weapons: combat aircraft, poison gas, tanks, submarines and ever more efficient machine guns, artillery pieces, rifles and bombs.

Science played an even larger role in the Second World War. By late 1944 Germany was losing the war and defeat was imminent. A year earlier, the Germans' main allies, the Italians, had toppled Mussolini and surrendered. But Germany kept fighting on, even though the British, American and Soviet armies were closing in. One reason German soldiers and civilians thought not all was lost was that they believed German scientists were about to turn the tide with so-called miracle weapons such as the V-2 rocket and jet-powered aircraft.

33. German V-2 rocket ready to launch. It didn't defeat the Allies, but it kept the Germans hoping for a technological miracle until the very last days of the war.

While the Germans were working on rockets and jets, the American Manhattan Project successfully developed atomic bombs. By the time the bomb was ready, in early August 1945, Germany had already surrendered, but Japan was fighting on. American forces were poised to invade its home islands. The Japanese vowed to resist the invasion and fight to the death, and there was every reason to believe that it was no idle threat. American generals told President Harry S. Truman that an invasion of Japan would cost the lives of a million American soldiers and would extend the war well into 1946. Truman decided to use the new bomb. Two weeks and two atom bombs later, Japan surrendered unconditionally and the war was over.

But science is not just about offensive weapons. It plays a major role in our defences as well. Today many Americans believe that the solution to terrorism is technological rather than political. Just give millions more to the nanotechnology industry, they believe, and the United States could send bionic spy-flies into every Afghan cave, Yemenite redoubt and North African encampment. Once that's done, Osama bin Laden's heirs will not be able to make a cup of coffee without a CIA spy-fly passing this vital information back to headquarters in Langley. Allocate millions more to brain research, and every airport could be equipped with ultra-sophisticated fMRI scanners that could immediately recognise angry and hateful thoughts in people's brains. Will it really work? Who knows. Is it wise to develop bionic flies and thought-reading scanners? Not necessarily. Be that as it may, as you read these lines, the US Department of Defense is transferring millions of dollars to nanotechnology and brain laboratories for work on these and other such ideas.

This obsession with military technology – from tanks to atom bombs to spy-flies – is a surprisingly recent phenomenon. Up until the nineteenth century, the vast majority of military revolutions were the product of organisational rather than technological changes. When alien civilisations met for the first time, technological gaps sometimes played an important role. But even in such cases, few thought of deliberately creating or enlarging such gaps. Most empires did not rise thanks to technological wizardry, and their rulers did not give much thought to technological improvement. The Arabs did not defeat the Sassanid Empire thanks to superior bows or swords, the Seljuks had no technological advantage over the Byzantines, and the Mongols did not conquer China with the help of some ingenious new weapon. In fact, in all these cases the vanquished enjoyed superior military and civilian technology.

The Roman army is a particularly good example. It was the best army of its day, yet technologically speaking, Rome had no edge over Carthage, Macedonia or the Seleucid Empire. Its advantage rested on efficient organisation, iron discipline and huge manpower reserves. The Roman army never set up a research and development department, and its weapons remained more or less the same for centuries on end. If the legions of Scipio Aemilianus – the general who levelled Carthage and defeated the Numantians in the second century BC – had suddenly popped up 500 years later in the age of Constantine the Great, Scipio would have had a fair chance of beating Constantine. Now imagine what would happen to a general from a few centuries back – say Napoleon – if he led his troops against a modern armoured brigade. Napoleon was a brilliant tactician, and his men were crack professionals, but their skills would be useless in the face of modern weaponry.

As in Rome, so also in ancient China: most generals and philosophers did not think it their duty to develop new weapons. The most important military invention in the history of China was gunpowder. Yet to the best of our knowledge, gunpowder was invented accidentally, by Daoist alchemists searching for the elixir of life. Gunpowder's subsequent career is even more telling. One might have thought that the Daoist alchemists would have made China master of the world. In fact, the Chinese used the new compound mainly for firecrackers. Even as the Song Empire collapsed in the face of a Mongol invasion, no emperor set up a medieval Manhattan Project to save the empire by inventing a doomsday weapon. Only in the fifteenth century – about 600 years after the invention of gunpowder – did cannons become a decisive factor on Afro-Asian battlefields. Why did it take so long for the deadly potential of this substance to be put to military use? Because it appeared at a time when neither kings, scholars, nor merchants thought that new military technology could save them or make them rich.

The situation began to change in the fifteenth and sixteenth centuries, but another 200 years went by before most rulers evinced any interest in financing the research and development of new weapons. Logistics and strategy continued to have far greater impact on the outcome of wars than technology. The Napoleonic military machine that crushed the armies of the European powers at Austerlitz

(1805) was armed with more or less the same weaponry that the army of Louis XVI had used. Napoleon himself, despite being an artilleryman, had little interest in new weapons, even though scientists and inventors tried to persuade him to fund the development of flying machines, submarines and rockets.

Science, industry and military technology intertwined only with the advent of the capitalist system and the Industrial Revolution. Once this relationship was established, however, it quickly transformed the world.

The Ideal of Progress

Until the Scientific Revolution most human cultures did not believe in progress. They thought the golden age was in the past, and that the world was stagnant, if not deteriorating. Strict adherence to the wisdom of the ages might perhaps bring back the good old times, and human ingenuity might conceivably improve this or that facet of daily life. However, it was considered impossible for human know-how to overcome the world's fundamental problems. If even Muhammad, Jesus, Buddha and Confucius – who knew everything there is to know – were unable to abolish famine, disease, poverty and war from the world, how could we expect to do so?

Many faiths believed that some day a messiah would appear and end all wars, famines and even death itself. But the notion that humankind could do so by discovering new knowledge and inventing new tools was worse than ludicrous – it was hubris. The story of the Tower of Babel, the story of Icarus, the story of the Golem and countless other myths taught people that any attempt to go beyond human limitations would inevitably lead to disappointment and disaster.

When modern culture admitted that there were many important things that it still did not know, and when that admission of

34. Benjamin Franklin disarming the gods.

ignorance was married to the idea that scientific discoveries could give us new powers, people began suspecting that real progress might be possible after all. As science began to solve one unsolvable problem after another, many became convinced that humankind could overcome any and every problem by acquiring and applying new knowledge. Poverty, sickness, wars, famines, old age and death itself were not the inevitable fate of humankind. They were simply the fruits of our ignorance.

A famous example is lightning. Many cultures believed that lightning was the hammer of an angry god, used to punish sinners. In the middle of the eighteenth century, in one of the most celebrated experiments in scientific history, Benjamin Franklin flew a kite during a lightning storm to test the hypothesis that lightning is simply an electric current. Franklin's empirical observations, coupled with his knowledge about the qualities of electrical energy, enabled him to invent the lightning rod and disarm the gods.

Poverty is another case in point. Many cultures have viewed poverty as an inescapable part of this imperfect world. According to the New Testament, shortly before the crucifixion a woman anointed Christ with precious oil worth 300 denarii. Jesus' disciples scolded the woman for wasting such a huge sum of money instead of giving it to the poor, but Jesus defended her, saying that 'The poor you will always have with you, and you can help them any time you want. But you will not always have me' (Mark 14:7). Today, fewer and fewer people, including fewer and fewer Christians, agree with Jesus on this matter. Poverty is increasingly seen as a technical problem amenable to intervention. It's common wisdom that policies based on the latest findings in agronomy, economics, medicine and sociology can eliminate poverty.

And indeed, many parts of the world have already been freed from the worst forms of deprivation. Throughout history, societies have suffered from two kinds of poverty: social poverty, which withholds from some people the opportunities available to others; and biological poverty, which puts the very lives of individuals at risk due to lack of food and shelter. Perhaps social poverty can never be eradicated, but in many countries around the world biological poverty is a thing of the past.

Until recently, most people hovered very close to the biological poverty line, below which a person lacks enough calories to sustain life for long. Even small miscalculations or misfortunes could easily push people below that line, into starvation. Natural disasters and man-made calamities often plunged entire populations over the abyss, causing the death of millions. Today most of the world's people have a safety net stretched below them. Individuals are protected from personal misfortune by insurance, state-sponsored social security and a plethora of local and international NGOs. When calamity strikes an entire region, worldwide relief efforts are usually successful in preventing the worst. People still suffer from numerous degradations, humiliations and poverty-related illnesses, but in most countries nobody is starving to death. In fact, in many societies more people are in danger of dying from obesity than from starvation.

The Gilgamesh Project

Of all mankind's ostensibly insoluble problems, one has remained the most vexing, interesting and important: the problem of death itself. Before the late modern era, most religions and ideologies took it for granted that death was our inevitable fate. Moreover, most faiths turned death into the main source of meaning in life. Try to imagine Islam, Christianity or the ancient Egyptian religion in a world without death. These creeds taught people that they must come to terms with death and pin their hopes on the afterlife, rather than seek to overcome death and live for ever here on earth. The best minds were busy giving meaning to death, not trying to escape it.

That is the theme of the most ancient myth to come down to us – the Gilgamesh myth of ancient Sumer. Its hero is the strongest and most capable man in the world, King Gilgamesh of Uruk, who could defeat anyone in battle. One day, Gilgamesh's best friend, Enkidu, died. Gilgamesh sat by the body and observed it for many days, until he saw a worm dropping out of his friend's nostril. At that moment Gilgamesh was gripped by a terrible horror, and he resolved that he himself would never die. He would somehow find a way to defeat death. Gilgamesh then undertook a journey to the end of the universe, killing lions, battling scorpion-men and finding his way into the underworld. There he shattered the mysterious stone 'things' of Urshanabi, the ferryman of the river of the dead, and found Utnapishtim, the last survivor of the primordial flood. Yet Gilgamesh failed in his quest. He returned home empty-handed, as mortal as ever, but with one new piece of wisdom. When the gods created man, Gilgamesh had learned, they set death as man's inevitable destiny, and man must learn to live with it.

Disciples of progress do not share this defeatist attitude. For men of science, death is not an inevitable destiny, but merely a technical problem. People die not because the gods decreed it, but due to various technical failures – a heart attack, cancer, an infection. And every technical problem has a technical solution. If the heart flutters, it can be stimulated by a pacemaker or replaced by a new heart. If cancer rampages, it can be killed with drugs or radiation. If bacteria proliferate, they can be subdued with antibiotics. True, at present we cannot solve all technical problems. But we are working on them. Our best minds are not wasting their time trying to give meaning to death. Instead, they are busy investigating the physiological, hormonal and genetic systems responsible for disease and old age. They are developing new medicines, revolutionary treatments and artificial organs that will lengthen our lives and might one day vanquish the Grim Reaper himself.

Until recently, you would not have heard scientists, or anyone else, speak so bluntly. 'Defeat death?! What nonsense! We are only trying to cure cancer, tuberculosis and Alzheimer's disease,' they insisted. People avoided the issue of death because the goal seemed too elusive. Why create unreasonable expectations? We're now at a point, however, where we can be frank about it. The leading project of the Scientific Revolution is to give humankind eternal life. Even if killing death seems a distant goal, we have already achieved things that were inconceivable a few centuries ago. In 1199, King Richard the Lionheart was struck by an arrow in his left shoulder. Today we'd say he incurred a minor injury. But in 1199, in the absence of antibiotics and effective sterilisation methods, this minor flesh wound turned infected and gangrene set in. The only way to stop the spread of gangrene in twelfth-century Europe was to cut off the infected limb, impossible when the infection was in a shoulder. The gangrene spread through the Lionheart's body and no one could help the king. He died in great agony two weeks later.

As recently as the nineteenth century, the best doctors still did not know how to prevent infection and stop the putrefaction of tissues. In field hospitals doctors routinely cut off the hands and legs of soldiers who received even minor limb injuries, fearing gangrene. These amputations, as well as all other medical procedures (such as tooth extraction), were done without any anaesthetics. The first anaesthetics – ether, chloroform and morphine – entered regular usage in Western medicine only in the middle of the nineteenth century. Before the advent of chloroform, four soldiers had to hold down a wounded comrade while the doctor sawed off the injured limb. On the morning after the Battle of Waterloo (1815), heaps of sawn-off hands and legs could be seen adjacent to the field hospitals. In those days, carpenters and butchers who enlisted to the army were often sent to serve in the medical corps, because surgery required little more than knowing your way around knives and saws.

In the two centuries since Waterloo, things have changed beyond recognition. Pills, injections and sophisticated operations save us from a spate of illnesses and injuries that once dealt an inescapable death sentence. They also protect us against countless daily aches and ailments, which premodern people simply accepted as part of life. The average life expectancy jumped from well below forty years, to around sixty-seven in the entire world, and to around eighty years in the developed world.[8]

Death suffered its worst setbacks in the arena of child mortality. Until the twentieth century, between a quarter and a third of the children of agricultural societies never reached adulthood. Most succumbed to childhood diseases such as diphtheria, measles and smallpox. In seventeenth-century England, 150 out of every 1,000 newborns died during their first year, and a third of all children were dead before they reached fifteen.[9] Today, only 5 out of 1,000 English babies die during their first year, and only 7 out of 1,000 die before age fifteen.[10]

We can better grasp the full impact of these figures by setting aside statistics and telling some stories. A good example is the family of King Edward I of England (1237–1307) and his wife, Queen Eleanor (1241–90). Their children enjoyed the best conditions and the most nurturing surroundings that could be provided in medieval Europe. They lived in palaces, ate as much food as they liked, had plenty of warm clothing, well-stocked fireplaces, the cleanest water available, an army of servants and the best doctors. The sources mention sixteen children that Queen Eleanor bore between 1255 and 1284:

1. An anonymous daughter, born in 1255, died at birth.
2. A daughter, Catherine, died either at age one or age three.
3. A daughter, Joan, died at six months.
4. A son, John, died at age five.
5. A son, Henry, died at age six.
6. A daughter, Eleanor, died at age twenty-nine.
7. An anonymous daughter died at five months.
8. A daughter, Joan, died at age thirty-five.
9. A son, Alphonso, died at age ten.
10. A daughter, Margaret, died at age fifty-eight.
11. A daughter, Berengeria, died at age two.
12. An anonymous daughter died shortly after birth.
13. A daughter, Mary, died at age fifty-three.
14. An anonymous son died shortly after birth.
15. A daughter, Elizabeth, died at age thirty-four.
16. A son, Edward.

The youngest, Edward, was the first of the boys to survive the dangerous years of childhood, and at his father's death he ascended the English throne as King Edward II. In other words, it took Eleanor sixteen tries to carry out the most fundamental mission of an English queen – to provide her husband with a male heir. Edward II's mother must have been a woman of exceptional patience and fortitude. Not so the woman Edward chose for his wife, Isabella of France. She had him murdered when he was forty-three.[11]

To the best of our knowledge, Eleanor and Edward I were a healthy couple and passed no fatal hereditary illnesses on to their children. Nevertheless, ten out of the sixteen – 62 per cent – died during childhood. Only six managed to live beyond the age of eleven, and only three – just 18 per cent – lived beyond the age of forty. In addition to these births, Eleanor most likely had a number of pregnancies that ended in miscarriage. On average, Edward and Eleanor lost a child every three years, ten children one after another. It's nearly impossible for a parent today to imagine such loss.

How long will the Gilgamesh Project – the quest for immortality – take to complete? A hundred years? Five hundred years? A thousand years? When we recall how little we knew about the human body in 1900, and how much knowledge we have gained in a single century, there is cause for optimism. Genetic engineers have recently managed to double the average life expectancy of *Caenorhabditis elegans* worms.[12] Could they do the same for *Homo sapiens*? Nanotechnology experts are developing a bionic immune system composed of millions of nano-robots, which would inhabit our bodies, open blocked blood vessels, fight viruses and bacteria, eliminate cancerous cells and even reverse ageing processes.[13] A few serious scholars suggest that by 2050, some humans will become a-mortal (not immortal, because they could still die of some accident, but a-mortal, meaning that in the absence of fatal trauma their lives could be extended indefinitely).

Whether or not Project Gilgamesh succeeds, from a historical perspective it is fascinating to see that most late modern religions and ideologies have already taken death and the afterlife out of the equation. Until the eighteenth century, religions considered death and its aftermath central to the meaning of life. Beginning in the eighteenth century, religions and ideologies such as liberalism, socialism and feminism lost all interest in the afterlife. What, exactly, happens to a Communist after he or she dies? What happens to a capitalist? What happens to a feminist? It is pointless to look for the answer in the writings of Marx, Adam Smith or Simone de Beauvoir. The only modern ideology that still awards death a central role is nationalism. In its more poetic and desperate moments, nationalism promises that whoever dies for the nation will for ever live in its collective memory. Yet this promise is so fuzzy that even most nationalists do not really know what to make of it.

The Sugar Daddy of Science

We are living in a technical age. Many are convinced that science and technology hold the answers to all our problems. We should just let the scientists and technicians go on with their work, and they will create heaven here on earth. But science is not an enterprise that takes place on some superior moral or spiritual plane above the rest of human activity. Like all other parts of our culture, it is shaped by economic, political and religious interests.

Science is a very expensive affair. A biologist seeking to understand the human immune system requires laboratories, test tubes, chemicals and electron microscopes, not to mention lab assistants, electricians, plumbers and cleaners. An economist seeking to model credit markets must buy computers, set up giant databanks and develop complicated data-processing programs. An archaeologist who wishes to understand the behaviour of archaic hunter-gatherers must travel to distant lands, excavate ancient ruins and date fossilised bones and artefacts. All of this costs money.

During the past 500 years modern science has achieved wonders thanks largely to the willingness of governments, businesses, foundations and private donors to channel billions of dollars into scientific research. These billions have done much more to chart the universe, map the planet and catalogue the animal kingdom than did Galileo Galilei, Christopher Columbus and Charles Darwin. If these particular geniuses had never been born, their insights would probably have occurred to others. But if the proper funding were unavailable, no intellectual brilliance could have compensated for that. If Darwin had never been born, for example, we'd today attribute the theory of evolution to Alfred Russel Wallace, who came up with the idea of evolution via natural selection independently of Darwin and just a few years later. But if the European powers had not financed geographical, zoological and botanical research around the world, neither Darwin nor Wallace would have had the necessary empirical data to develop the theory of evolution. It is likely that they would not even have tried.

Why did the billions start flowing from government and business coffers into labs and universities? In academic circles, many are naive enough to believe in pure science. They believe that government and business altruistically give them money to pursue whatever research projects strike their fancy. But this hardly describes the realities of science funding.

Most scientific studies are funded because somebody believes they can help attain some political, economic or religious goal. For example, in the sixteenth century, kings and bankers channelled enormous resources to finance geographical expeditions around the world but not a penny for studying child psychology. This is because kings and bankers surmised that the discovery of new geographical knowledge would enable them to conquer new lands and set up trade empires, whereas they couldn't see any profit in understanding child psychology.

In the 1940s the governments of America and the Soviet Union channelled enormous resources to the study of nuclear physics rather than underwater archaeology. They surmised that studying nuclear physics would enable them to develop nuclear weapons, whereas underwater archaeology was unlikely to help win wars. Scientists themselves are not always aware of the political, economic and religious interests that control the flow of money; many scientists do, in fact, act out of pure intellectual curiosity. However, only rarely do scientists dictate the scientific agenda.

Even if we wanted to finance pure science unaffected by political, economic or religious interests, it would probably be impossible. Our resources are limited, after all. Ask a congressman to allocate an additional million dollars to the National Science Foundation for basic research, and he'll justifiably ask whether that money wouldn't be better used to fund teacher training or to give a needed tax break to a troubled factory in his district. To channel limited resources we must answer questions such as 'What is more important?' and 'What is good?' And these are not scientific questions. Science can explain what exists in the world, how things work, and what might be in the future. By definition, it has no pretensions to knowing what *should* be in the future. Only religions and ideologies seek to answer such questions.

Consider the following quandary: two biologists from the same department, possessing the same professional skills, have both applied for a million-dollar grant to finance their current research projects. Professor Slughorn wants to study a disease that infects the udders of cows, causing a 10 per cent decrease in their milk production. Professor Sprout wants to study whether cows suffer mentally when they are separated from their calves. Assuming that the amount of money is limited, and that it is impossible to finance both research projects, which one should be funded?

There is no scientific answer to this question. There are only political, economic and religious answers. In today's world, it is obvious that Slughorn has a better chance of getting the money. Not because udder diseases are scientifically more interesting than bovine mentality, but because the dairy industry, which stands to benefit from the research, has more political and economic clout than the animal-rights lobby.

Perhaps in a strict Hindu society, where cows are sacred, or in a society committed to animal rights, Professor Sprout would have a better shot. But as long as she lives in a society that values the commercial potential of milk and the health of its human citizens over the feelings of cows, she'd best write up her research proposal so as to appeal to those assumptions. For example, she might write that 'Depression leads to a decrease in milk production. If we understand the mental world of dairy cows, we could develop psychiatric medication that will improve their mood, thus raising milk production by up to 10 per cent. I estimate that there is a global annual market of $250 million for bovine psychiatric medications.'

Science is unable to set its own priorities. It is also incapable of determining what to do with its discoveries. For example, from a purely scientific viewpoint it is unclear what we should do with our increasing understanding of genetics. Should we use this knowledge to cure cancer, to create a race of genetically engineered supermen, or to engineer dairy cows with super-sized udders? It is obvious that a liberal government, a Communist government, a Nazi government and a capitalist business corporation would use the very same scientific discovery for completely different purposes, and there is no *scientific* reason to prefer one usage over others.

In short, scientific research can flourish only in alliance with some religion or ideology. The ideology justifies the costs of the research. In exchange, the ideology influences the scientific agenda and determines what to do with the discoveries. Hence in order to comprehend how humankind has reached Alamogordo and the moon – rather than any number of alternative destinations – it is not enough to survey the achievements of physicists, biologists and sociologists. We have to take into account the ideological, political and economic forces that shaped physics, biology and sociology, pushing them in certain directions while neglecting others.

Two forces in particular deserve our attention: imperialism and capitalism. The feedback loop between science, empire and capital has arguably been history's chief engine for the past 500 years. The following chapters analyse its workings. First we'll look at how the twin turbines of science and empire were latched to one another, and then learn how both were hitched up to the money pump of capitalism.

15

The Marriage of Science and Empire

HOW FAR IS THE SUN FROM THE EARTH? It's a question that intrigued many early modern astronomers, particularly after Copernicus argued that the sun, rather than the earth, is located at the centre of the universe. A number of astronomers and mathematicians tried to calculate the distance, but their methods provided widely varying results. A reliable means of making the measurement was finally proposed in the middle of the eighteenth century. Every few years, the planet Venus passes directly between the sun and the earth. The duration of the transit differs when seen from distant points on the earth's surface because of the tiny difference in the angle at which the observer sees it. If several observations of the same transit were made from different continents, simple trigonometry was all it would take to calculate our exact distance from the sun.

Astronomers predicted that the next Venus transits would occur in 1761 and 1769. So expeditions were sent from Europe to the four corners of the world in order to observe the transits from as many distant points as possible. In 1761 scientists observed the transit from Siberia, North America, Madagascar and South Africa. As the 1769 transit approached, the European scientific community mounted a supreme effort, and scientists were dispatched as far as northern Canada and California (which was then a wilderness). The Royal Society of London for the Improvement of Natural Knowledge concluded that this was not enough. To obtain the most accurate results it was imperative to send an astronomer all the way to the south-western Pacific Ocean.

The Royal Society resolved to send an eminent astronomer, Charles Green, to Tahiti, and spared neither effort nor money. But, since it was funding such an expensive expedition, it hardly made sense to use it to make just a single astronomical observation. Green was therefore accompanied by a team of eight other scientists from several disciplines, headed by botanists Joseph Banks and Daniel Solander. The team also included artists assigned to produce drawings of the new lands, plants, animals and peoples that the scientists would no doubt encounter. Equipped with the most advanced scientific instruments that Banks and the Royal Society could buy, the expedition was placed under the command of Captain James Cook, an experienced seaman as well as an accomplished geographer and ethnographer.

The expedition left England in 1768, observed the Venus transit from Tahiti in 1769, reconnoitred several Pacific islands, visited Australia and New Zealand, and returned to England in 1771. It brought back enormous quantities of astronomical, geographical, meteorological, botanical, zoological and anthropological data. Its findings made major contributions to a number of disciplines, sparked the imagination of Europeans with astonishing tales of the South Pacific, and inspired future generations of naturalists and astronomers.

One of the fields that benefited from the Cook expedition was medicine. At the time, ships that set sail to distant shores knew that more than half their crew members would die on the journey. The nemesis was not angry natives, enemy warships or homesickness. It was a mysterious ailment called scurvy. Men who came down with the disease grew lethargic and depressed, and their gums and other soft tissues bled. As the disease progressed, their teeth fell out, open sores appeared, and they grew feverish and jaundiced and lost control of their limbs. Between the sixteenth and eighteenth centuries, scurvy is estimated to have claimed the lives of about 2 million sailors. No one knew what caused it, and no matter what remedy was tried, sailors continued to die in droves. The turning point came in 1747, when a British physician, James Lind, conducted a controlled experiment on sailors who suffered from the disease. He separated them into several groups and gave each group a different treatment. One of the test groups was instructed to eat citrus fruits, a common folk remedy for scurvy. The patients in this group promptly recovered. Lind did not know what the citrus fruits had that the sailors' bodies lacked, but we now know that it is vitamin C. A typical shipboard diet at that time was notably lacking in foods that are rich in this essential nutrient. On long-range voyages sailors usually subsisted on biscuits and beef jerky, and ate almost no fruits or vegetables.

The Royal Navy was not convinced by Lind's experiments, but James Cook was. He resolved to prove the doctor right. He loaded his ship with a large quantity of sauerkraut and ordered his sailors to eat lots of fresh fruits and vegetables whenever the expedition made landfall. Cook did not lose a single sailor to scurvy. In the following decades, all the world's navies adopted Cook's nautical diet, and the lives of countless sailors and passengers were saved.[1]

However, the Cook expedition had another, far less benign result. Cook was not only an experienced seaman and geographer, but also a naval officer. The Royal Society financed a large part of the expedition's expenses, but the ship itself was provided by the Royal Navy. The navy also seconded eighty-five well-armed sailors and marines, and equipped the ship with artillery, muskets, gunpowder and other weaponry. Much of the information collected by the expedition – particularly the astronomical, geographical, meteorological and anthropological data – was of obvious political and military value. The discovery of an effective treatment for scurvy greatly contributed to British control of the world's oceans and its ability to send armies to the other side of the world. Cook claimed for Britain many of the islands and lands he 'discovered', most notably Australia. The

Cook expedition laid the foundation for the British occupation of the south-western Pacific Ocean; for the conquest of Australia, Tasmania and New Zealand; for the settlement of millions of Europeans in the new colonies; and for the extermination of their native cultures and most of their native populations.[2]

In the century following the Cook expedition, the most fertile lands of Australia and New Zealand were taken from their previous inhabitants by European settlers. The native population dropped by up to 90 per cent and the survivors were subjected to a harsh regime of racial oppression. For the Aborigines of Australia, and to a lesser extent for the Maori of New Zealand, the Cook expedition was the beginning of a catastrophe from which they have never fully recovered.

An even worse fate befell the natives of Tasmania. Having survived for 10,000 years in splendid isolation, they were almost exterminated within a century of Cook's arrival. European settlers first drove them off the richest parts of the island, and then, coveting even the remaining wilderness, hunted them down and killed them systematically. Some of the last survivors were hounded into an evangelical concentration camp, where well-meaning but not particularly open-minded missionaries tried to indoctrinate them in the ways of the modern world. The Tasmanians were instructed in reading and writing, Christianity and various 'productive skills' such as sewing clothes and farming. But they refused to learn. They became ever more melancholic, stopped having children, lost all interest in life, and finally chose the only escape route from the modern world of science and progress – death.

Alas, science and progress pursued them even to the afterlife. The corpses of dead Tasmanians were seized in the name of science by anthropologists and curators. They were dissected, weighed and measured, and analysed in learned articles. The skulls and skeletons were then put on display in museums and anthropological collections. Only in 1976 did the Tasmanian Museum give up for burial the skeleton of Truganini, often thought to be the last full-blooded native Tasmanian, who had died a hundred years earlier. The English Royal College of Surgeons held on to samples of her skin and hair until 2002.

Was Cook's ship a scientific expedition protected by a military force or a military expedition with a few scientists tagging along? That's like asking whether your petrol tank is half empty or half full. It was both. The Scientific Revolution and modern imperialism were inseparable. People such as Captain James Cook and the botanist Joseph Banks could hardly distinguish science from empire. Nor could luckless Truganini.

Why Europe?

The fact that people from a large island in the northern Atlantic conquered a large island south of Australia is one of history's more bizarre occurrences. Not long before Cook's expedition, the British Isles and western Europe in general were but distant backwaters of the Mediterranean world. Little of importance ever happened there. Even the Roman Empire – the only important premodern European empire – derived most of its wealth from its North African, Balkan and Middle Eastern provinces. Rome's western European provinces were a poor Wild West, which contributed little aside from minerals and slaves. Northern Europe was so desolate and barbarous that it wasn't even worth conquering.

Only at the end of the fifteenth century did Europe become a hothouse of important military, political, economic and cultural developments. Between 1500 and 1750, western Europe gained momentum and became master of the 'Outer World', meaning the two American continents and the oceans. Yet even then Europe was no match for the great powers of Asia. Europeans managed to conquer America and gain supremacy at sea mainly because the Asiatic powers showed little interest in them. The early modern era was a golden age for the Ottoman Empire in the Mediterranean,

35. Truganini, the last native Tasmanian.

the Safavid Empire in Persia, the Mughal Empire in India, and the Chinese Ming and Qing dynasties. They expanded their territories significantly and enjoyed unprecedented demographic and economic growth. In 1775 Asia accounted for 80 per cent of the world economy. The combined economies of India and China alone represented two-thirds of global production. In comparison, Europe was an economic dwarf.[3]

The global centre of power shifted to Europe only between 1750 and 1850, when Europeans humiliated the Asian powers in a series of wars and conquered large parts of Asia. By 1900 Europeans firmly controlled the world's economy and most of its territory. In 1950 western Europe and the United States together accounted for more than half of global production, whereas China's portion had been reduced to 5 per cent.[4] Under the European aegis a new global order and global culture emerged. Today all humans are, to a much greater extent than they usually want to admit, European in dress, thought and taste. They may be fiercely anti-European in their rhetoric, but almost everyone on the planet views politics, medicine, war and economics through European eyes, and listens to music written in European modes with words in European languages. Even today's burgeoning Chinese economy, which may soon regain its global primacy, is built on a European model of production and finance.

How did the people of this frigid finger of Eurasia manage to break out of their remote corner of the globe and conquer the entire world? Europe's scientists are often given much of the credit. It's unquestionable that from 1850 onward European domination rested to a large extent on the military–industrial–scientific complex and technological wizardry. All successful late modern empires cultivated scientific research in the hope of harvesting technological innovations, and many scientists spent most of their time working on arms, medicines and machines for their imperial masters. A common saying among European soldiers facing African enemies was 'Come what may, we have machine guns, and they don't.' Civilian technologies were no less important. Canned food fed soldiers, railroads and steamships transported soldiers and their provisions, while a new arsenal of medicines cured soldiers, sailors and locomotive engineers. These logistical advances played a more significant role in the European conquest of Africa than did the machine gun.

But that wasn't the case before 1850. The military–industrial–scientific complex was still in its infancy; the technological fruits of the Scientific Revolution were unripe; and the technological gap between European, Asiatic and African powers was small. In 1770, James Cook certainly had far better technology than the Australian Aborigines, but so did the Chinese and the Ottomans. Why then was Australia explored and colonised by Captain James Cook and not by Captain Wan Zhengse or Captain Hussein Pasha? More importantly, if in 1770 Europeans had no significant technological advantage over Muslims, Indians and Chinese, how did they manage in the following century to open such a gap between themselves and the rest of the world?

Why did the military–industrial–scientific complex blossom in Europe rather than India? When Britain leaped forward, why were France, Germany and the United States quick to follow, whereas China lagged behind? When the gap between industrial and non-industrial nations became an obvious economic and political factor, why did Russia, Italy and Austria succeed in closing it, whereas Persia, Egypt and the Ottoman Empire failed? After all, the technology of the first industrial wave was relatively simple. Was it so hard for Chinese or Ottomans to engineer steam engines, manufacture machine guns and lay down railroads?

The world's first commercial railroad opened for business in 1830, in Britain. By 1850, Western nations were criss-crossed by almost 40,000 kilometres of railroads – but in the whole of Asia, Africa and Latin America there were only 4,000 kilometres of tracks. In 1880, the West boasted more than 350,000 kilometres of railroads, whereas in the rest of the world there were but 35,000 kilometres of train lines (and most of these were laid by the British in India).[5] The first railroad in China opened only in 1876. It was twenty-five kilometres long and built by Europeans – the Chinese government destroyed it the following year. In 1880 the Chinese Empire did not operate a single railroad. The first railroad in Persia was built only in 1888, and it connected Tehran with a Muslim holy site about ten kilometres south of the capital. It was constructed and operated by a Belgian company. In 1950, the total railway network of Persia still amounted to a meagre 2,500 kilometres, in a country seven times the size of Britain.[6]

The Chinese and Persians did not lack technological inventions such as steam engines (which could be freely copied or bought). They lacked the values, myths, judicial apparatus and sociopolitical structures that took centuries to form and mature in the West and which could not be copied and internalised rapidly. France and the

United States quickly followed in Britain's footsteps because the French and Americans already shared the most important British myths and social structures. The Chinese and Persians could not catch up as quickly because they thought and organised their societies differently.

This explanation sheds new light on the period from 1500 to 1850. During this era Europe did not enjoy any obvious technological, political, military or economic advantage over the Asian powers, yet the continent built up a unique potential, whose importance suddenly became obvious around 1850. The apparent equality between Europe, China and the Muslim world in 1750 was a mirage. Imagine two builders, each busy constructing very tall towers. One builder uses wood and mud bricks, whereas the other uses steel and concrete. At first it seems that there is not much of a difference between the two methods, since both towers grow at a similar pace and reach a similar height. However, once a critical threshold is crossed, the wood and mud tower cannot stand the strain and collapses, whereas the steel and concrete tower grows storey by storey, as far as the eye can see.

What potential did Europe develop in the early modern period that enabled it to dominate the late modern world? There are two complementary answers to this question: modern science and capitalism. Europeans were used to thinking and behaving in a scientific and capitalist way even before they enjoyed any significant technological advantages. When the technological bonanza began, Europeans could harness it far better than anybody else. So it is hardly coincidental that science and capitalism form the most important legacy that European imperialism has bequeathed the post-European world of the twenty-first century. Europe and Europeans no longer rule the world, but science and capital are growing ever stronger. The victories of capitalism are examined in the following chapter. This chapter is dedicated to the love story between European imperialism and modern science.

The Mentality of Conquest

Modern science flourished in and thanks to European empires. The discipline obviously owes a huge debt to ancient scientific traditions, such as those of classical Greece, China, India and Islam, yet its unique character began to take shape only in the early modern period, hand in hand with the imperial expansion of Spain, Portugal, Britain, France, Russia and the Netherlands. During the early modern period, Chinese, Indians, Muslims, Native Americans and Polynesians continued to make important contributions to the Scientific Revolution. The insights of Muslim economists were studied by Adam Smith and Karl Marx, treatments pioneered by Native American doctors found their way into English medical texts, and data extracted from Polynesian informants revolutionised Western anthropology. But until the mid-twentieth century, the people who collated these myriad scientific discoveries, creating scientific disciplines in the process, were the intellectual elites of the global European empires. The Far East and the Islamic world produced minds as intelligent and curious as those of Europe. However, between 1500 and 1950 they did not produce anything that comes even close to Newtonian physics or Darwinian biology.

This does not mean that Europeans have a unique gene for science, or that they will for ever dominate the study of physics and biology. Just as Islam began as an Arab monopoly but was subsequently taken over by Turks and Persians, so modern science began as a European speciality, but is today becoming a multi-ethnic enterprise.

What forged the historical bond between modern science and European imperialism? Technology was an important factor in the nineteenth and twentieth centuries, but in the early modern era it was of limited importance. The key factor was that the plant-seeking botanist and the colony-seeking naval officer shared a similar mindset.

Both scientist and conqueror began by admitting ignorance – they both said, 'I don't know what's out there.' They both felt compelled to go out and make new discoveries. And they both hoped the new knowledge thus acquired would make them masters of the world.

European imperialism was entirely unlike all other imperial projects in history. Previous seekers of empire tended to assume that they already understood the world. Conquest merely utilised and spread *their* view of the world. The Arabs, to name one example, did not conquer Egypt, Spain or India in order to discover something they did not know. The Romans, Mongols and Aztecs voraciously conquered new lands in search of power and wealth – not of knowledge. In contrast, European imperialists set out to distant shores in the hope of obtaining new knowledge along with new territories.

James Cook was not the first explorer to think this way. The Portuguese and Spanish voyagers of the fifteenth and sixteenth centuries already did. Prince Henry the Navigator and Vasco da Gama explored the coasts of Africa and, while doing so, seized control of islands and harbours. Christopher Columbus 'discovered' America and immediately claimed sovereignty over the new lands for the kings of Spain. Ferdinand Magellan found a way around the world, and simultaneously laid the foundation for the Spanish conquest of the Philippines.

As time went by, the conquest of knowledge and the conquest of territory became ever more tightly intertwined. In the eighteenth and nineteenth centuries, almost every important military expedition that left Europe for distant lands had on board scientists who set out not to fight but to make scientific discoveries. When Napoleon invaded Egypt in 1798, he took 165 scholars with him. Among other things, they founded an entirely new discipline, Egyptology, and made important contributions to the study of religion, linguistics and botany.

In 1831, the Royal Navy sent the ship HMS *Beagle* to map the coasts of South America, the Falklands Islands and the Galapagos Islands. The navy needed this knowledge in order to tighten Britain's imperial grip over South America. The ship's captain, who was an amateur scientist, decided to add a geologist to the expedition to study geological formations they might encounter on the way. After several professional geologists refused his invitation, the captain offered the job to a twenty-two-year-old Cambridge graduate, Charles Darwin. Darwin had studied to become an Anglican parson but was far more interested in geology and natural sciences than in the Bible. He jumped at the opportunity, and the rest is history. The captain spent his time on the voyage drawing military maps while Darwin collected the empirical data and formulated the insights that would eventually become the theory of evolution.

On 20 July 1969, Neil Armstrong and Buzz Aldrin landed on the surface of the moon. In the months leading up to their expedition, the Apollo 11 astronauts trained in a remote moon-like desert in the western United States. The area is home to several Native American communities, and there is a story – or legend – describing an encounter between the astronauts and one of the locals.

One day as they were training, the astronauts came across an old Native American. The man asked them what they were doing there. They replied that they were part of a research expedition that would shortly travel to explore the moon. When the old man heard that, he fell silent for a few moments, and then asked the astronauts if they could do him a favour.

'What do you want?' they asked.

'Well,' said the old man, 'the people of my tribe believe that holy spirits live on the moon. I was wondering if you could pass an important message to them from my people.'

'What's the message?' asked the astronauts.

The man uttered something in his tribal language, and then asked the astronauts to repeat it again and again until they had memorised it correctly.

'What does it mean?' asked the astronauts.

'Oh, I cannot tell you. It's a secret that only our tribe and the moon spirits are allowed to know.'

When they returned to their base, the astronauts searched and searched until they found someone who could speak the tribal language, and asked him to translate the secret message. When they repeated what they had memorised, the translator started to laugh uproariously. When he calmed down, the astronauts asked him what it meant. The man explained that the sentence they had memorised so carefully meant 'Don't believe a single word these people are telling you. They have come to steal your lands.'

Empty Maps

The modern 'explore and conquer' mentality is nicely illustrated by the development of world maps. Many cultures drew world maps long before the modern age. Obviously, none of them really knew the whole of the world. No Afro-Asian culture knew about America, and no American culture knew about Afro-Asia. But unfamiliar areas were simply left out, or filled with imaginary monsters and wonders. These maps had no empty spaces. They gave the impression of a familiarity with the entire world.

During the fifteenth and sixteenth centuries, Europeans began to draw world maps with lots of empty spaces – one indication of the development of the scientific mindset, as well as of the European imperial drive. The empty maps were a psychological and ideological breakthrough, a clear admission that Europeans were ignorant of large parts of the world.

The crucial turning point came in 1492, when Christopher Columbus sailed westward from Spain, seeking a new route to East

Asia. Columbus still believed in the old 'complete' world maps. Using them, he calculated that Japan should have been located about 7,000 kilometres west of Spain. In fact, more than 20,000 kilometres and an entire unknown continent separate East Asia from Spain. On 12 October 1492, at about 2 a.m., Columbus' expedition collided with the unknown continent. Juan Rodriguez Bermejo, watching from the mast of the ship *Pinta*, spotted an island in what we now call the Bahamas, and shouted 'Land! Land!'

Columbus believed he had reached a small island off the East Asian coast. He called the people he found there 'Indians' because he thought he had landed in the Indies – what we now call the East Indies or the Indonesian archipelago. Columbus stuck to this error for the rest of his life. The idea that he had discovered a completely unknown continent was inconceivable for him and for many of his generation. For thousands of years, not only the greatest thinkers and scholars but also the infallible Scriptures had known only Europe, Africa and Asia. Could they all have been wrong? Could the Bible have missed half the world? It would be as if in 1969, on its way to the moon, Apollo 11 had crashed into a hitherto unknown moon circling the earth, which all previous observations had somehow failed to spot. In his refusal to admit ignorance, Columbus was still a medieval man. He was convinced he knew the whole world, and even his momentous discovery failed to convince him otherwise.

The first modern man was Amerigo Vespucci, an Italian sailor who took part in several expeditions to America in the years 1499–1504. Between 1502 and 1504, two texts describing these expeditions were published in Europe. They were attributed to Vespucci. These texts argued that the new lands discovered by Columbus were not islands off the East Asian coast, but rather an entire continent unknown to the Scriptures, classical geographers and contemporary Europeans. In 1507, convinced by these arguments, a respected mapmaker named Martin Waldseemüller published an updated world

36. A European world map from 1459 (Europe is in the top left corner). The map is filled with details, even when depicting areas that were completely unfamiliar to Europeans, such as southern Africa.

map, the first to show the place where Europe's westward-sailing fleets had landed as a separate continent. Having drawn it, Waldseemüller had to give it a name. Erroneously believing that Amerigo Vespucci had been the person who discovered it, Waldseemüller named the continent in his honour – America. The Waldseemüller map became very popular and was copied by many other cartographers, spreading the name he had given the new land. There is poetic justice in the fact that a quarter of the world, and two of its seven continents, are named after a little-known Italian whose sole claim to fame is that he had the courage to say, 'We don't know.'

The discovery of America was the foundational event of the Scientific Revolution. It not only taught Europeans to favour present observations over past traditions, but the desire to conquer America also obliged Europeans to search for new knowledge at breakneck speed. If they really wanted to control the vast new territories, they had to gather enormous amounts of new data about the geography, climate, flora, fauna, languages, cultures and history of the new continent. Christian Scriptures, old geography books and ancient oral traditions were of little help.

Henceforth not only European geographers, but European scholars in almost all other fields of knowledge began to draw maps with spaces left to fill in. They began to admit that their theories were not perfect and that there were important things that they did not know.

The Europeans were drawn to the blank spots on the map as if they were magnets, and promptly started filling them in. During the fifteenth and sixteenth centuries, European expeditions circumnavigated Africa, explored America, crossed the Pacific and Indian Oceans, and created a network of bases and colonies all over the world. They established the first truly global empires and knitted together the first global trade network. The European imperial expeditions transformed the history of the world: from being a series of histories of isolated peoples and cultures, it became the history of a single integrated human society.

These European explore-and-conquer expeditions are so familiar to us that we tend to overlook just how extraordinary they were. Nothing like them had ever happened before. Long-distance campaigns of conquest are not a natural undertaking. Throughout history most human societies were so busy with local conflicts and neighbourhood quarrels that they never considered exploring and conquering distant lands. Most great empires extended their control only over their immediate neighbourhood – they reached far-flung

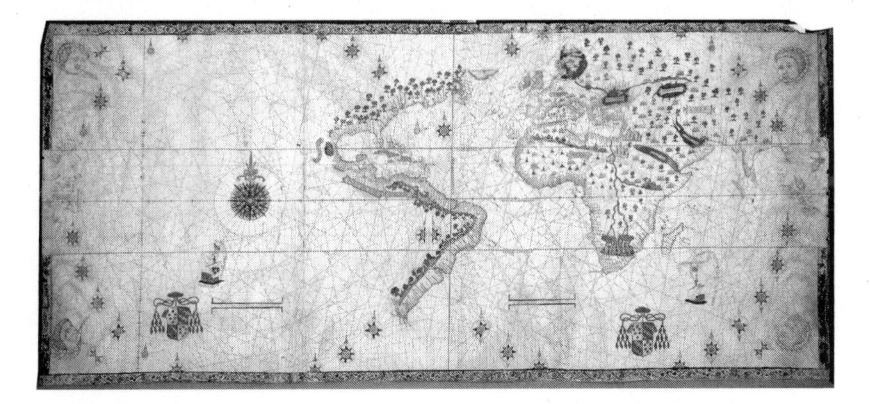

37. The Salviati World Map, 1525. While the 1459 world map is full of continents, islands and detailed explanations, the Salviati map is mostly empty. The eye wanders south along the American coastline, until it peters into emptiness. Anyone looking at the map and possessing even minimal curiosity is tempted to ask, 'What's beyond this point?' The map gives no answers. It invites the observer to set sail and find out.

lands simply because their neighbourhood kept expanding. Thus the Romans conquered Etruria in order to defend Rome (*c.*350–300 BC). They then conquered the Po Valley in order to defend Etruria (*c.*200 BC). They subsequently conquered Provence to defend the Po Valley (*c.*120 BC), Gaul to defend Provence (*c.*50 BC), and Britain in order to defend Gaul (*c.* AD 50). It took them 400 years to get from Rome to London. In 350 BC, no Roman would have conceived of sailing directly to Britain and conquering it.

Occasionally an ambitious ruler or adventurer would embark on a long-range campaign of conquest, but such campaigns usually followed well-beaten imperial or commercial paths. The campaigns of Alexander the Great, for example, did not result in the establishment of a new empire, but rather in the usurpation of an existing empire – that of the Persians. The closest precedents to the modern

European empires were the ancient naval empires of Athens and Carthage, and the medieval naval empire of Majapahit, which held sway over much of Indonesia in the fourteenth century. Yet even these empires rarely ventured into unknown seas – their naval exploits were local undertakings when compared to the global ventures of the modern Europeans.

Many scholars argue that the voyages of Admiral Zheng He of the Chinese Ming dynasty heralded and eclipsed the European voyages of discovery. Between 1405 and 1433, Zheng led seven huge armadas from China to the far reaches of the Indian Ocean. The largest of these comprised almost 300 ships and carried close to 30,000 people.[7] They visited Indonesia, Sri Lanka, India, the Persian Gulf, the Red Sea and East Africa. Chinese ships anchored in Jedda, the main harbour of the Hejaz, and in Malindi, on the Kenyan coast. Columbus' fleet of 1492 – which consisted of three small ships manned by 120 sailors – was like a trio of mosquitoes compared to Zheng He's drove of dragons.[8]

Yet there was a crucial difference. Zheng He explored the oceans, and assisted pro-Chinese rulers, but he did not try to conquer or colonise the countries he visited. Moreover, the expeditions of Zheng He were not deeply rooted in Chinese politics and culture. When the ruling faction in Beijing changed during the 1430s, the new overlords abruptly terminated the operation. The great fleet was dismantled, crucial technical and geographical knowledge was lost, and no explorer of such stature and means ever set out again from a Chinese port. Chinese rulers in the coming centuries, like most Chinese rulers in previous centuries, restricted their interests and ambitions to the Middle Kingdom's immediate environs.

The Zheng He expeditions prove that Europe did not enjoy an outstanding technological edge. What made Europeans exceptional was their unparalleled and insatiable ambition to explore and conquer. Although they might have had the ability, the Romans never attempted to conquer India or Scandinavia, the Persians never

38. Zheng He's flagship next to that of Columbus.

attempted to conquer Madagascar or Spain, and the Chinese never attempted to conquer Indonesia or Africa. Most Chinese rulers left even nearby Japan to its own devices. There was nothing peculiar about that. The oddity is that early modern Europeans caught a fever that drove them to sail to distant and completely unknown lands full of alien cultures, take one step on to their beaches, and immediately declare, 'I claim all these territories for my king!'

Invasion from Outer Space

Around 1517, Spanish colonists in the Caribbean islands began to hear vague rumours about a powerful empire somewhere in the centre of the Mexican mainland. A mere four years later, the Aztec capital was a smouldering ruin, the Aztec Empire was a thing of the past, and Hernán Cortés lorded over a vast new Spanish Empire in Mexico.

The Spaniards did not stop to congratulate themselves or even to catch their breath. They immediately commenced explore-and-conquer operations in all directions. The previous rulers of Central America – the Aztecs, the Toltecs, the Maya – barely knew South America existed, and never made any attempt to subjugate it, over the course of 2,000 years. Yet within little more than ten years of the Spanish conquest of Mexico, Francisco Pizarro had discovered the Inca Empire in South America, vanquishing it in 1532.

Had the Aztecs and Incas shown a bit more interest in the world surrounding them – and had they known what the Spaniards had done to their neighbours – they might have resisted the Spanish conquest more keenly and successfully. In the years separating Columbus' first journey to America (1492) from the landing of Cortés in Mexico (1519), the Spaniards conquered most of the Caribbean islands, setting up a chain of new colonies. For the subjugated natives, these colonies were hell on earth. They were ruled with an iron fist by greedy and unscrupulous colonists who enslaved them and set them to work in mines and plantations, killing anyone who offered the slightest resistance. Most of the native population soon died, either because of the harsh working conditions or the virulence of the diseases that hitch-hiked to America on the conquerors' sailing ships. Within twenty years, almost the entire native Caribbean population was wiped out. The Spanish colonists began importing African slaves to fill the vacuum.

This genocide took place on the very doorstep of the Aztec Empire, yet when Cortés landed on the empire's eastern coast, the Aztecs knew nothing about it. The coming of the Spaniards was the equivalent of an alien invasion from outer space. The Aztecs were convinced that they knew the entire world and that they ruled most of it. To them it was unimaginable that outside their domain could exist anything like these Spaniards. When Cortés and his men

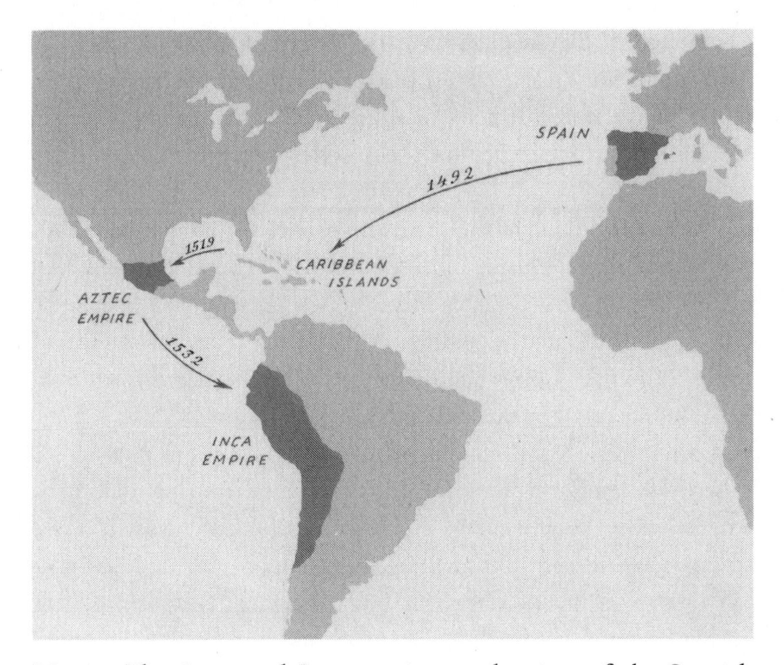

Map 7. The Aztec and Inca empires at the time of the Spanish conquest.

landed on the sunny beaches of today's Veracruz, it was the first time the Aztecs encountered a completely unknown people.

The Aztecs did not know how to react. They had trouble deciding what these strangers were. Unlike all known humans, the aliens had white skins. They also had lots of facial hair. Some had hair the colour of the sun. They stank horribly. (Native hygiene was far better than Spanish hygiene. When the Spaniards first arrived in Mexico, natives bearing incense burners were assigned to accompany them wherever they went. The Spaniards thought it was a mark of divine honour. We know from native sources that they found the newcomers' smell unbearable.)

The aliens' material culture was even more bewildering. They came in giant ships, the like of which the Aztecs had never imagined, let alone seen. They rode on the back of huge and terrifying animals, swift as the wind. They could produce lightning and thunder out of shiny metal sticks. They had flashing long swords and impenetrable armour, against which the natives' wooden swords and flint spears were useless.

Some Aztecs thought these must be gods. Others argued that they were demons, or the ghosts of the dead, or powerful sorcerers. Instead of concentrating all available forces and wiping out the Spaniards, the Aztecs deliberated, dawdled and negotiated. They saw no reason to rush. After all, Cortés had no more than 550 Spaniards with him. What could 550 men do to an empire of millions?

Cortés was equally ignorant about the Aztecs, but he and his men held significant advantages over their adversaries. While the Aztecs had no experience to prepare them for the arrival of these strange-looking and foul-smelling aliens, the Spaniards knew that the earth was full of unknown human realms, and no one had greater expertise in invading alien lands and dealing with situations about which they were utterly ignorant. For the modern European conqueror, like the modern European scientist, plunging into the unknown was exhilarating.

So when Cortés anchored off that sunny beach in July 1519, he did not hesitate to act. Like a science-fiction alien emerging from his spaceship, he declared to the awestruck locals: 'We come in peace. Take us to your leader.' Cortés explained that he was a peaceful emissary from the great king of Spain, and asked for a diplomatic interview with the Aztec ruler, Montezuma II. (This was a shameless lie. Cortés led an independent expedition of greedy adventurers. The king of Spain had never heard of Cortés, nor of the Aztecs.) Cortés was given guides, food and some military assistance by local enemies of the Aztecs. He then marched towards the Aztec capital, the great metropolis of Tenochtitlan.

The Aztecs allowed the aliens to march all the way to the capital, then respectfully led the aliens' leader to meet Emperor Montezuma.

In the middle of the interview, Cortés gave a signal, and steel-armed Spaniards butchered Montezuma's bodyguards (who were armed only with wooden clubs, and stone blades). The honoured guest took his host prisoner.

Cortés was now in a very delicate situation. He had captured the emperor, but was surrounded by tens of thousands of furious enemy warriors, millions of hostile civilians, and an entire continent about which he knew practically nothing. He had at his disposal only a few hundred Spaniards, and the closest Spanish reinforcements were in Cuba, more than 1,500 kilometres away.

Cortés kept Montezuma captive in the palace, making it look as if the king remained free and in charge and as if the 'Spanish ambassador' were no more than a guest. The Aztec Empire was an extremely centralised polity, and this unprecedented situation paralysed it. Montezuma continued to behave as if he ruled the empire, and the Aztec elite continued to obey him, which meant they obeyed Cortés. This situation lasted for several months, during which time Cortés interrogated Montezuma and his attendants, trained translators in a variety of local languages, and sent small Spanish expeditions in all directions to become familiar with the Aztec Empire and the various tribes, peoples and cities that it ruled.

The Aztec elite eventually revolted against Cortés and Montezuma, elected a new emperor, and drove the Spaniards from Tenochtitlan. However, by now numerous cracks had appeared in the imperial edifice. Cortés used the knowledge he had gained to prise the cracks open wider and split the empire from within. He convinced many of the empire's subject peoples to join him against the ruling Aztec elite. The subject peoples miscalculated badly. They hated the Aztecs, but knew nothing of Spain or the Caribbean genocide. They assumed that with Spanish help they could shake off the Aztec yoke. The idea that the Spanish would take over never occurred to them. They were sure that if Cortés and his few hundred henchmen caused any trouble, they could easily be overwhelmed. The rebellious peoples provided Cortés with an army of tens of thousands of local troops, and with its help Cortés besieged Tenochtitlan and conquered the city.

At this stage more and more Spanish soldiers and settlers began arriving in Mexico, some from Cuba, others all the way from Spain. When the local peoples realised what was happening, it was too late. Within a century of the landing at Veracruz, the native population of the Americas had shrunk by about 90 per cent, due mainly to unfamiliar diseases that reached America with the invaders. The survivors found themselves under the thumb of a greedy and racist regime that was far worse than that of the Aztecs.

Ten years after Cortés landed in Mexico, Pizarro arrived on the shore of the Inca Empire. He had far fewer soldiers than Cortés – his expedition numbered just 168 men! Yet Pizarro benefited from all the knowledge and experience gained in previous invasions. The Inca, in contrast, knew nothing about the fate of the Aztecs. Pizarro plagiarised Cortés. He declared himself a peaceful emissary from the king of Spain, invited the Inca ruler, Atahualpa, to a diplomatic interview, and then kidnapped him. Pizarro proceeded to conquer the paralysed empire with the help of local allies. If the subject peoples of the Inca Empire had known the fate of the inhabitants of Mexico, they would not have thrown in their lot with the invaders. But they did not know.

The native peoples of America were not the only ones to pay a heavy price for their parochial outlook. The great empires of Asia – the Ottoman, the Safavid, the Mughal and the Chinese – very quickly heard that the Europeans had discovered something big. Yet they displayed little interest in these discoveries. They continued to believe that the world revolved around Asia, and made no attempt to compete with the Europeans for control of America or of the new ocean lanes in the Atlantic and the Pacific. Even puny European kingdoms such as Scotland and Denmark sent a few explore-and-conquer expeditions to America, but not one expedition of either exploration or conquest was ever sent to America from the Islamic world, India or China. The first non-European power that tried to send a military expedition to America was Japan. That happened in June 1942, when a Japanese expedition conquered Kiska and Attu, two small islands off the Alaskan coast, capturing in the process ten US soldiers and a dog. The Japanese never got any closer to the mainland.

It is hard to argue that the Ottomans or Chinese were too far away, or that they lacked the technological, economic or military wherewithal. The resources that sent Zheng He from China to East Africa in the 1420s should have been enough to reach America. The Chinese just weren't interested. The first Chinese world map to show America was not issued until 1602 – and then by a European missionary!

For 300 years, Europeans enjoyed undisputed mastery in America and Oceania, in the Atlantic and the Pacific. The only significant struggles in those regions were between different European powers. The wealth and resources accumulated by the Europeans eventually enabled them to invade Asia too, defeat its empires, and divide it among themselves. When the Ottomans, Persians, Indians and Chinese woke up and began paying attention, it was too late.

Only in the twentieth century did non-European cultures adopt a truly global vision. This was one of the crucial factors that led to the collapse of European hegemony. Thus in the Algerian War of Independence (1954–62), Algerian guerrillas defeated a French army with an overwhelming numerical, technological and economic advantage. The Algerians prevailed because they were supported by a global anti-colonial network, and because they worked out how to harness the world's media to their cause – as well as public opinion in France itself. The defeat that little North Vietnam inflicted on the American colossus was based on a similar strategy. These guerrilla forces showed that even superpowers could be defeated if a local struggle became a global cause. It is interesting to contemplate what might have happened had Montezuma been able to manipulate public opinion in Spain and gain assistance from one of Spain's rivals – Portugal, France or the Ottoman Empire.

Rare Spiders and Forgotten Scripts

Modern science and modern empires were motivated by the restless feeling that perhaps something important awaited beyond the horizon – something they had better explore and master. Yet the connection between science and empire went much deeper. Not just the motivation, but also the practices of empire-builders were entangled with those of scientists. For modern Europeans, building an empire was a scientific project, while setting up a scientific discipline was an imperial project.

When the Muslims conquered India, they did not bring along archaeologists to systematically study Indian history, anthropologists to study Indian cultures, geologists to study Indian soils, or zoologists to study Indian fauna. When the British conquered India, they did all of these things. On 10 April 1802 the Great Survey of India was launched. It lasted sixty years. With the help of tens of thousands of native labourers, scholars and guides, the British carefully mapped the whole of India, marking borders, measuring distances, and even calculating for the first time the exact height of Mount Everest and the other Himalayan peaks. The British explored the military resources of Indian provinces and the location of their gold mines, but they also took the trouble to collect information about rare Indian spiders, to catalogue colourful butterflies, to trace the ancient origins of extinct Indian languages, and to dig up forgotten ruins.

Mohenjo-daro was one of the chief cities of the Indus Valley civilisation, which flourished in the third millennium BC and was destroyed around 1900 BC. None of India's pre-British rulers –

neither the Mauryas, nor the Guptas, nor the Delhi sultans, nor the great Mughals – had given the ruins a second glance. But a British archaeological survey took notice of the site in 1922. A British team then excavated it, and discovered the first great civilisation of India, which no Indian had been aware of.

Another telling example of British scientific curiosity was the deciphering of cuneiform script. This was the main script used throughout the Middle East for close to 3,000 years, but the last person able to read it probably died sometime in the early first millennium AD. Since then, inhabitants of the region frequently encountered cuneiform inscriptions on monuments, steles, ancient ruins and broken pots. But they had no idea how to read the weird, angular scratches and, as far as we know, they never tried. Cuneiform came to the attention of Europeans in 1618, when the Spanish ambassador in Persia went sightseeing in the ruins of ancient Persepolis, where he saw inscriptions that nobody could explain to him. News of the unknown script spread among European savants and piqued their curiosity. In 1657 European scholars published the first transcription of a cuneiform text from Persepolis. More and more transcriptions followed, and for close to two centuries scholars in the West tried to decipher them. None succeeded.

In the 1830s, a British officer named Henry Rawlinson was sent to Persia to help the shah train his army in the European style. In his spare time Rawlinson travelled around Persia and one day he was led by local guides to a cliff in the Zagros Mountains and shown the huge Behistun Inscription. About fifteen metres high and twenty-five metres wide, it had been etched high up on the cliff face on the command of King Darius I sometime around 500 BC. It was written in cuneiform script in three languages: Old Persian, Elamite and Babylonian. The inscription was well known to the local population, but nobody could read it. Rawlinson became convinced that if he could decipher the writing it would enable him and other scholars to read the numerous inscriptions and texts that were at the time being discovered all over the Middle East, opening a door into an ancient and forgotten world.

The first step in deciphering the lettering was to produce an accurate transcription that could be sent back to Europe. Rawlinson defied death to do so, scaling the steep cliff to copy the strange letters. He hired several locals to help him, most notably a Kurdish boy who climbed to the most inaccessible parts of the cliff in order to copy the upper portion of the inscription. In 1847 the project was completed, and a full and accurate copy was sent to Europe.

Rawlinson did not rest on his laurels. As an army officer, he had military and political missions to carry out, but whenever he had a spare moment he puzzled over the secret script. He tried one method after another and finally managed to decipher the Old Persian part of the inscription. This was easiest, since Old Persian was not that different from modern Persian, which Rawlinson knew well. An understanding of the Old Persian section gave him the key he needed to unlock the secrets of the Elamite and Babylonian sections. The great door swung open, and out came a rush of ancient but lively voices – the bustle of Sumerian bazaars, the proclamations of Assyrian kings, the arguments of Babylonian bureaucrats. Without the efforts of modern European imperialists such as Rawlinson, we would not have known much about the fate of the ancient Middle Eastern empires.

Another notable imperialist scholar was William Jones. Jones arrived in India in September 1783 to serve as a judge in the Supreme Court of Bengal. He was so captivated by the wonders of India that within less than six months of his arrival he had founded the Asiatic Society. This academic organisation was devoted to studying the cultures, histories and societies of Asia, and in particular those of India. Within two years Jones published his observations on the Sanskrit language, which pioneered the science of comparative linguistics.

In his publications Jones pointed out surprising similarities between Sanskrit, an ancient Indian language that became the sacred tongue of Hindu ritual, and the Greek and Latin languages, as well as similarities between all these languages and Gothic, Celtic, Old Persian, German, French and English. Thus in Sanskrit, 'mother' is '*matar*', in Latin it is '*mater*', and in Old Celtic it is '*mathir*'. Jones surmised that all these languages must share a common origin, developing from a now-forgotten ancient ancestor. He was thus the first to identify what later came to be called the Indo-European family of languages.

Jones' study was an important milestone not merely due to his bold (and accurate) hypotheses, but also because of the orderly methodology that he developed to compare languages. It was adopted by other scholars, enabling them systematically to study the development of all the world's languages.

Linguistics received enthusiastic imperial support. The European empires believed that in order to govern effectively they must know the languages and cultures of their subjects. British officers arriving in India were supposed to spend up to three years in a Calcutta college, where they studied Hindu and Muslim law alongside English law; Sanskrit, Urdu and Persian alongside Greek and Latin; and Tamil, Bengali and Hindustani culture alongside mathematics, economics and geography. The study of linguistics provided invaluable help in understanding the structure and grammar of local languages.

Thanks to the work of people like William Jones and Henry Rawlinson, the European conquerors knew their empires very well. Far better, indeed, than any previous conquerors, or even than the native population itself. Their superior knowledge had obvious practical advantages. Without such knowledge, it is unlikely that a ridiculously small number of Britons could have succeeded in governing, oppressing and exploiting so many hundreds of millions of Indians for two centuries. Throughout the nineteenth and early twentieth centuries, fewer than 5,000 British officials, about 40,000–70,000 British soldiers, and perhaps another 100,000 British business people, hangers-on, wives and children were sufficient to conquer and rule up to 300 million Indians.[9]

Yet these practical advantages were not the only reason why empires financed the study of linguistics, botany, geography and history. No less important was the fact that science gave the empires ideological justification. Modern Europeans came to believe that acquiring new knowledge was always good. The fact that the empires produced a constant stream of new knowledge branded them as progressive and positive enterprises. Even today, histories of sciences such as geography, archaeology and botany cannot avoid crediting the European empires, at least indirectly. Histories of botany have little to say about the suffering of the Aboriginal Australians, but they usually find some kind words for James Cook and Joseph Banks.

Furthermore, the new knowledge accumulated by the empires made it possible, at least in theory, to benefit the conquered populations and bring them the benefits of 'progress' – to provide them with medicine and education, to build railroads and canals, to ensure justice and prosperity. Imperialists claimed that their empires were not vast enterprises of exploitation but rather altruistic projects conducted for the sake of the non-European races – in Rudyard Kipling's words, 'the White Man's burden':

> Take up the White Man's burden –
> Send forth the best ye breed –
> Go bind your sons to exile
> To serve your captives' need;
> To wait in heavy harness,
> On fluttered folk and wild –
> Your new-caught, sullen peoples,
> Half-devil and half-child.

Of course, the facts often belied this myth. The British conquered Bengal, the richest province of India, in 1764. The new rulers were interested in little except enriching themselves. They adopted a disastrous economic policy that a few years later led to the outbreak of the Great Bengal Famine. It began in 1769, reached catastrophic levels in 1770, and lasted until 1773. About 10 million Bengalis, a third of the province's population, died in the calamity.[10]

In truth, neither the narrative of oppression and exploitation nor that of 'the White Man's burden' completely matches the facts. The European empires did so many different things on such a large scale, that you can find plenty of examples to support whatever you want to say about them. You think that these empires were evil monstrosities that spread death, oppression and injustice around the world? You could easily fill an encyclopedia with their crimes. You want to argue that they in fact improved the conditions of their subjects with new medicines, better economic conditions and greater security? You could fill another encyclopedia with their achievements. Due to their close cooperation with science, these empires wielded so much power and changed the world to such an extent that perhaps they cannot be simply labelled as good or evil. They created the world as we know it, including the ideologies we use in order to judge them.

But science was also used by imperialists to more sinister ends. Biologists, anthropologists and even linguists provided scientific proof that Europeans are superior to all other races, and consequently have the right (if not perhaps the duty) to rule over them. After William Jones argued that all Indo-European languages descend from a single ancient language many scholars were eager to discover who the speakers of that language had been. They noticed that the earliest Sanskrit speakers, who had invaded India from Central Asia more than 3,000 years ago, had called themselves Arya. The speakers of the earliest Persian language called themselves Airiia. European scholars consequently surmised that the people who spoke the primordial language that gave birth to both Sanskrit and Persian (as well as to Greek, Latin, Gothic and Celtic) must have called themselves Aryans. Could it be a coincidence that those who founded the magnificent Indian, Persian, Greek and Roman civilisations were all Aryans?

Next, British, French and German scholars wedded the linguistic theory about the industrious Aryans to Darwin's theory of natural selection and posited that the Aryans were not just a linguistic group but a biological entity – a race. And not just any race, but a master race of tall, light-haired, blue-eyed, hard-working, and super-rational humans who emerged from the mists of the north to lay the foundations of culture throughout the world. Regrettably, the Aryans who invaded India and Persia intermarried with the local natives they found in these lands, losing their light complexions and blond hair, and with them their rationality and diligence. The civilisations of India and Persia consequently declined. In Europe, on the other hand, the Aryans preserved their racial purity. This is why Europeans had managed to conquer the world, and why they were fit to rule it – provided they took precautions not to mix with inferior races.

Such racist theories enjoyed prominence and respectability for many generations, justifying the Western conquest of the world. Eventually, in the late twentieth century, just as the Western empires crumbled, racism became anathema among scientists and politicians alike. But the belief in Western superiority did not vanish. Instead, it took on new forms. Racism was replaced by culturism. Today's elites usually justify superiority in terms of historical differences between cultures rather than biological differences between races. We no longer say, 'It's in their blood.' We say, 'It's in their culture.'

Thus European right-wing parties which oppose Muslim immigration usually take care to avoid racial terminology. Marine Le Pen's speechwriters would have been shown the door on the spot had they suggested that the leader of France's Front national go on television to declare that 'We don't want those inferior Semites to dilute our Aryan blood and spoil our Aryan civilisation.' Instead, the French Front national, the Dutch Party for Freedom, the Alliance for the

Future of Austria and their like tend to argue that Western culture, as it has evolved in Europe, is characterised by democratic values, tolerance and gender equality, whereas Muslim culture, which evolved in the Middle East, is characterised by hierarchical politics, fanaticism and misogyny. Since the two cultures are so different, and since many Muslim immigrants are unwilling (and perhaps unable) to adopt Western values, they should not be allowed to enter, lest they foment internal conflicts and corrode European democracy and liberalism.

Such culturist arguments are fed by scientific studies in the humanities and social sciences that highlight the so-called clash of civilisations and the fundamental differences between different cultures. Not all historians and anthropologists accept these theories or support their political usages. But whereas biologists today have an easy time disavowing racism, simply explaining that the biological differences between present-day human populations are trivial, it is harder for historians and anthropologists to disavow culturism. After all, if the differences between human cultures are trivial, why should we pay historians and anthropologists to study them?

Scientists have provided the imperial project with practical knowledge, ideological justification and technological gadgets. Without this contribution it is highly questionable whether Europeans could have conquered the world. The conquerors returned the favour by providing scientists with information and protection, supporting all kinds of strange and fascinating projects and spreading the scientific way of thinking to the far corners of the earth. Without imperial support, it is doubtful whether modern science would have progressed very far. There are very few scientific disciplines that did not begin their lives as servants to imperial growth and that do not owe a large proportion of their discoveries, collections, buildings and scholarships to the generous help of army officers, navy captains and imperial governors.

This is obviously not the whole story. Science was supported by other institutions, not just by empires. And the European empires rose and flourished thanks also to factors other than science. Behind the meteoric rise of both science and empire lurks one particularly important force: capitalism. Were it not for businessmen seeking to make money, Columbus would not have reached America, James Cook would not have reached Australia, and Neil Armstrong would never have taken that small step on the surface of the moon.

16

The Capitalist Creed

MONEY HAS BEEN ESSENTIAL BOTH FOR building empires and for promoting science. Neither modern armies nor university laboratories can be sustained without banks.

It is not easy to grasp the true role of economics in modern history. Whole volumes have been written about how money founded states and ruined them, opened new horizons and enslaved millions, moved the wheels of industry and drove hundreds of species into extinction. Yet to understand modern economic history, you really need to understand just a single word. The word is growth. For better or worse, in sickness and in health, the modern economy has been growing like a hormone-soused teenager. It eats up everything it can find and puts on inches faster than you can count.

For most of history the economy stayed much the same size. Yes, global production increased, but this was due mostly to demographic expansion and the settlement of new lands. Per capita production remained static. But all that changed in the modern age. In 1500, global production of goods and services was equal to about $250 billion; today it hovers around $60 trillion. More importantly, in 1500, annual per capita production averaged $550, while today every man, woman and child produces, on average, $8,800 a year.[1] What accounts for this stupendous growth?

Economics is a notoriously complicated subject. To make things easier, let's imagine a simple example.

Samuel Greedy, a shrewd financier, founds a bank in El Dorado, California.

A. A. Stone, an up-and-coming contractor in El Dorado, finishes his first big job, receiving payment in cash to the tune of $1 million. He deposits this sum in Mr Greedy's bank. The bank now has $1 million in capital.

In the meantime, Jane McDoughnut, an experienced but impecunious El Dorado chef, thinks she sees a business opportunity – there's no really good bakery in her part of town. But she doesn't have enough money of her own to buy a proper facility complete with industrial ovens, sinks, knives and pots. She goes to the bank, presents her business plan to Greedy, and persuades him that it's a worthwhile investment. He issues her a $1 million loan, by crediting her account in the bank with that sum.

McDoughnut now hires Stone, the contractor, to build and furnish her bakery. His price is $1,000,000.

When she pays him, with a cheque drawn on her account, Stone deposits it in his account in the Greedy bank.

So how much money does Stone have in his bank account? Right, $2 million.

How much money – cash – is actually located in the bank's safe? Yes, $1 million.

It doesn't stop there. As contractors are wont to do, two months into the job Stone informs McDoughnut that, due to unforeseen problems and expenses, the bill for constructing the bakery will actually be $2 million. Mrs McDoughnut is not pleased, but she can hardly stop the job in the middle. So she pays another visit to the bank, convinces Mr Greedy to give her an additional loan, and he puts another $1 million in her account. She transfers the money to the contractor's account.

How much money does Stone have in his account now? He's got $3 million.

But how much money is actually sitting in the bank? Still just

$1 million. In fact, the same $1 million that's been in the bank all along.

Current US banking law permits the bank to repeat this exercise seven more times. The contractor would eventually have $10 million in his account, even though the bank still has but $1 million in its vaults. Banks are allowed to loan $10 for every dollar they actually possess, which means that 90 per cent of all the money in our bank accounts is not covered by actual coins and notes.[2] If all of the account holders at Barclays Bank suddenly demand their money, Barclays will promptly collapse (unless the government steps in to save it). The same is true of Lloyds, Deutsche Bank, Citibank, and all other banks in the world.

It sounds like a giant Ponzi scheme, doesn't it? But if it's a fraud, then the entire modern economy is a fraud. The fact is, it's not a deception, but rather a tribute to the amazing abilities of the human imagination. What enables banks – and the entire economy – to survive and flourish is our trust in the future. This trust is the sole backing for most of the money in the world.

In the bakery example, the discrepancy between the contractor's account statement and the amount of money actually in the bank is Mrs McDoughnut's bakery. Mr Greedy has put the bank's money into the asset, trusting that one day it would be profitable. The bakery hasn't baked a loaf of bread yet, but McDoughnut and Greedy anticipate that a year hence it will be selling thousands of loaves, rolls, cakes and cookies each day, at a handsome profit. Mrs McDoughnut will then be able to repay her loan, with interest. If at that point Mr Stone decides to withdraw his savings, Greedy will be able to come up with the cash. The entire enterprise is thus founded on trust in an imaginary future – the trust that the entrepreneur and the banker have in the bakery of their dreams, along with the contractor's trust in the future solvency of the bank.

We've already seen that money is an astounding thing because it can represent myriad different objects and convert anything into almost anything else. However, before the modern era this ability was limited. In most cases, money could represent and convert only things that actually existed in the present. This imposed a severe limitation on growth, since it made it very hard to finance new enterprises.

Consider our bakery again. Could McDoughnut get it built if money could represent only tangible objects? No. In the present, she has a lot of dreams, but no tangible resources. The only way she could get her bakery built would be to find a contractor willing to work today and receive payment in a few years' time, if and when the bakery starts making money. Alas, such contractors are rare breeds. So our entrepreneur is in a bind. Without a bakery, she can't bake cakes. Without cakes, she can't make money. Without money, she can't hire a contractor. Without a contractor, she has no bakery.

Humankind was trapped in this predicament for thousands of years. As a result, economies remained frozen. The way out of the trap was discovered only in the modern era, with the appearance of a new system based on trust in the future. In it, people agreed to represent imaginary goods – goods that do not exist in the present – with a special kind of money they called 'credit'. Credit enables us to build the present at the expense of the future. It's founded on the assumption that our future resources are sure to be far more abundant than our present resources. A host of new and wonderful opportunities open up if we can build things in the present using future income.

If credit is such a wonderful thing, why did nobody think of it earlier? Of course they did. Credit arrangements of one kind or another have existed in all known human cultures, going back at least to ancient Sumer. The problem in previous eras was not that no one had the idea or knew how to use it. It was that people seldom wanted to extend much credit because they didn't trust that the future would be better than the present. They generally believed that times past had been better than their own times and that the future would be worse, or at best much the same. To put that in economic terms, they believed that the total amount of wealth was limited, if not dwindling. People therefore considered it a bad bet to assume that they personally, or their kingdom, or the entire world, would be producing more wealth ten years down the line. Business looked like a zero-sum game. Of course, the profits of one particular bakery might rise, but only at the expense of the bakery next door. Venice might flourish, but only by impoverishing Genoa. The king of England might enrich himself, but only by robbing the king of France. You could cut the pie in many different ways, but it never got any bigger.

That's why many cultures concluded that making bundles of money was sinful. As Jesus said, 'It is easier for a camel to pass through the eye of a needle than for a rich man to enter into the kingdom of God' (Matthew 19:24). If the pie is static, and I have a big part of it, then I must have taken somebody else's slice. The rich were obliged to do penance for their evil deeds by giving some of their surplus wealth to charity.

If the global pie stayed the same size, there was no margin for credit. Credit is the difference between today's pie and tomorrow's pie. If the pie stays the same, why extend credit? It would be an unacceptable risk unless you believed that the baker or king asking for your money might be able to steal a slice from a competitor. So it was hard to get a loan in the premodern world, and when you got one it was usually *small, short-term, and subject to high interest rates.* Upstart entrepreneurs thus found it difficult to open new bakeries and great kings who wanted to build palaces or wage wars had no choice but to raise the necessary funds through high taxes and tariffs. That was fine for kings (as long as their subjects remained docile), but a scullery maid who had a great idea for a bakery and wanted to move up in the world generally could only dream of wealth while scrubbing down the royal kitchen's floors.

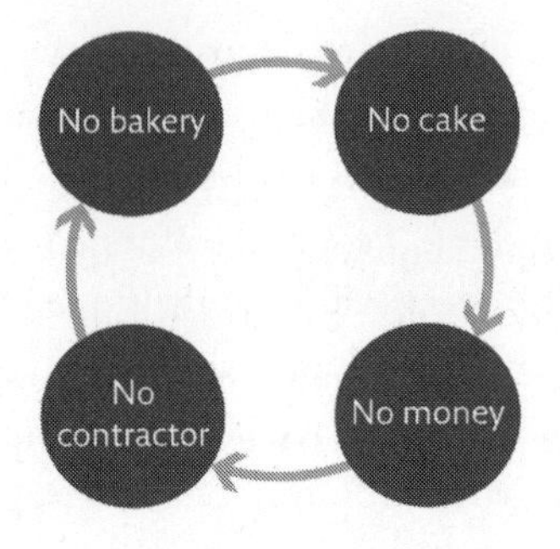

The Entrepreneur's Dilemma

It was lose–lose. Because credit was limited, people had trouble financing new businesses. Because there were few new businesses, the economy did not grow. Because it did not grow, people assumed it never would, and those who had capital were wary of extending credit. The expectation of stagnation fulfilled itself.

The Magic Circle of the Modern Economy

A Growing Pie

Then came the Scientific Revolution and the idea of progress. The idea of progress is built on the notion that if we admit our ignorance and invest resources in research, things can improve. This idea was soon translated into economic terms. Whoever believes in progress believes that geographical discoveries, technological inventions and organisational developments can increase the sum total of human production, trade and wealth. New trade routes in the Atlantic could flourish without ruining old routes in the Indian Ocean. New goods could be produced without reducing the production of old ones. For instance, one could open a new bakery specialising in chocolate cakes and croissants without causing bakeries specialising in bread to go bust. Everybody would simply develop new tastes and eat more. I can be wealthy without your becoming poor; I can be obese without your dying of hunger. The entire global pie can grow.

Over the last 500 years the idea of progress convinced people to put more and more trust in the future. This trust created credit; credit brought real economic growth; and growth strengthened the trust in the future and opened the way for even more credit. It didn't happen overnight – the economy behaved more like a roller coaster than a balloon. But over the long run, with the bumps evened out, the general direction was unmistakable. Today, there is so much credit in the world that governments, business corporations

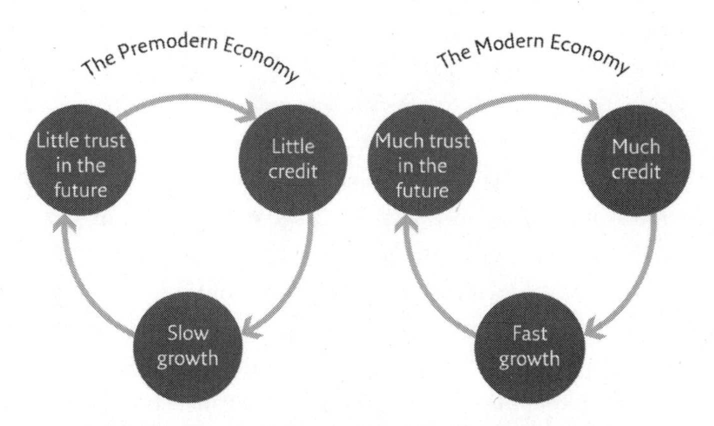

The Economic History of the World in a Nutshell

and private individuals easily obtain *large, long-term and low-interest loans* that far exceed current income.

The belief in the growing global pie eventually turned revolutionary. In 1776 the Scottish economist Adam Smith published *The Wealth of Nations*, probably the most important economics manifesto of all time. In the eighth chapter of its first volume, Smith made the following novel argument: when a landlord, a weaver or a shoemaker has greater profits than he needs to maintain his own family, he uses the surplus to employ more assistants, in order to further increase his profits. The more profits he has, the more assistants he can employ. It follows that an increase in the profits of private entrepreneurs is the basis for the increase in collective wealth and prosperity.

This may not strike you as very original, because we all live in a capitalist world that takes Smith's argument for granted. We hear variations on this theme every day in the news. Yet Smith's claim that the selfish human urge to increase private profits is the basis for collective wealth is one of the most revolutionary ideas in human history – revolutionary not just from an economic perspective, but even more so from a moral and political perspective. What Smith says is, in fact, that greed is good, and that by becoming richer I benefit everybody, not just myself. *Egoism is altruism.*

Smith taught people to think about the economy as a 'win–win situation', in which my profits are also your profits. Not only can we both enjoy a bigger slice of pie at the same time, but the increase in your slice depends upon the increase in my slice. If I am poor, you too will be poor since I cannot buy your products or services. If I am rich, you too will be enriched since you can now sell me something. Smith denied the traditional contradiction between wealth and morality, and threw open the gates of heaven for the rich. Being rich meant being moral. In Smith's story, people become rich not by despoiling their neighbours, but by increasing the overall size of the pie. And when the pie grows, everyone benefits. The rich are accordingly the most useful and benevolent people in society, because they turn the wheels of growth for everyone's advantage.

All this depends, however, on the rich using their profits to open new factories and hire new employees, rather than wasting them on non-productive activities. Smith therefore repeated like a mantra the maxim that 'When profits increase, the landlord or weaver will employ more assistants' and not 'When profits increase, Scrooge will hoard his money in a chest and take it out only to count his coins.' A crucial part of the modern capitalist economy was the emergence of a new ethic, according to which profits ought to be reinvested in production. This brings about more profits, which are again reinvested in production, which brings more profits, et cetera ad infinitum. Investments can be made in many ways: enlarging the factory, conducting scientific research, developing new products. Yet all these investments must somehow increase production and translate into larger profits. In the new capitalist creed, the first and most sacred commandment is: 'The profits of production must be reinvested in increasing production.'

That's why capitalism is called 'capitalism'. Capitalism distinguishes 'capital' from mere 'wealth'. Capital consists of money, goods and resources that are invested in production. Wealth, on the other hand, is buried in the ground or wasted on unproductive activities. A pharaoh who pours resources into a non-productive pyramid is not a capitalist. A pirate who loots a Spanish treasure fleet and buries a chest full of glittering coins on the beach of some Caribbean island is not a capitalist. But a hard-working factory hand who reinvests part of his income in the stock market is.

The idea that 'The profits of production must be reinvested in increasing production' sounds trivial. Yet it was alien to most people throughout history. In premodern times, people believed that the level of production was more or less constant. So why reinvest your profits if production won't increase by much, no matter what you do? Thus medieval noblemen espoused an ethic of generosity and

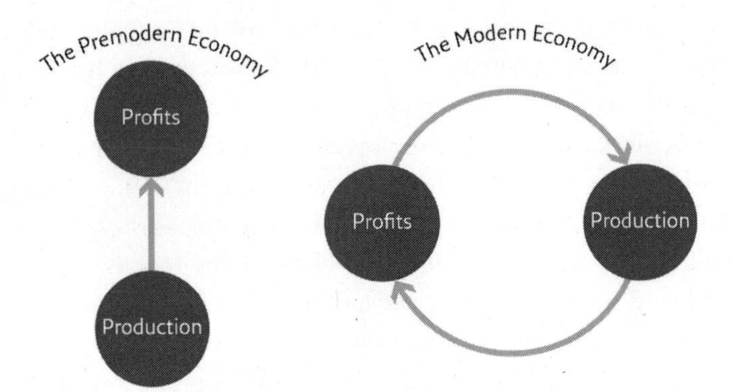

conspicuous consumption. They spent their revenues on tournaments, banquets, palaces and wars, and on charity and monumental cathedrals. Few tried to reinvest profits in increasing their manors' output, developing better kinds of wheat, or looking for new markets.

In the modern era, the nobility has been overtaken by a new elite whose members are true believers in the capitalist creed. The new capitalist elite is made up not of dukes and marquises, but of board chairmen, stock traders and industrialists. These magnates are far richer than the medieval nobility, but they are far less interested in extravagant consumption, and they spend a much smaller part of their profits on non-productive activities.

Medieval noblemen wore colourful robes of gold and silk, and devoted much of their time to attending banquets, carnivals and glamorous tournaments. In comparison, modern CEOs don dreary uniforms called suits that afford them all the panache of a flock of crows, and they have little time for festivities. The typical venture capitalist rushes from one business meeting to another, trying to figure out where to invest his capital and following the ups and downs of the stocks and bonds he owns. True, his suits might be Versace and he might get to travel in a private jet, but these expenses are nothing compared to what he invests in increasing human production.

It's not just Versace-clad business moguls who invest to increase productivity. Ordinary folk and government agencies think along similar lines. How many dinner conversations in modest neighbourhoods sooner or later bog down in interminable debate about whether it is better to invest one's savings in the stock market, bonds or property? Governments too strive to invest their tax revenues in productive enterprises that will increase future income – for example, building a new port could make it easier for factories to export their products, enabling them to make more taxable income, thereby increasing the government's future revenues. Another government might prefer to invest in education, on the grounds that educated people form the basis for the lucrative high-tech industries, which pay lots of taxes without needing extensive port facilities.

Capitalism began as a theory about how the economy functions. It was both descriptive and prescriptive – it offered an account of how money worked and promoted the idea that reinvesting profits in production leads to fast economic growth. But capitalism gradually became far more than just an economic doctrine. It now encompasses an ethic – a set of teachings about how people should behave, educate their children and even think. Its principal tenet is that economic growth is the supreme good, or at least a proxy for the supreme good, because justice, freedom and even happiness all depend on economic growth. Ask a capitalist how to bring justice and political freedom to a place like Zimbabwe or Afghanistan, and you are likely to get a lecture on how economic affluence and a thriving middle class are essential for stable democratic institutions, and about the need therefore to inculcate Afghan tribesmen in the values of free enterprise, thrift and self-reliance.

This new religion has had a decisive influence on the development of modern science, too. Scientific research is usually funded by either governments or private businesses. When capitalist governments and businesses consider investing in a particular scientific project, the first questions are usually 'Will this project enable us to increase production and profits? Will it produce economic growth?' A project that can't clear these hurdles has little chance of finding a sponsor. No history of modern science can leave capitalism out of the picture.

Conversely, the history of capitalism is unintelligible without taking science into account. Capitalism's belief in perpetual economic growth flies in the face of almost everything we know about the universe. A society of wolves would be extremely foolish to believe that the supply of sheep would keep on growing indefinitely. The human economy has nevertheless managed to keep on growing throughout the modern era, thanks only to the fact that scientists come up with another discovery or gadget every few years – such as the continent of America, the internal combustion engine, or genetically engineered sheep. Banks and governments print money, but ultimately, it is the scientists who foot the bill.

Over the last few years, banks and governments have been frenziedly printing money. Everybody is terrified that the current economic crisis may stop the growth of the economy. So they are creating trillions of dollars, euros and yen out of thin air, pumping cheap credit into the system, and hoping that the scientists, technicians and engineers will manage to come up with something really big, before the bubble bursts. Everything depends on the people in the labs. New discoveries in fields such as biotechnology and artificial intelligence could create entire new industries, whose profits could back the trillions of make-believe money that the banks and governments have created since 2008. If the labs do not fulfil these expectations before the bubble bursts, we are heading towards very rough times.

Columbus Searches for an Investor

Capitalism played a decisive role not only in the rise of modern science, but also in the emergence of European imperialism. And it was European imperialism that created the capitalist credit system in the first place. Of course, credit was not invented in modern Europe. It existed in almost all agricultural societies, and in the early modern period the emergence of European capitalism was closely linked to economic developments in Asia. Remember, too, that until the late eighteenth century, Asia was the world's economic powerhouse, meaning that Europeans had far less capital at their disposal than the Chinese, Muslims or Indians.

However, in the sociopolitical systems of China, India and the Muslim world, credit played only a secondary role. Merchants and bankers in the markets of Istanbul, Isfahan, Delhi and Beijing may have thought along capitalist lines, but the kings and generals in the palaces and forts tended to despise merchants and mercantile thinking. Most non-European empires of the early modern era were established by great conquerors such as Nurhaci and Nader Shah, or by bureaucratic and military elites as in the Qing and Ottoman empires. Financing wars through taxes and plunder (without making fine distinctions between the two), they owed little to credit systems, and they cared even less about the interests of bankers and investors.

In Europe, on the other hand, kings and generals gradually adopted the mercantile way of thinking, until merchants and bankers became the ruling elite. The European conquest of the world was increasingly financed through credit rather than taxes, and was increasingly directed by capitalists whose main ambition was to receive maximum returns on their investments. The empires built by bankers and merchants in frock coats and top hats defeated the empires built by kings and noblemen in gold clothes and shining armour. The mercantile empires were simply much shrewder in financing their conquests. Nobody wants to pay taxes, but everyone is happy to invest.

In 1484 Christopher Columbus approached the king of Portugal with the proposal that he finance a fleet that would sail westward to find a new trade route to East Asia. Such explorations were a very risky and costly business. A lot of money was needed in order to build ships, buy supplies, and pay sailors and soldiers – and there was no guarantee that the investment would yield a return. The king of Portugal declined.

Like a present-day start-up entrepreneur, Columbus did not give up. He pitched his idea to other potential investors in Italy, France, England, and again in Portugal. Each time he was rejected. He then tried his luck with Ferdinand and Isabella, rulers of newly united Spain. He took on some experienced lobbyists, and with their help he managed to convince Queen Isabella to invest. As every schoolchild knows, Isabella hit the jackpot. Columbus' discoveries enabled the Spaniards to conquer America, where they established gold and silver mines as well as sugar and tobacco plantations that enriched the Spanish kings, bankers and merchants beyond their wildest dreams.

A hundred years later, princes and bankers were willing to extend far more credit to Columbus' successors, and they had more capital at their disposal, thanks to the treasures reaped from America. Equally important, princes and bankers had far more trust in the potential of exploration, and were more willing to part with their money. This was the magic circle of imperial capitalism: credit financed new discoveries; discoveries led to colonies; colonies provided profits; profits built trust; and trust translated into more credit. Nurhaci and Nader Shah ran out of fuel after a few thousand kilometres. Capitalist entrepreneurs only increased their financial momentum from conquest to conquest.

But these expeditions remained chancy affairs, so credit markets nevertheless remained quite cautious. Many expeditions returned to Europe empty-handed, having discovered nothing of value. The English, for instance, wasted a lot of capital in fruitless attempts to discover a north-western passage to Asia through the Arctic. Many other expeditions didn't return at all. Ships hit icebergs, foundered in tropical storms, or fell victim to pirates. In order to increase the number of potential investors and reduce the risk they incurred,

Europeans turned to limited liability joint-stock companies. Instead of a single investor betting all his money on a single rickety ship, the joint-stock company collected money from a large number of investors, each risking only a small portion of his capital. The risks were thereby curtailed, but no cap was placed on the profits. Even a small investment in the right ship could turn you into a millionaire.

Decade by decade, western Europe witnessed the development of a sophisticated financial system that could raise large amounts of credit on short notice and put it at the disposal of private entrepreneurs and governments. This system could finance explorations and conquests far more efficiently than any kingdom or empire. The new-found power of credit can be seen in the bitter struggle between Spain and the Netherlands. In the sixteenth century, Spain was the most powerful state in Europe, holding sway over a vast global empire. It ruled much of Europe, huge chunks of North and South America, the Philippine Islands, and a string of bases along the coasts of Africa and Asia. Every year, fleets heavy with American and Asian treasures returned to the ports of Seville and Cadiz. The Netherlands was a small and windy swamp, devoid of natural resources, a tiny corner of the king of Spain's dominions.

In 1568 the Dutch, who were mainly Protestant, revolted against their Catholic Spanish overlord. At first the rebels seemed to play the role of Don Quixote, courageously tilting at invincible windmills. Yet within eighty years the Dutch had not only secured their independence from Spain, but had managed to replace the Spaniards and their Portuguese allies as masters of the ocean highways, build a global Dutch empire, and become the richest state in Europe.

The secret of Dutch success was credit. The Dutch burghers, who had little taste for combat on land, hired mercenary armies to fight the Spanish for them. The Dutch themselves meanwhile took to the sea in ever-larger fleets. Mercenary armies and cannon-brandishing fleets cost a fortune, but the Dutch were able to finance their military expeditions more easily than the mighty Spanish Empire because they secured the trust of the burgeoning European financial system at a time when the Spanish king was carelessly eroding its trust in him. Financiers extended the Dutch enough credit to set up armies and fleets, and these armies and fleets gave the Dutch control of world trade routes, which in turn yielded handsome profits. The profits allowed the Dutch to repay the loans, which strengthened the trust of the financiers. Amsterdam was fast becoming not only one of the most important ports of Europe, but also the continent's financial Mecca.

How exactly did the Dutch win the trust of the financial system? Firstly, they were sticklers about repaying their loans on time and in full, making the extension of credit less risky for lenders. Secondly, their country's judicial system enjoyed independence and protected private rights – in particular private property rights. Capital trickles away from dictatorial states that fail to defend private individuals and their property. Instead, it flows into states upholding the rule of law and private property.

Imagine that you are the son of a solid family of German financiers. Your father sees an opportunity to expand the business by opening branches in major European cities. He sends you to Amsterdam and your younger brother to Madrid, giving you each 10,000 gold coins to invest. Your brother lends his start-up capital at interest to the king of Spain, who needs it to raise an army to fight the king of France. You decide to lend yours to a Dutch merchant, who wants to invest in scrubland on the southern end of a desolate island called Manhattan, certain that property values there will skyrocket as the Hudson River turns into a major trade artery. Both loans are to be repaid within a year.

The year passes. The Dutch merchant sells the land he's bought at a handsome markup and repays your money with the interest he promised. Your father is pleased. But your little brother in Madrid is getting nervous. The war with France ended well for the king of

Spain, but he has now embroiled himself in a conflict with the Turks. He needs every penny to finance the new war, and thinks this is far more important than repaying old debts. Your brother sends letters to the palace and asks friends with connections at court to intercede, but to no avail. Not only has your brother not earned the promised interest – he's lost the principal. Your father is not pleased.

Now, to make matters worse, the king sends a treasury official to your brother to tell him, in no uncertain terms, that he expects to receive another loan of the same size, forthwith. Your brother has no money to lend. He writes home to dad, trying to persuade him that this time the king will come through. The paterfamilias has a soft spot for his youngest, and agrees with a heavy heart. Another 10,000 gold coins disappear into the Spanish treasury, never to be seen again. Meanwhile in Amsterdam, things are looking bright. You make more and more loans to enterprising Dutch merchants, who repay them promptly and in full. But your luck does not hold indefinitely. One of your usual clients has a hunch that wooden clogs are going to be the next fashion craze in Paris, and asks you for a loan to set up a footwear emporium in the French capital. You lend him the money, but unfortunately the clogs don't catch on with the French ladies, and the disgruntled merchant refuses to repay the loan.

Your father is furious, and tells both of you it is time to unleash the lawyers. Your brother files suit in Madrid against the Spanish monarch, while you file suit in Amsterdam against the erstwhile wooden-shoe wizard. In Spain, the law courts are subservient to the king – the judges serve at his pleasure and fear punishment if they do not do his will. In the Netherlands, the courts are a separate branch of government, not dependent on the country's burghers and princes. The court in Madrid throws out your brother's suit, while the court in Amsterdam finds in your favour and puts a lien on the clog merchant's assets to force him to pay up. Your father has learned his lesson. Better to do business with merchants than with kings, and better to do it in Holland than in Madrid.

And your brother's travails are not over. The king of Spain desperately needs more money to pay his army. He's sure that your father has cash to spare. So he brings trumped-up treason charges against your brother. If he doesn't come up with 20,000 gold coins forthwith, he'll get cast into a dungeon and rot there until he dies.

Your father has had enough. He pays the ransom for his beloved son, but swears never to do business in Spain again. He closes his Madrid branch and relocates your brother to Rotterdam. Two branches in Holland now look like a really good idea. He hears that even Spanish capitalists are smuggling their fortunes out of their country. They, too, realise that if they want to keep their money and use it to gain more wealth, they are better off investing it where the rule of law prevails and where private property is respected – in the Netherlands, for example.

In such ways did the king of Spain squander the trust of investors at the same time that Dutch merchants gained their confidence. And it was the Dutch merchants – not the Dutch state – who built the Dutch Empire. The king of Spain kept on trying to finance and maintain his conquests by raising unpopular taxes from a disgruntled populace. The Dutch merchants financed conquest by getting loans, and increasingly also by selling shares in their companies that entitled their holders to receive a portion of the company's profits. Cautious investors who would never have given their money to the king of Spain, and who would have thought twice before extending credit to the Dutch government, happily invested fortunes in the Dutch joint-stock companies that were the mainstay of the new empire.

If you thought a company was going to make a big profit but it had already sold all its shares, you could buy some from people who owned them, probably for a higher price than they originally paid. If you bought shares and later discovered that the company was in dire straits, you could try to unload your stock for a lower price. The resulting trade in company shares led to the establishment in most major European cities of stock exchanges, places where the shares of companies were traded.

The most famous Dutch joint-stock company, the Vereenigde Oostindische Compagnie, or VOC for short, was chartered in 1602, just as the Dutch were throwing off Spanish rule and the boom of Spanish artillery could still be heard not far from Amsterdam's ramparts. VOC used the money it raised from selling shares to build ships, send them to Asia, and bring back Chinese, Indian and Indonesian goods. It also financed military actions taken by company ships against competitors and pirates. Eventually VOC money financed the conquest of Indonesia.

Indonesia is the world's biggest archipelago. Its thousands upon thousands of islands were ruled in the early seventeenth century by hundreds of kingdoms, principalities, sultanates and tribes. When VOC merchants first arrived in Indonesia in 1603, their aims were strictly commercial. However, in order to secure their commercial interests and maximise the profits of the shareholders, VOC merchants began to fight against local potentates who charged inflated tariffs, as well as against European competitors. VOC armed its merchant ships with cannons; it recruited European, Japanese, Indian and Indonesian mercenaries; and it built forts and conducted full-scale battles and sieges. This enterprise may sound a little strange to us, but in the early modern age it was common for private companies to hire not only soldiers, but also generals and admirals, cannons and ships, and even entire off-the-shelf armies. The international community took this for granted and didn't raise an eyebrow when a private company established an empire.

Island after island fell to VOC mercenaries and a large part of Indonesia became a VOC colony. VOC ruled Indonesia for close to 200 years. Only in 1800 did the Dutch state assume control of Indonesia, making it a Dutch national colony for the following 150 years. Today some people warn that twenty-first-century corporations are accumulating too much power. Early modern history shows just how far that can go if businesses are allowed to pursue their self-interest unchecked.

While VOC operated in the Indian Ocean, the Dutch West Indies Company, or WIC, plied the Atlantic. In order to control trade on the important Hudson River, WIC built a settlement called New Amsterdam on an island at the river's mouth. The colony was threatened by Native Americans and repeatedly attacked by the British, who eventually captured it in 1664. The British changed its name to New York. The remains of the wall built by WIC to defend its colony against Native Americans and British are today paved over by the world's most famous street – Wall Street.

As the seventeenth century wound to an end, complacency and costly continental wars caused the Dutch to lose not only New York, but also their place as Europe's financial and imperial engine. The vacancy was hotly contested by France and Britain. At first France seemed to be in a far stronger position. It was bigger than Britain, richer, more populous, and it possessed a larger and more experienced army. Yet Britain managed to win the trust of the financial system whereas France proved itself unworthy. The behaviour of the French Crown was particularly notorious during what was called the Mississippi Bubble, the largest financial crisis of eighteenth-century Europe. That story also begins with an empire-building joint-stock company.

In 1717 the Mississippi Company, chartered in France, set out to colonise the lower Mississippi valley, establishing the city of New Orleans in the process. To finance its ambitious plans, the company, which had good connections at the court of King Louis XV, sold shares on the Paris stock exchange. John Law, the company's director, was also the governor of the central bank of France. Furthermore, the king had appointed him controller-general of finances, an office roughly equivalent to that of a modern finance minister. In 1717

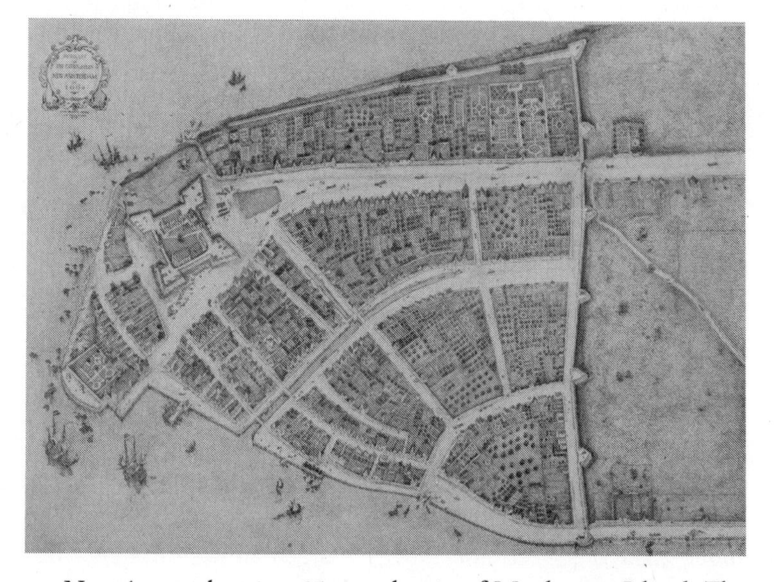

39. New Amsterdam in 1660, at the tip of Manhattan Island. The settlement's protective wall is today paved over by Wall Street.

the lower Mississippi valley offered few attractions besides swamps and alligators, yet the Mississippi Company spread tales of fabulous riches and boundless opportunities. French aristocrats, businessmen and the stolid members of the urban bourgeoisie fell for these fantasies, and Mississippi share prices skyrocketed. Initially, shares were offered at 500 livres apiece. On 1 August 1719, shares traded at 2,750 livres. By 30 August, they were worth 4,100 livres, and on 4 September, they reached 5,000 livres. On 2 December the price of a Mississippi share crossed the threshold of 10,000 livres. Euphoria swept the streets of Paris. People sold all their possessions and took huge loans in order to buy Mississippi shares. Everybody believed they'd discovered the easy way to riches.

A few days later, the panic began. Some speculators realised that the share prices were totally unrealistic and unsustainable. They figured that they had better sell while stock prices were at their peak. As the supply of shares available rose, their price declined. When other investors saw the price going down, they also wanted to get out quick. The stock price plummeted further, setting off an avalanche. In order to stabilise prices, the central bank of France – at the direction of its governor, John Law – bought up Mississippi shares, but it could not do so for ever. Eventually it ran out of money. When this happened, the controller-general of finances, the same John Law, authorised the printing of more money in order to buy additional shares. This placed the entire French financial system inside the bubble. And not even this financial wizardry could save the day. The price of Mississippi shares dropped from 10,000 livres back to 1,000 livres, and then collapsed completely, and the shares lost every sou of their worth. By now, the central bank and the royal treasury owned a huge amount of worthless stock and had no money. The big speculators emerged largely unscathed – they had sold in time. Small investors lost everything, and many committed suicide.

The Mississippi Bubble was one of history's most spectacular financial crashes. The royal French financial system never recuperated fully from the blow. The way in which the Mississippi Company used its political clout to manipulate share prices and fuel the buying frenzy caused the public to lose faith in the French banking system and in the financial wisdom of the French king. Louis XV found it more and more difficult to raise credit. This became one of the chief reasons that the overseas French Empire fell into British hands. While the British could borrow money easily and at low interest rates, France had difficulties securing loans, and had to pay high interest on them. In order to finance his growing debts, the king of France borrowed more and more money at higher and higher interest rates. Eventually, in the 1780s, Louis XVI, who had ascended to the throne on his grandfather's death, realised that half his annual budget was tied to servicing the interest on his loans, and that he was heading towards bankruptcy. Reluctantly, in 1789, Louis XVI convened the Estates General, the French parliament that had not met for a century and a half, in order to find a solution to the crisis. Thus began the French Revolution.

While the French overseas empire was crumbling, the British Empire was expanding rapidly. Like the Dutch Empire before it, the British Empire was established and run largely by private joint-stock companies based in the London stock exchange. The first English settlements in North America were established in the early seventeenth century by joint-stock companies such as the London Company, the Plymouth Company, the Dorchester Company and the Massachusetts Company.

The Indian subcontinent too was conquered not by the British state, but by the mercenary army of the British East India Company. This company outperformed even the VOC. From its headquarters in Leadenhall Street, London, it ruled a mighty Indian empire for about a century, maintaining a huge military force of up to 350,000 soldiers, considerably outnumbering the armed forces of the British monarchy. Only in 1858 did the British Crown nationalise India along with the company's private army. Napoleon made fun of the British, calling them a nation of shopkeepers. Yet these shopkeepers defeated Napoleon himself, and their empire was the largest the world has ever seen.

In the Name of Capital

The nationalisation of Indonesia by the Dutch Crown (1800) and of India by the British Crown (1858) hardly ended the embrace of capitalism and empire. On the contrary, the connection only grew stronger during the nineteenth century. Joint-stock companies no longer needed to establish and govern private colonies – their managers and large shareholders now pulled the strings of power in London, Amsterdam and Paris, and they could count on the state to look after their interests. As Marx and other social critics quipped, Western governments were becoming a capitalist trade union.

The most notorious example of how governments did the bidding of big money was the First Opium War, fought between Britain and China (1840–42). In the first half of the nineteenth century, the British East India Company and sundry British business people made fortunes by exporting drugs, particularly opium, to China. Millions of Chinese became addicts, debilitating the country both economically and socially. In the late 1830s the Chinese government issued a ban on drug trafficking, but British drug merchants simply ignored the law. Chinese authorities began to confiscate and destroy drug cargos. The drug cartels had close connections in Westminster and Downing Street – many MPs and Cabinet ministers in fact held stock in the drug companies – so they pressured the government to take action.

In 1840 Britain duly declared war on China in the name of 'free trade'. It was a walkover. The overconfident Chinese were no match for Britain's new wonder weapons – steamboats, heavy artillery, rockets and rapid-fire rifles. Under the subsequent peace treaty, China agreed not to constrain the activities of British drug merchants and to compensate them for damages inflicted by the Chinese police. Furthermore, the British demanded and received control of Hong Kong, which they proceeded to use as a secure base for drug trafficking (Hong Kong remained in British hands until 1997). In the late nineteenth century, about 40 million Chinese, a tenth of the country's population, were opium addicts.[3]

Egypt, too, learned to respect the long arm of British capitalism. During the nineteenth century, French and British investors lent huge sums to the rulers of Egypt, first in order to finance the Suez Canal project, and later to fund far less successful enterprises. Egyptian debt swelled, and European creditors increasingly meddled in Egyptian affairs. In 1881 Egyptian nationalists had had enough and rebelled. They declared a unilateral abrogation of all foreign debt. Queen Victoria was not amused. A year later she dispatched her army and navy to the Nile, and Egypt remained a British protectorate until after the Second World War.

These were hardly the only wars fought in the interests of investors. In fact, war itself could become a commodity, just like opium. In 1821 the Greeks rebelled against the Ottoman Empire. The uprising aroused great sympathy in liberal and Romantic circles in Britain – Lord Byron, the poet, even went to Greece to fight alongside the insurgents. But London financiers saw an opportunity as well. They proposed to the rebel leaders the issue of tradable Greek Rebellion Bonds on the London stock exchange. The Greeks would promise to repay the bonds, plus interest, if and when they won their independence. Private investors bought bonds to make a profit, or out of sympathy for the Greek cause, or both. The value of Greek Rebellion Bonds rose and fell on the London stock exchange in tempo with military successes and failures on the battlefields of Hellas. The Turks gradually gained the upper hand. With a rebel defeat imminent, the bondholders faced the prospect of losing their trousers. The bondholders' interest was the national interest, so the

40. The Battle of Navarino (1827).

British organised an international fleet that, in 1827, sank the main Ottoman flotilla in the Battle of Navarino. After centuries of subjugation, Greece was finally free. But freedom came with a huge debt that the new country had no way of repaying. The Greek economy was mortgaged to British creditors for decades to come.

The bear hug between capital and politics has had far-reaching implications for the credit market. The amount of credit in an economy is determined not only by purely economic factors such as the discovery of a new oil field or the invention of a new machine, but also by political events such as regime changes or more ambitious foreign policies. After the Battle of Navarino, British capitalists were more willing to invest their money in risky overseas deals. They had seen that if a foreign debtor refused to repay loans, the British army would get their money back.

This is why today a country's credit rating is far more important to its economic well-being than are its natural resources. Credit ratings indicate the probability that a country will pay its debts. In addition to purely economic data, they take into account political, social and even cultural factors. An oil-rich country cursed with a despotic government, endemic warfare and a corrupt judicial system will usually receive a low credit rating. As a result, it is likely to remain relatively poor since it will not be able to raise the necessary capital to make the most of its oil bounty. A country devoid of natural resources, but which enjoys peace, a fair judicial system and a free government is likely to receive a high credit rating. As such, it may be able to raise enough cheap capital to support a good education system and foster a flourishing high-tech industry.

The Cult of the Free Market

Capital and politics influence each other to such an extent that their relations are hotly debated by economists, politicians and the general public alike. Ardent capitalists tend to argue that capital should be free to influence politics, but politics should not be allowed to influence capital. They argue that when governments interfere in the markets, political interests cause them to make unwise investments that result in slower growth. For example, a government may impose heavy taxation on industrialists and use the money to give lavish unemployment benefits, which are popular with voters. In the view of many business people, it would be far better if the government left the money with them. They would use it, they claim, to open new factories and hire the unemployed.

In this view, the wisest economic policy is to keep politics out of the economy, reduce taxation and government regulation to a minimum, and allow market forces free rein to take their course. Private investors, unencumbered by political considerations, will invest their money where they can get the most profit, so the way to ensure the most economic growth – which will benefit everyone, industrialists and workers – is for the government to do as little as possible. This free-market doctrine is today the most common and influential variant of the capitalist creed. The most enthusiastic advocates of the free market criticise military adventures abroad with as much zeal as welfare programmes at home. They offer governments the same advice that Zen masters offer initiates: just do nothing.

But in its extreme form, belief in the free market is as naive as belief in Santa Claus. There simply is no such thing as a market free of all political bias. The most important economic resource is trust in the future, and this resource is constantly threatened by thieves and charlatans. Markets by themselves offer no protection against fraud, theft and violence. It is the job of political systems to ensure trust by legislating sanctions against cheats and to establish and support police forces, courts and jails which will enforce the law. When kings fail to do their jobs and regulate the markets properly, it leads to loss of trust, dwindling credit and economic depression. That was the lesson taught by the Mississippi Bubble of 1719, and anyone who forgot it was reminded by the US housing bubble of 2007, and the ensuing credit crunch and recession.

The Capitalist Hell

There is an even more fundamental reason why it's dangerous to give markets a completely free rein. Adam Smith taught that the shoemaker would use his surplus to employ more assistants. This implies that egoistic greed is beneficial for all, since profits are utilised to expand production and hire more employees.

Yet what happens if the greedy shoemaker increases his profits by paying employees less and increasing their work hours? The standard answer is that the free market would protect the employees. If our shoemaker pays too little and demands too much, the best employees would naturally abandon him and go to work for his competitors. The tyrant shoemaker would find himself left with the worst labourers, or with no labourers at all. He would have to mend his ways or go out of business. His own greed would compel him to treat his employees well.

This sounds bulletproof in theory, but in practice the bullets get through all too easily. In a completely free market, unsupervised by kings and priests, avaricious capitalists can establish monopolies or collude against their workforces. If there is a single corporation controlling all shoe factories in a country, or if all factory owners conspire to reduce wages simultaneously, then the labourers are no longer able to protect themselves by switching jobs.

Even worse, greedy bosses might curtail the workers' freedom of movement through debt peonage or slavery. At the end of the Middle Ages, slavery was almost unknown in Christian Europe. During the early modern period, the rise of European capitalism went hand in hand with the rise of the Atlantic slave trade. Unrestrained market forces, rather than tyrannical kings or racist ideologues, were responsible for this calamity.

When the Europeans conquered America, they opened gold and silver mines and established sugar, tobacco and cotton plantations. These mines and plantations became the mainstay of American production and export. The sugar plantations were particularly important. In the Middle Ages, sugar was a rare luxury in Europe. It was imported from the Middle East at prohibitive prices and used sparingly as a secret ingredient in delicacies and snake-oil medicines. After large sugar plantations were established in America, ever-increasing amounts of sugar began to reach Europe. The price of sugar dropped and Europe developed an insatiable sweet tooth. Entrepreneurs met this need by producing huge quantities of sweets: cakes, cookies, chocolate, candy, and sweetened beverages such as cocoa, coffee and tea. The annual sugar intake of the average Englishman rose from near zero in the early seventeenth century to around eight kilograms in the early nineteenth century.

However, growing cane and extracting its sugar was a labour-intensive business. Few people wanted to work long hours in malaria-infested sugar fields under a tropical sun. Contract labourers would have produced a commodity too expensive for mass consumption. Sensitive to market forces, and greedy for profits and economic growth, European plantation owners switched to slaves.

From the sixteenth to the nineteenth centuries, about 10 million African slaves were imported to America. About 70 per cent of them worked on the sugar plantations. Labour conditions were abominable. Most slaves lived a short and miserable life, and millions more died during wars waged to capture slaves or during the long voyage from inner Africa to the shores of America. All this so that Europeans could enjoy their sweet tea and candy – and sugar barons could enjoy huge profits.

The slave trade was not controlled by any state or government. It was a purely economic enterprise, organised and financed by the free market according to the laws of supply and demand. Private slave-trading companies sold shares on the Amsterdam, London and Paris stock exchanges. Middle-class Europeans looking for a good investment bought these shares. Relying on this money, the companies bought ships, hired sailors and soldiers, purchased slaves in Africa, and transported them to America. There they sold the slaves to the plantation owners, using the proceeds to purchase plantation products such as sugar, cocoa, coffee, tobacco, cotton and rum. They returned to Europe, sold the sugar and cotton for a good price, and then sailed to Africa to begin another round. The shareholders were very pleased with this arrangement. Throughout the eighteenth century the yield on slave-trade investments was about 6 per cent a year – they were extremely profitable, as any modern consultant would be quick to admit.

This is the fly in the ointment of free-market capitalism. It cannot ensure that profits are gained in a fair way, or distributed in a fair manner. On the contrary, the craving to increase profits and production blinds people to anything that might stand in the way. When growth becomes a supreme good, unrestricted by any other ethical considerations, it can easily lead to catastrophe. Some religions, such as Christianity and Nazism, have killed millions out of burning hatred. Capitalism has killed millions out of cold indifference coupled with greed. The Atlantic slave trade did not stem from racist hatred towards Africans. The individuals who bought the shares, the brokers who sold them, and the managers of the slave-trade companies rarely thought about the Africans. Nor did the owners of the sugar plantations. Many owners lived far from their plantations, and the only information they demanded were neat ledgers of profits and losses.

It is important to remember that the Atlantic slave trade was not a single aberration in an otherwise spotless record. The Great Bengal Famine, discussed in the previous chapter, was caused by a similar dynamic – the British East India Company cared more about its profits than about the lives of 10 million Bengalis. VOC's military campaigns in Indonesia were financed by upstanding Dutch burghers who loved their children, gave to charity, and enjoyed good music and fine art, but had no regard for the suffering of the inhabitants of Java, Sumatra and Malacca. Countless other crimes and misdemeanours accompanied the growth of the modern economy in other parts of the planet.

The nineteenth century brought no improvement in the ethics of capitalism. The Industrial Revolution that swept through Europe enriched the bankers and capital-owners, but condemned millions of workers to a life of abject poverty. In the European colonies things were even worse. In 1876, King Leopold II of Belgium set up a non-governmental humanitarian organisation with the declared aim of exploring Central Africa and fighting the slave trade along the Congo River. It was also charged with improving conditions for the inhabitants of the region by building roads, schools and hospitals. In 1885 the European powers agreed to give this organisation control of 2.3 million square kilometres in the Congo basin. This territory, seventy-five times the size of Belgium, was henceforth known as the Congo Free State. Nobody asked the opinion of the territory's 20–30 million inhabitants.

Within a short time the humanitarian organisation became a business enterprise whose real aim was growth and profit. The schools and hospitals were forgotten, and the Congo basin was instead filled with mines and plantations, run by mostly Belgian officials who ruthlessly exploited the local population. The rubber industry was particularly notorious. Rubber was fast becoming an industrial staple, and rubber export was the Congo's most important source of income. The African villagers who collected the rubber were required to provide higher and higher quotas. Those who failed to deliver their quota were punished brutally for their 'laziness'. Their arms were chopped off and occasionally entire villages were massacred. According to the most moderate estimates, between 1885 and 1908 the pursuit of growth and profits cost the lives of 6 million individuals (at least

20 per cent of the Congo's population). Some estimates reach up to 10 million deaths.[4]

In recent decades, and especially after 1945, capitalist greed was somewhat reined in, not least due to the fear of Communism. Yet inequities are still rampant. The economic pie of 2014 is far larger than the pie of 1500, but it is distributed so unevenly that many African peasants and Indonesian labourers return home after a hard day's work with less food than did their ancestors 500 years ago. Much like the Agricultural Revolution, so too the growth of the modern economy might turn out to be a colossal fraud. The human species and the global economy may well keep growing, but many more individuals may live in hunger and want.

Capitalism has two answers to this criticism. First, capitalism has created a world that nobody but a capitalist is capable of running. The only serious attempt to manage the world differently – Communism – was so much worse in almost every conceivable way that nobody has the stomach to try again. In 8500 BC one could cry bitter tears over the Agricultural Revolution, but it was too late to give up agriculture. Similarly, we may not like capitalism, but we cannot live without it.

The second answer is that we just need more patience – paradise, the capitalists promise, is right around the corner. True, mistakes have been made, such as the Atlantic slave trade and the exploitation of the European working class. But we have learned our lesson, and if we just wait a little longer and allow the pie to grow a little bigger, everybody will receive a fatter slice. The division of spoils will never be equitable, but there will be enough to satisfy every man, woman and child – even in the Congo.

There are, indeed, some positive signs. At least when we use purely material criteria – such as life expectancy, child mortality and calorie intake – the standard of living of the average human in 2014 is significantly higher than it was in 1914, despite the exponential growth in the number of humans.

Yet can the economic pie grow indefinitely? Every pie requires raw materials and energy. Prophets of doom warn that sooner or later *Homo sapiens* will exhaust the raw materials and energy of planet Earth. And what will happen then?

17

The Wheels of Industry

THE MODERN ECONOMY GROWS THANKS to our trust in the future and to the willingness of capitalists to reinvest their profits in production. Yet that does not suffice. Economic growth also requires energy and raw materials, and these are finite. When and if they run out, the entire system will collapse.

But the evidence provided by the past is that they are finite only in theory. Counter-intuitively, while humankind's use of energy and raw materials has mushroomed in the last few centuries, the amounts available for our exploitation have actually *increased*. Whenever a shortage of either has threatened to slow economic growth, investments have flowed into scientific and technological research. These have invariably produced not only more efficient ways of exploiting existing resources, but also completely new types of energy and materials.

Consider the vehicle industry. Over the last 300 years, humankind has manufactured billions of vehicles – from carts and wheelbarrows, to trains, cars, supersonic jets and space shuttles. One might have expected that such a prodigious effort would have exhausted the energy sources and raw materials available for vehicle production, and that today we would be scraping the bottom of the barrel. Yet the opposite is the case. Whereas in 1700 the global vehicle industry relied overwhelmingly on wood and iron, today it has at its disposal a cornucopia of new-found materials such as plastic, rubber, aluminium and titanium, none of which our ancestors even knew about. Whereas in 1700 carts were built mainly by the muscle power of carpenters and smiths, today the machines in Toyota and Boeing factories are powered by petroleum combustion engines and nuclear power stations. A similar revolution has swept almost all other fields of industry. We call it the Industrial Revolution.

For millennia prior to the Industrial Revolution, humans already knew how to make use of a large variety of energy sources. They burned wood in order to smelt iron, heat houses and bake cakes. Sailing ships harnessed wind power to move around, and watermills captured the flow of rivers to grind grain. Yet all these had clear limits and problems. Trees were not available everywhere, the wind didn't always blow when you needed it, and water power was only useful if you lived near a river.

An even bigger problem was that people didn't know how to convert one type of energy into another. They could harness the movement of wind and water to sail ships and push millstones, but not to heat water or smelt iron. Conversely, they could not use the heat energy produced by burning wood to make a millstone move. Humans had only one machine capable of performing such energy conversion tricks: the body. In the natural process of metabolism, the bodies of humans and other animals burn organic fuels known as food and convert the released energy into the movement of muscles. Men, women and beasts could consume grain and meat, burn up their carbohydrates and fats, and use the energy to haul a rickshaw or pull a plough.

Since human and animal bodies were the only energy conversion device available, muscle power was the key to almost all human activities. Human muscles built carts and houses, ox muscles ploughed fields, and horse muscles transported goods. The energy that fuelled these organic muscle-machines came ultimately from a single source – plants. Plants in their turn obtained their energy from the sun. By the process of photosynthesis, they captured solar energy and packed it into organic compounds. Almost everything people did throughout history was fuelled by solar energy that was captured by plants and converted into muscle power.

Human history was consequently dominated by two main cycles: the growth cycles of plants and the changing cycles of solar energy (day and night, summer and winter). When sunlight was scarce and when wheat fields were still green, humans had little energy. Granaries were empty, tax collectors were idle, soldiers found it difficult to move and fight, and kings tended to keep the peace. When the sun shone brightly and the wheat ripened, peasants harvested the crops and filled the granaries. Tax collectors hurried to take their share. Soldiers flexed their muscles and sharpened their swords. Kings convened councils and planned their next campaigns. Everyone was fuelled by solar energy – captured and packaged in wheat, rice and potatoes.

The Secret in the Kitchen

Throughout these long millennia, day in and day out, people stood face to face with the most important invention in the history of energy production – and failed to notice it. It stared them in the eye every time a housewife or servant put up a kettle to boil water for tea or put a pot full of potatoes on the stove. The minute the water boiled, the lid of the kettle or the pot jumped. Heat was being converted to movement. But jumping pot lids were an annoyance, especially if you forgot the pot on the stove and the water boiled over. Nobody saw their real potential.

A partial breakthrough in converting heat into movement followed the invention of gunpowder in ninth-century China. At first, the idea of using gunpowder to propel projectiles was so counter-intuitive that for centuries gunpowder was used primarily to produce fire bombs. But eventually – perhaps after some bomb expert ground gunpowder in a mortar only to have the pestle shoot out with force – guns made their appearance. About 600 years passed between the invention of gunpowder and the development of effective artillery.

Even then, the idea of converting heat into motion remained so counter-intuitive that another three centuries went by before people invented the next machine that used heat to move things around. The new technology was born in British coal mines. As the British population swelled, forests were cut down to fuel the growing economy and make way for houses and fields. Britain suffered from an increasing shortage of firewood. It began burning coal as a substitute. Many coal seams were located in waterlogged areas, and flooding prevented miners from accessing the lower strata of the mines. It was a problem looking for a solution. Around 1700, a strange noise began reverberating around British mineshafts. That noise – harbinger of the Industrial Revolution – was subtle at first, but it grew louder and louder with each passing decade until it enveloped the entire world in a deafening cacophony. It emanated from a steam engine.

There are many types of steam engines, but they all share one common principle. You burn some kind of fuel, such as coal, and use the resulting heat to boil water, producing steam. As the steam expands it pushes a piston. The piston moves, and anything that is connected to the piston moves with it. You have converted heat into movement! In eighteenth-century British coal mines, the piston was connected to a pump that extracted water from the bottom of the mineshafts. The earliest engines were incredibly inefficient. You needed to burn a huge load of coal in order to pump out even a tiny amount of water. But in the mines coal was plentiful and close at hand, so nobody cared.

In the decades that followed, British entrepreneurs improved the efficiency of the steam engine, brought it out of the mineshafts, and connected it to looms and gins. This revolutionised textile production, making it possible to produce ever-larger quantities of cheap textiles. In the blink of an eye, Britain became the workshop of the world. But even more importantly, getting the steam engine out of the mines broke an important psychological barrier. If you could burn coal in order to move textile looms, why not use the same method to move other things, such as vehicles?

In 1825, a British engineer connected a steam engine to a train of mine wagons full of coal. The engine drew the wagons along an iron rail some twenty kilometres long from the mine to the nearest harbour. This was the first steam-powered locomotive in history. Clearly, if steam could be used to transport coal, why not other goods? And why not even people? On 15 September 1830, the first commercial railway line was opened, connecting Liverpool with Manchester. The trains moved under the same steam power that had previously pumped water and moved textile looms. A mere twenty years later, Britain had tens of thousands of kilometres of railway tracks.[1]

Henceforth, people became obsessed with the idea that machines and engines could be used to convert one type of energy into another. Any type of energy, anywhere in the world, might be harnessed to whatever need we had, if we could just invent the right machine. For example, when physicists realised that an immense amount of energy is stored within atoms, they immediately started thinking about how this energy could be released and used to make electricity, power submarines and annihilate cities. Six hundred years passed between the moment Chinese alchemists discovered gunpowder and the moment Turkish cannon pulverised the walls of Constantinople. Only forty years passed between the moment Einstein determined that any kind of mass could be converted into energy – that's what $E = mc^2$ means – and the moment atom bombs obliterated Hiroshima and Nagasaki and nuclear power stations mushroomed all over the globe.

Another crucial discovery was the internal combustion engine, which took little more than a generation to revolutionise human transportation and turn petroleum into liquid political power.

Petroleum had been known for thousands of years, and was used to waterproof roofs and lubricate axles. Yet until just a century ago nobody thought it was useful for much more than that. The idea of spilling blood for the sake of oil would have seemed ludicrous. You might fight a war over land, gold, pepper or slaves, but not oil.

The career of electricity was more startling yet. Two centuries ago electricity played no role in the economy, and was used at most for arcane scientific experiments and cheap magic tricks. A series of inventions turned it into our universal genie in a lamp. We flick our fingers and it prints books and sews clothes, keeps our vegetables fresh and our ice cream frozen, cooks our dinners and executes our criminals, registers our thoughts and records our smiles, lights up our nights and entertains us with countless television shows. Few of us understand how electricity does all these things, but even fewer can imagine life without it.

An Ocean of Energy

At heart, the Industrial Revolution has been a revolution in energy conversion. It has demonstrated again and again that there is no limit to the amount of energy at our disposal. Or, more precisely, that the only limit is set by our ignorance. Every few decades we discover a new energy source, so that the sum total of energy at our disposal just keeps growing.

Why are so many people afraid that we are running out of energy? Why do they warn of disaster if we exhaust all available fossil fuels? Clearly the world does not lack energy. All we lack is the knowledge necessary to harness and convert it to our needs. The amount of energy stored in all the fossil fuel on earth is negligible compared to the amount that the sun dispenses every day, free of charge. Only a tiny proportion of the sun's energy reaches us, yet it amounts to 3,766,800 exajoules of energy each year (a joule is a unit of energy in the metric system, about the amount you expend to lift a small apple one metre straight up; an exajoule is a billion billion joules – that's a lot of apples).[2] All the world's plants capture only about 3,000 of those solar exajoules through the process of photosynthesis.[3] All human activities and industries put together consume about 500 exajoules annually, equivalent to the amount of energy earth receives from the sun in just ninety minutes.[4] And that's only solar energy. In addition, we are surrounded by other enormous sources of energy, such as nuclear energy and gravitational energy, the latter most evident in the power of the ocean tides caused by the moon's pull on the earth.

Prior to the Industrial Revolution, the human energy market was almost completely dependent on plants. People lived alongside a green energy reservoir carrying 3,000 exajoules a year, and tried to pump as much of its energy as they could. Yet there was a clear limit to how much they could extract. During the Industrial Revolution, we came to realise that we are actually living alongside an enormous ocean of energy, one holding billions upon billions of exajoules of potential power. All we need to do is invent better pumps.

Learning how to harness and convert energy effectively solved the other problem that slows economic growth – the scarcity of raw materials. As humans worked out how to harness large quantities of cheap energy, they could begin exploiting previously inaccessible deposits of raw materials (for example, mining iron in the Siberian wastelands), or transporting raw materials from ever more distant locations (for example, supplying a British textile mill with Australian wool). Simultaneously, scientific breakthroughs enabled humankind to invent completely new raw materials, such as plastic, and discover previously unknown natural materials, such as silicon and aluminium.

Chemists discovered aluminium only in the 1820s, but separating the metal from its ore was extremely difficult and costly. For decades, aluminium was much more expensive than gold. In the 1860s,

Emperor Napoleon III of France commissioned aluminium cutlery to be laid out for his most distinguished guests. Less important visitors had to make do with the gold knives and forks.[5] But at the end of the nineteenth century chemists discovered a way to extract immense amounts of cheap aluminium, and current global production stands at 30 million tons per year. Napoleon III would be surprised to hear that his subjects' descendants use cheap disposable aluminium foil to wrap their sandwiches and put away their leftovers.

Two thousand years ago, when people in the Mediterranean basin suffered from dry skin they smeared olive oil on their hands. Today, they open a tube of hand cream. Below is the list of ingredients of a simple modern hand cream that I bought at a local store:

deionised water, stearic acid, glycerin, caprylic/caprictiglyceride, propylene glycol, isopropyl myristate, panax ginseng root extract, fragrance, cetyl alcohol, triethanolamine, dimeticone, arctostaphylos uva-ursi leaf extract, magnesium ascorbyl phosphate, imidazolidinyl urea, methyl paraben, camphor, propyl paraben, hydroxyisohexyl 3-cyclohexene carboxaldehyde, hydroxycitronellal, linalool, butylphenyl methylproplonal, citronnellol, limonene, geraniol.

Almost all of these ingredients were invented or discovered in the last two centuries.

During the First World War, Germany was placed under blockade and suffered severe shortages of raw materials, in particular saltpetre, an essential ingredient in gunpowder and other explosives. The most important saltpetre deposits were in Chile and India; there were none at all in Germany. True, saltpetre could be replaced by ammonia, but that was expensive to produce as well. Luckily for the Germans, one of their fellow citizens, a Jewish chemist named Fritz Haber, had discovered in 1908 a process for producing ammonia literally out of thin air. When war broke out, the Germans used Haber's discovery to commence industrial production of explosives using air as a raw material. Some scholars believe that if it hadn't been for Haber's discovery, Germany would have been forced to surrender long before November 1918.[6] The discovery won Haber (who during the war also pioneered the use of poison gas in battle) a Nobel Prize in 1918. In chemistry, not in peace.

Life on the Conveyor Belt

The Industrial Revolution yielded an unprecedented combination of cheap and abundant energy and cheap and abundant raw materials. The result was an explosion in human productivity. The explosion was felt first and foremost in agriculture. Usually, when we think of the Industrial Revolution, we think of an urban landscape of smoking chimneys, or the plight of exploited coal miners sweating in the bowels of the earth. Yet the Industrial Revolution was above all else the Second Agricultural Revolution.

During the last 200 years, industrial production methods became the mainstay of agriculture. Machines such as tractors began to undertake tasks that were previously performed by muscle power, or not performed at all. Fields and animals became vastly more productive thanks to artificial fertilisers, industrial insecticides and an entire arsenal of hormones and medications. Refrigerators, ships and aeroplanes have made it possible to store produce for months, and transport it quickly and cheaply to the other side of the world. Europeans began to dine on fresh Argentinian beef and Japanese sushi.

Even plants and animals were mechanised. Around the time that *Homo sapiens* was elevated to divine status by humanist religions, farm animals stopped being viewed as living creatures that could feel pain and distress, and instead came to be treated as machines. Today these animals are often mass-produced in factory-like facilities, their bodies shaped in accordance with industrial needs. They pass their entire lives as cogs in a giant production line, and the length and quality of their existence is determined by the profits and losses of business corporations. Even when the industry takes care to keep them alive, reasonably healthy and well fed, it has no intrinsic interest in the animals' social and psychological needs (except when these have a direct impact on production).

Egg-laying hens, for example, have a complex world of behavioural needs and drives. They feel strong urges to scout their environment, forage and peck around, determine social hierarchies, build nests and groom themselves. But the egg industry often locks the hens inside tiny coops, and it is not uncommon for it to squeeze four hens to a cage, each given a floor space of about twenty-five by twenty-two centimetres. The hens receive sufficient food, but they are unable to claim a territory, build a nest or engage in other natural activities. Indeed, the cage is so small that hens are often unable even to flap their wings or stand fully erect.

Pigs are among the most intelligent and inquisitive of mammals, second perhaps only to the great apes. Yet industrialised pig farms routinely confine nursing sows inside such small crates that they are literally unable to turn around (not to mention walk or forage). The sows are kept in these crates day and night for four weeks after giving birth. Their offspring are then taken away to be fattened up and the sows are impregnated with the next litter of piglets.

Many dairy cows live almost all their allotted years inside a small enclosure; standing, sitting and sleeping in their own urine and excrement. They receive their measure of food, hormones and medications from one set of machines, and get milked every few hours by another set of machines. The cow in the middle is treated as little more than a mouth that takes in raw materials and an udder that produces a commodity. Treating living creatures possessing complex emotional worlds as if they were machines is likely to cause them not only physical discomfort, but also much social stress and psychological frustration.[7]

41. Chicks on a conveyor belt in a commercial hatchery. Male chicks and imperfect female chicks are picked off the conveyor belt and are then asphyxiated in gas chambers, dropped into automatic shredders, or simply thrown into the rubbish, where they are crushed to death. Hundreds of millions of chicks die each year in such hatcheries.

Just as the Atlantic slave trade did not stem from hatred towards Africans, so the modern animal industry is not motivated by animosity. Again, it is fuelled by indifference. Most people who produce and consume eggs, milk and meat rarely stop to think about the fate of the chickens, cows or pigs whose flesh and emissions they are eating. Those who do think often argue that such animals are really little different from machines, devoid of sensations and emotions, incapable of suffering. Ironically, the same scientific disciplines which shape our milk machines and egg machines have lately demonstrated beyond reasonable doubt that mammals and birds have a complex sensory and emotional make-up. They not only feel physical pain, but can also suffer from emotional distress.

Evolutionary psychology maintains that the emotional and social needs of farm animals evolved in the wild, when they were essential for survival and reproduction. For example, a wild cow had to know how to form close relations with other cows and bulls, or else she could not survive and reproduce. In order to learn the necessary skills, evolution implanted in calves – as in the young of all other social mammals – a strong desire to play (playing is the mammalian way of learning social behaviour). And it implanted in them an even stronger desire to bond with their mothers, whose milk and care were essential for survival.

What happens if farmers now take a young calf, separate her from her mother, put her in a closed cage, give her food, water and inoculations against diseases, and then, when she is old enough, inseminate her with bull sperm? From an objective perspective, this calf no longer needs either maternal bonding or playmates in order to survive and reproduce. But from a subjective perspective, the calf still feels a very strong urge to bond with her mother and to play with other calves. If these urges are not fulfilled, the calf suffers greatly. This is the basic lesson of evolutionary psychology: a need shaped in the wild continues to be felt subjectively even if it is no longer really necessary for survival and reproduction. The tragedy of industrial agriculture is that it takes great care of the objective needs of animals, while neglecting their subjective needs.

The truth of this theory has been known at least since the 1950s, when the American psychologist Harry Harlow studied the development of monkeys. Harlow separated infant monkeys from their mothers several hours after birth. The monkeys were isolated inside cages, and then raised by dummy mothers. In each cage, Harlow placed two dummy mothers. One was made of metal wires, and was fitted with a milk bottle from which the infant monkey could suck. The other was made of wood covered with cloth, which made it resemble a real monkey mother, but it provided the infant monkey with no material sustenance whatsoever. It was assumed that the infants would cling to the nourishing metal mother rather than to the barren cloth one.

To Harlow's surprise, the infant monkeys showed a marked

42. One of Harlow's orphaned monkeys clings to the cloth mother even while sucking milk from the metal mother.

preference for the cloth mother, spending most of their time with her. When the two mothers were placed in close proximity, the infants held on to the cloth mother even while they reached over to suck milk from the metal mother. Harlow suspected that perhaps the infants did so because they were cold. So he fitted an electric bulb inside the wire mother, which now radiated heat. Most of the monkeys, except for the very young ones, continued to prefer the cloth mother.

Follow-up research showed that Harlow's orphaned monkeys grew up to be emotionally disturbed even though they had received all the nourishment they required. They never fitted into monkey society, had difficulties communicating with other monkeys, and suffered from high levels of anxiety and aggression. The conclusion was inescapable: monkeys must have psychological needs and desires that go beyond their material requirements, and if these are not fulfilled, they will suffer greatly. Harlow's infant monkeys preferred to spend their time in the hands of the barren cloth mother because they were looking for an emotional bond and not only for milk. In the following decades, numerous studies showed that this conclusion applies not only to monkeys, but to other mammals, as well as birds. At present, millions of farm animals are subjected to the same conditions as Harlow's monkeys, as farmers routinely separate calves, kids and other youngsters from their mothers, to be raised in isolation.[8]

Altogether, tens of billions of farm animals live today as part of a mechanised assembly line, and about 50 billion of them are slaughtered annually. These industrial livestock methods have led to a sharp increase in agricultural production and in human food reserves. Together with the mechanisation of plant cultivation, industrial animal husbandry is the basis for the entire modern socio-economic order. Before the industrialisation of agriculture, most of the food produced in fields and farms was 'wasted' feeding peasants and farmyard animals. Only a small percentage was available to feed artisans, teachers, priests and bureaucrats. Consequently, in almost all societies peasants comprised more than 90 per cent of the population. Following the industrialisation of agriculture, a shrinking number of farmers was enough to feed a growing number of clerks and factory hands. Today in the United States, only 2 per cent of the population makes a living from agriculture, yet this 2 per cent produces enough not only to feed the entire US population, but also to export surpluses to the rest of the world.[9] Without the industrialisation of agriculture the urban Industrial Revolution could never have taken place – there would not have been enough hands and brains to staff factories and offices.

As those factories and offices absorbed the billions of hands and brains that were released from fieldwork, they began pouring out an unprecedented avalanche of products. Humans now produce far more steel, manufacture much more clothing, and build many more structures than ever before. In addition, they produce a mind-boggling array of previously unimaginable goods, such as light bulbs, mobile phones, cameras and dishwashers. For the first time in human history, supply began to outstrip demand. And an entirely new problem was born: who is going to buy all this stuff?

The Age of Shopping

The modern capitalist economy must constantly increase production if it is to survive, like a shark that must swim or suffocate. Yet it's not enough just to produce. Somebody must also buy the products, or industrialists and investors alike will go bust. To prevent this catastrophe and to make sure that people will always buy whatever new stuff industry produces, a new kind of ethic appeared: consumerism.

Most people throughout history lived under conditions of scarcity. Frugality was thus their watchword. The austere ethics of the Puritans and Spartans are but two famous examples. A good person avoided luxuries, never threw food away, and patched up torn trousers instead of buying a new pair. Only kings and nobles allowed themselves to renounce such values publicly and conspicuously flaunt their riches.

Consumerism sees the consumption of ever more products and services as a positive thing. It encourages people to treat themselves, spoil themselves, and even kill themselves slowly by overconsumption. Frugality is a disease to be cured. You don't have to look far to see the consumer ethic in action – just read the back of a cereal box. Here's a quote from a box of one of my favourite breakfast cereals, produced by an Israeli firm, Telma:

Sometimes you need a treat. Sometimes you need a little extra energy. There are times to watch your weight and times when you've just got to have something . . . right now! Telma offers a variety of tasty cereals just for you – treats without remorse.

The same package sports an ad for another brand of cereal called Health Treats:

Health Treats offers lots of grains, fruits and nuts for an experience that combines taste, pleasure and health. For an enjoyable treat in the middle of the day, suitable for a healthy lifestyle. *A real treat with the wonderful taste of more* [emphasis in the original].

Throughout most of history, people were likely to be repelled rather than attracted by such a text. They would have branded it as selfish, decadent and morally corrupt. Consumerism has worked very hard, with the help of popular psychology ('Just do it'), to convince people that indulgence is good for you, whereas frugality is self-oppression.

It has succeeded. We are all good consumers. We buy countless products that we don't really need, and that until yesterday we didn't know existed. Manufacturers deliberately design short-term goods and invent new and unnecessary models of perfectly satisfactory products that we must purchase in order to stay 'in'. Shopping has become a favourite pastime, and consumer goods have become essential mediators in relationships between family members, spouses and friends. Religious holidays such as Christmas have become shopping festivals. In the United States, even Memorial Day – originally a solemn day for remembering fallen soldiers – is now an occasion for special sales. Most people mark this day by going shopping, perhaps to prove that the defenders of freedom did not die in vain.

The flowering of the consumerist ethic is manifested most clearly in the food market. Traditional agricultural societies lived in the awful shade of starvation. In the affluent world of today one of the leading health problems is obesity, which strikes the poor (who stuff themselves with hamburgers and pizzas) even more severely than the rich (who eat organic salads and fruit smoothies). Each year the US population spends more money on diets than the amount needed to feed all the hungry people in the rest of the world. Obesity is a double victory for consumerism. Instead of eating little, which will lead to economic contraction, people eat too much and then buy diet products – contributing to economic growth twice over.

How can we square the consumerist ethic with the capitalist ethic of the business person, according to which profits should not be wasted, and should instead be reinvested in production? It's simple. As in previous eras, there is today a division of labour between the elite and the masses. In medieval Europe, aristocrats spent their money carelessly on extravagant luxuries, whereas peasants lived frugally, minding every penny. Today, the tables have turned. The rich take great care managing their assets and investments, while the less well-heeled go into debt buying cars and televisions they don't really need.

The capitalist and consumerist ethics are two sides of the same coin, a merger of two commandments. The supreme commandment of the rich is 'Invest!' The supreme commandment of the rest of us is 'Buy!'

The capitalist–consumerist ethic is revolutionary in another respect. Most previous ethical systems presented people with a pretty tough deal. They were promised paradise, but only if they cultivated compassion and tolerance, overcame craving and anger, and restrained their selfish interests. This was too tough for most. The history of ethics is a sad tale of wonderful ideals that nobody can live up to. Most Christians did not imitate Christ, most Buddhists failed to follow Buddha, and most Confucians would have caused Confucius a temper tantrum.

In contrast, most people today successfully live up to the capitalist–consumerist ideal. The new ethic promises paradise on condition that the rich remain greedy and spend their time making more money, and that the masses give free rein to their cravings and passions – and buy more and more. This is the first religion in history whose followers actually do what they are asked to do. How, though, do we know that we'll really get paradise in return? We've seen it on television.

18

A Permanent Revolution

THE INDUSTRIAL REVOLUTION OPENED up new ways to convert energy and to produce goods, largely liberating humankind from its dependence on the surrounding ecosystem. Humans cut down forests, drained swamps, dammed rivers, flooded plains, laid down tens of thousands of kilometres of railroad tracks, and built skyscraping metropolises. As the world was moulded to fit the needs of *Homo sapiens*, habitats were destroyed and species went extinct. Our once green and blue planet is becoming a concrete and plastic shopping centre.

Today, the earth's continents are home to almost 7 billion Sapiens. If you took all these people and put them on a large set of scales, their combined mass would be about 300 million tons. If you then took all our domesticated farmyard animals – cows, pigs, sheep and chickens – and placed them on an even larger set of scales, their mass would amount to about 700 million tons. In contrast, the combined mass of all surviving large wild animals – from porcupines and penguins to elephants and whales – is less than 100 million tons. Our children's books, our iconography and our TV screens are still full of giraffes, wolves and chimpanzees, but the real world has very few of them left. There are about 80,000 giraffes in the world, compared to 1.5 billion cattle; only 200,000 wolves, compared to 400 million domesticated dogs; only 250,000 chimpanzees – in contrast to billions of humans. Humankind really has taken over the world.[1]

Ecological degradation is not the same as resource scarcity. As we saw in the previous chapter, the resources available to humankind are constantly increasing, and are likely to continue to do so. That's why doomsday prophesies of resource scarcity are probably misplaced. In contrast, the fear of ecological degradation is only too well founded. The future may see Sapiens gaining control of a cornucopia of new materials and energy sources, while simultaneously destroying what remains of the natural habitat and driving most other species to extinction.

In fact, ecological turmoil might endanger the survival of *Homo sapiens* itself. Global warming, rising oceans and widespread pollution could make the earth less hospitable to our kind, and the future might consequently see a spiralling race between human power and human-induced natural disasters. As humans use their power to counter the forces of nature and subjugate the ecosystem to their needs and whims, they might cause more and more unanticipated and dangerous side effects. These are likely to be controllable only by even more drastic manipulations of the ecosystem, which would result in even worse chaos.

Many call this process 'the destruction of nature'. But it's not really destruction, it's change. Nature cannot be destroyed. Sixty-five million years ago, an asteroid wiped out the dinosaurs, but in so doing opened the way forward for mammals. Today, humankind is driving many species into extinction and might even annihilate itself. But other organisms are doing quite well. Rats and cockroaches, for example, are in their heyday. These tenacious creatures would probably creep out from beneath the smoking rubble of a nuclear Armageddon, ready and able to spread their DNA. Perhaps 65 million years from now, intelligent rats will look back gratefully on the decimation wrought by humankind, just as we today can thank that dinosaur-busting asteroid.

Still, the rumours of our own extinction are premature. Since the Industrial Revolution, the world's human population has burgeoned as never before. In 1700 the world was home to some 700 million humans. In 1800 there were 950 million of us. By 1900 we almost doubled our numbers to 1.6 billion. And by 2000 that quadrupled to 6 billion. Today there are just shy of 7 billion Sapiens.

Modern Time

While all these Sapiens have grown increasingly impervious to the whims of nature, they have become ever more subject to the dictates of modern industry and government. The Industrial Revolution opened the way to a long line of experiments in social engineering and an even longer series of unpremeditated changes in daily life and human mentality. One example among many is the replacement of the rhythms of traditional agriculture with the uniform and precise schedule of industry.

Traditional agriculture depended on cycles of natural time and organic growth. Most societies were unable to make precise time measurements, nor were they terribly interested in doing so. The world went about its business without clocks and timetables, subject only to the movements of the sun and the growth cycles of plants. There was no uniform working day, and all routines changed drastically from season to season. People knew where the sun was, and watched anxiously for portents of the rainy season and harvest time, but they did not know the hour and hardly cared about the year. If a lost time traveller popped up in a medieval village and asked a passerby, 'What year is this?' the villager would be as bewildered by the question as by the stranger's ridiculous clothing.

In contrast to medieval peasants and shoemakers, modern industry cares little about the sun or the season. It sanctifies precision and uniformity. For example, in a medieval workshop each shoemaker made an entire shoe, from sole to buckle. If one shoemaker was late for work, it did not stall the others. However, in a

43. Charlie Chaplin as a simple worker caught in the wheels of the industrial assembly line, from the film *Modern Times* (1936).

modern footwear-factory assembly line, every worker mans a machine that produces just a small part of a shoe, which is then passed on to the next machine. If the worker who operates machine no. 5 has overslept, it stalls all the other machines. In order to prevent such calamities, everybody must adhere to a precise timetable. Each worker arrives at work at exactly the same time. Everybody takes their lunch break together, whether they are hungry or not. Everybody goes home when a whistle announces that the shift is over – not when they have finished their project.

The Industrial Revolution turned the timetable and the assembly line into a template for almost all human activities. Shortly after factories imposed their time frames on human behaviour, schools too adopted precise timetables, followed by hospitals, government offices and grocery stores. Even in places devoid of assembly lines and machines, the timetable became king. If the shift at the factory ends at 5 p.m., the local pub had better be open for business by 5:02.

A crucial link in the spreading timetable system was public transportation. If workers needed to start their shift by 08:00, the train or bus had to reach the factory gate by 07:55. A few minutes' delay would lower production and perhaps even lead to the lay-offs of the unfortunate latecomers. In 1784 a carriage service with a published schedule began operating in Britain. Its timetable specified only the hour of departure, not arrival. Back then, each British city and town had its own local time, which could differ from London time by up to half an hour. When it was 12:00 in London, it was perhaps 12:20 in Liverpool and 11:50 in Canterbury. Since there were no telephones, no radio or television, and no fast trains – who could know, and who cared?[2]

The first commercial train service began operating between Liverpool and Manchester in 1830. Ten years later, the first train timetable was issued. The trains were much faster than the old carriages, so the quirky differences in local hours became a severe nuisance. In 1847, British train companies put their heads together and agreed that henceforth all train timetables would be calibrated to Greenwich Observatory time, rather than the local times of Liverpool, Manchester or Glasgow. More and more institutions followed the lead of the train companies. Finally, in 1880, the British government took the unprecedented step of legislating that all timetables in Britain must follow Greenwich. For the first time in history, a country adopted a national time and obliged its population to live according to an artificial clock rather than local ones or sunrise-to-sunset cycles.

This modest beginning spawned a global network of timetables, synchronised down to the tiniest fractions of a second. When the broadcast media – first radio, then television – made their debut, they entered a world of timetables and became its main enforcers and evangelists. Among the first things radio stations broadcast were time signals, beeps that enabled far-flung settlements and ships at sea to set their clocks. Later, radio stations adopted the custom of broadcasting the news every hour. Nowadays, the first item of every news broadcast – more important even than the outbreak of war – is the time. During the Second World War, BBC News was broadcast to Nazi-occupied Europe. Each news programme opened with a live broadcast of Big Ben tolling the hour – the magical sound of freedom. Ingenious German physicists found a way to determine the weather conditions in London based on tiny differences in the tone of the broadcast ding-dongs. This information offered invaluable help to the Luftwaffe. When the British Secret Service discovered this, they replaced the live broadcast with a set recording of the famous clock.

In order to run the timetable network, cheap but precise portable clocks became ubiquitous. In Assyrian, Sassanid or Inca cities there might have been at most a few sundials. In European medieval cities there was usually a single clock – a giant machine mounted on top of a high tower in the town square. These tower clocks were notoriously inaccurate, but since there were no other clocks in town to contradict them, it hardly made any difference. Today, a single affluent family generally has more timepieces at home than an entire medieval country. You can tell the time by looking at your wristwatch, glancing at your Android, peering at the alarm clock by your bed, gazing at the clock on the kitchen wall, staring at the microwave, catching a glimpse of the TV or DVD, or taking in the taskbar on your computer out of the corner of your eye. You need to make a conscious effort *not* to know what time it is.

The typical person consults these clocks several dozen times a day, because almost everything we do has to be done on time. An alarm clock wakes us up at 7 a.m., we heat our frozen bagel for exactly fifty seconds in the microwave, brush our teeth for three minutes until the electric toothbrush beeps, catch the 07:40 train to work, run on the treadmill at the gym until the beeper announces that half an hour is over, sit down in front of the TV at 7 p.m. to watch our favourite show, get interrupted at preordained moments by commercials that cost $1,000 per second, and eventually unload all our angst on a therapist who restricts our prattle to the now standard fifty-minute therapy hour.

The Industrial Revolution brought about dozens of major upheavals in human society. Adapting to industrial time is just one of them. Other notable examples include urbanisation, the disappearance of the peasantry, the rise of the industrial proletariat, the empowerment of the common person, democratisation, youth culture and the disintegration of patriarchy.

Yet all of these upheavals are dwarfed by the most momentous social revolution that ever befell humankind: the collapse of the family and the local community and their replacement by the state and the market. As best we can tell, from the earliest times, more than a million years ago, humans lived in small, intimate communities, most of whose members were kin. The Cognitive Revolution and the Agricultural Revolution did not change that. They glued together families and communities to create tribes, cities, kingdoms and empires, but families and communities remained the basic building blocks of all human societies. The Industrial Revolution, on the other hand, managed within little more than two centuries to break these building blocks into atoms. Most of the traditional functions of families and communities were handed over to states and markets.

The Collapse of the Family and the Community

Prior to the Industrial Revolution, the daily life of most humans ran its course within three ancient frames: the nuclear family, the extended family and the local intimate community.* Most people worked in the family business – the family farm or the family workshop, for example – or they worked in their neighbours' family businesses. The family was also the welfare system, the health system, the education system, the construction industry, the trade union, the pension fund, the insurance company, the radio, the television, the newspapers, the bank and even the police.

When a person fell sick, the family took care of her. When a person grew old, the family supported her, and her children were her pension fund. When a person died, the family took care of the orphans. If a person wanted to build a hut, the family lent a hand. If a person wanted to open a business, the family raised the necessary money. If a person wanted to marry, the family chose, or at least vetted, the prospective spouse. If conflict arose with a neighbour, the family muscled in. But if a person's illness was too grave for the family to manage, or a new business demanded too large an investment, or the neighbourhood quarrel escalated to the point of violence, the local community came to the rescue.

The community offered help on the basis of local traditions and an economy of favours, which often differed greatly from the supply and demand laws of the free market. In an old-fashioned medieval community, when my neighbour was in need, I helped build his hut and guard his sheep, without expecting any payment in return. When I was in need, my neighbour returned the favour. At the same time, the local potentate might have drafted all of us villagers to construct his castle without paying us a penny. In exchange, we counted on him to defend us against brigands and barbarians. Village life involved many transactions but few payments. There were some markets, of course, but their roles were limited. You could buy rare spices, cloth and tools, and hire the services of lawyers and doctors. Yet less than 10 per cent of commonly used products and services

* An 'intimate community' is a group of people who know one another well and depend on each other for survival.

were bought in the market. Most human needs were taken care of by the family and the community.

There were also kingdoms and empires that performed important tasks such as waging wars, building roads and constructing palaces. For these purposes kings raised taxes and occasionally enlisted soldiers and labourers. Yet, with few exceptions, they tended to stay out of the daily affairs of families and communities. Even if they wanted to intervene, most kings could do so only with difficulty. Traditional agricultural economies had few surpluses with which to feed crowds of government officials, policemen, social workers, teachers and doctors. Consequently, most rulers did not develop mass welfare systems, health-care systems or educational systems. They left such matters in the hands of families and communities. Even on rare occasions when rulers tried to intervene more intensively in the daily lives of the peasantry (as happened, for example, in the Qin Empire in China), they did so by converting family heads and community elders into government agents.

Often enough, transportation and communication difficulties made it so complicated to intervene in the affairs of remote communities that many kingdoms preferred to cede even the most basic royal prerogatives – such as taxation and violence – to communities. The Ottoman Empire, for instance, allowed family vendettas to mete out justice, rather than supporting a large imperial police force. If my cousin killed somebody, the victim's brother might kill me in sanctioned revenge. The sultan in Istanbul or even the provincial pasha did not intervene in such clashes, as long as violence remained within acceptable limits.

In the Chinese Ming Empire (1368–1644), the population was organised into the *baojia* system. Ten families were grouped to form a *jia*, and ten *jia* constituted a *bao*. When a member of a *bao* commited a crime, other *bao* members could be punished for it, in particular the *bao* elders. Taxes too were levied on the *bao*, and it was the responsibility of the *bao* elders rather than of the state officials to assess the situation of each family and determine the amount of tax it should pay. From the empire's perspective, this system had a huge advantage. Instead of maintaining thousands of revenue officials and tax collectors, who would have to monitor the earnings and expenses of every family, these tasks were left to the community elders. The elders knew how much each villager was worth and they could usually enforce tax payments without involving the imperial army.

Many kingdoms and empires were in truth little more than large protection rackets. The king was the *capo di tutti capi* who collected protection money, and in return made sure that neighbouring crime syndicates and local small fry did not harm those under his protection. He did little else.

Life in the bosom of family and community was far from ideal. Families and communities could oppress their members no less brutally than do modern states and markets, and their internal dynamics were often fraught with tension and violence – yet people had little choice. A person who lost her family and community around 1750 was as good as dead. She had no job, no education and no support in times of sickness and distress. Nobody would loan her money or defend her if she got into trouble. There were no policemen, no social workers and no compulsory education. In order to survive, such a person quickly had to find an alternative family or community. Boys and girls who ran away from home could expect, at best, to become servants in some new family. At worst, there was the army or the brothel.

All this changed dramatically over the last two centuries. The Industrial Revolution gave the market immense new powers, provided the state with new means of communication and transportation, and placed at the government's disposal an army of clerks, teachers, policemen and social workers. At first the market and the state discovered their path blocked by traditional families and communities who had little love for outside intervention. Parents and community elders were reluctant to let the younger generation be indoctrinated by nationalist education systems, conscripted into armies or turned into a rootless urban proletariat.

Over time, states and markets used their growing power to weaken the traditional bonds of family and community. The state sent its policemen to stop family vendettas and replace them with court decisions. The market sent its hawkers to wipe out long-standing local traditions and replace them with ever-changing commercial fashions. Yet this was not enough. In order really to break the power of family and community, they needed the help of a fifth column.

The state and the market approached people with an offer that could not be refused. 'Become individuals,' they said. 'Marry whomever you desire, without asking permission from your parents. Take up whatever job suits you, even if community elders frown. Live wherever you wish, even if you cannot make it every week to the family dinner. You are no longer dependent on your family or your community. We, the state and the market, will take care of you instead. We will provide food, shelter, education, health, welfare and employment. We will provide pensions, insurance and protection.'

Romantic literature often presents the individual as somebody caught in a struggle against the state and the market. Nothing could be further from the truth. The state and the market are the mother and father of the individual, and the individual can survive only thanks to them. The market provides us with work, insurance and a pension. If we want to study a profession, the government's schools are there to teach us. If we want to open a business, the bank loans us money. If we want to build a house, a construction company builds it and the bank gives us a mortgage, in some cases subsidised or insured by the state. If violence flares up, the police protect us. If we are sick for a few days, our health insurance takes care of us. If we are debilitated for months, social security steps in. If we need around-the-clock assistance, we can go to the market and hire a nurse – usually some stranger from the other side of the world who takes care of us with the kind of devotion that we no longer expect from our own children. If we have the means, we can spend our golden years at a senior citizens' home. The tax authorities treat us as individuals, and do not expect us to pay the neighbours' taxes. The courts, too, see us as individuals, and never punish us for the crimes of our cousins.

Not only adult men, but also women and children, are recognised as individuals. Throughout most of history, women were often seen as the property of family or community. Modern states, on the other hand, see women as individuals, enjoying economic and legal rights independently of their family and community. They may hold their own bank accounts, decide whom to marry, and even choose to divorce or live on their own.

But the liberation of the individual comes at a cost. Many of us now bewail the loss of strong families and communities and feel alienated and threatened by the power the impersonal state and market wield over our lives. States and markets composed of alienated individuals can intervene in the lives of their members much more easily than states and markets composed of strong families and communities. When neighbours in a high-rise apartment building cannot even agree on how much to pay their janitor, how can we expect them to resist the state?

The deal between states, markets and individuals is an uneasy one. The state and the market disagree about their mutual rights and obligations, and individuals complain that both demand too much and provide too little. In many cases individuals are exploited by markets, and states employ their armies, police forces and bureaucracies to persecute individuals instead of defending them. Yet it is amazing that this deal works at all – however imperfectly. For it breaches countless generations of human social arrangements. Millions of years of evolution have designed us to live and think as community members. Within a mere two centuries we have become alienated individuals. Nothing testifies better to the awesome power of culture.

The nuclear family did not disappear completely from the modern landscape. When states and markets took from the family most of its economic and political roles, they left it some important emotional functions. The modern family is still supposed to provide for intimate needs, which state and market are (so far) incapable of providing. Yet even here the family is subject to increasing interventions. The market shapes to an ever-greater degree the way people conduct their romantic and sexual lives. Whereas traditionally the family was the main matchmaker, today it's the market that tailors our romantic and sexual preferences, and then lends a hand in providing for them – for a fat fee. Previously bride and groom met in the family living room, and money passed from the hands of one father to another. Today courting is done at bars and cafés, and money passes from the hands of lovers to waitresses. Even more money is transferred to the bank accounts of fashion designers, gym managers, dieticians, cosmeticians and plastic surgeons, who help us arrive at the café looking as similar as possible to the market's ideal of beauty.

The state, too, keeps a sharper eye on family relations, especially between parents and children. Parents are obliged to send their children to be educated by the state. Parents who are especially abusive or violent with their children may be restrained by the state. If need be, the state may even imprison the parents or transfer their children to foster families. Until not long ago, the suggestion that the state ought to prevent parents from beating or humiliating their children would have been rejected out of hand as ludicrous and unworkable. In most societies parental authority was sacred. Respect of and obedience to one's parents were among the most hallowed values, and parents could do almost anything they wanted, including killing newborn babies, selling children into slavery and marrying

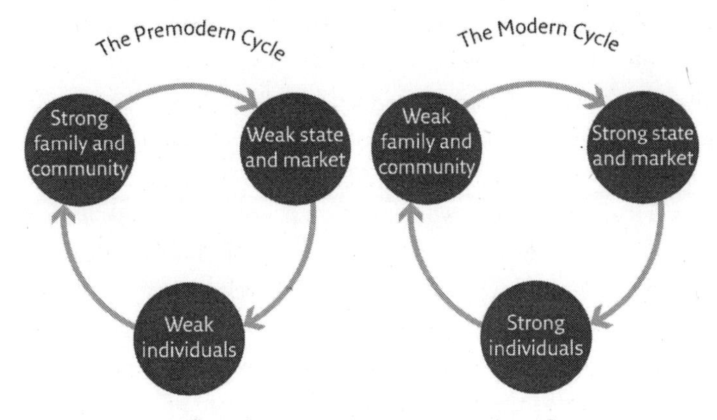

Family and community vs state and market

off daughters to men more than twice their age. Today, parental authority is in full retreat. Youngsters are increasingly excused from obeying their elders, whereas parents are blamed for anything that goes wrong in the life of their child. Mum and Dad are about as likely to get off in the Freudian courtroom as were defendants in a Stalinist show trial.

Imagined Communities

Like the nuclear family, the community could not completely disappear from our world without any emotional replacement. Markets and states today provide most of the material needs once provided by communities, but they must also supply tribal bonds.

Markets and states do so by fostering 'imagined communities' that contain millions of strangers, and which are tailored to national and commercial needs. An imagined community is a community of people who don't really know each other, but imagine that they do. Such communities are not a novel invention. Kingdoms, empires and churches functioned for millennia as imagined communities. In ancient China, tens of millions of people saw themselves as members of a single family, with the emperor as its father. In the Middle Ages, millions of devout Muslims imagined that they were all brothers and sisters in the great community of Islam. Yet throughout history, such imagined communities played second fiddle to intimate communities of several dozen people who knew each other well. The intimate communities fulfilled the emotional needs of their members and were essential for everyone's survival and welfare. In the last two centuries, the intimate communities have withered, leaving imagined communities to fill in the emotional vacuum.

The two most important examples for the rise of such imagined communities are the nation and the consumer tribe. The nation is the imagined community of the state. The consumer tribe is the imagined community of the market. Both are *imagined* communities because it is impossible for all customers in a market or for all members of a nation really to know one another the way villagers knew one another in the past. No German can intimately know the other 80 million members of the German nation, or the other 500 million customers inhabiting the European Common Market (which evolved first into the European Community and finally became the European Union).

Consumerism and nationalism work extra hours to make us imagine that millions of strangers belong to the same community as ourselves, that we all have a common past, common interests and a common future. This isn't a lie. It's imagination. Like money, limited liability companies and human rights, nations and consumer tribes are inter-subjective realities. They exist only in our collective imagination, yet their power is immense. As long as millions of Germans believe in the existence of a German nation, get excited at the sight of German national symbols, retell German national myths, and are willing to sacrifice money, time and limbs for the German nation, Germany will remain one of the strongest powers in the world.

The nation does its best to hide its imagined character. Most nations argue that they are a natural and eternal entity, created in some primordial epoch by mixing the soil of the motherland with the blood of the people. Yet such claims are usually exaggerated. Nations existed in the distant past, but their importance was much smaller than today because the importance of the state was much smaller. A resident of medieval Nuremberg might have felt some loyalty towards the German nation, but she felt far more loyalty towards her family and local community, which took care of most of her needs. Moreover, whatever importance ancient nations may have had, few of them survived. Many of today's nations coalesced only in the last few centuries.

The Middle East provides ample examples. The Syrian, Lebanese, Jordanian and Iraqi nations are the product of haphazard borders drawn in the sand by French and British diplomats who ignored local history, geography and economy. These diplomats determined in 1918 that the people of Kurdistan, Baghdad and Basra would henceforth be 'Iraqis'. It was primarily the French who decided who would be Syrian and who Lebanese. Saddam Hussein and Hafez al-Assad tried their best to promote and reinforce their Anglo-French-manufactured national consciousnesses, but their bombastic speeches about the allegedly eternal Iraqi and Syrian nations had a hollow ring.

It goes without saying that nations cannot be created from thin air. Those who worked hard to construct Iraq or Syria made use of real historical, geographical and cultural raw materials – some of which are centuries and millennia old. Saddam Hussein co-opted the heritage of the Abbasid caliphate and the Babylonian Empire, even calling one of his crack armoured units the Hammurabi Division. Yet that does not turn the Iraqi nation into an ancient entity. If I bake a cake from flour, oil and sugar, all of which have been sitting in my pantry for the past two months, it does not mean that the cake itself is two months old.

In the battle for human loyalty, national communities have to compete with tribes of customers. People who do not know one another intimately but share the same consumption habits and interests often feel part of the same consumer tribe – and define themselves as such. Madonna fans, for example, constitute a consumer tribe. They define themselves largely by shopping. They buy Madonna concert tickets, CDs, posters, shirts and ring tones, and thereby define who they are. Manchester United fans, vegetarians and environmentalists are other examples. True, few people are willing to die for the environment or for Manchester United. But most people nowadays spend far more time in the supermarket than on the battlefield, and in the supermarket the consumer tribe is often mightier than the nation.

Perpetuum Mobile

The revolutions of the last two centuries have been so swift and radical that they have changed the most fundamental characteristic of the social order. Traditionally, the social order was hard and rigid. 'Order' implied stability and continuity. Swift social revolutions were exceptional, and most social transformations resulted from the accumulation of numerous small steps. Humans tended to assume that the social structure was inflexible and eternal. Families and communities might struggle to change their place within the order, but the idea that you could change the fundamental structure of the order was alien. People tended to reconcile themselves to the status quo, declaring that 'this is how it always was, and this is how it always will be'.

Over the last two centuries, the pace of change became so quick that the social order acquired a dynamic and malleable nature. It now exists in a state of permanent flux. When we speak of modern revolutions we tend to think of 1789 (the French Revolution), 1848 (the liberal revolutions) or 1917 (the Russian Revolution). But the fact is that, these days, every year is revolutionary. Today, even a thirty-year-old can honestly tell disbelieving teenagers, 'When I was young, the world was completely different.' The Internet, for example, came into wide usage only in the early 1990s, hardly twenty years ago. Today we cannot imagine the world without it.

Hence any attempt to define the characteristics of modern society is akin to defining the colour of a chameleon. The only characteristic of which we can be certain is the incessant change. People have become used to this, and most of us think about the social order as something flexible, which we can engineer and improve at will. The main promise of premodern rulers was to safeguard the traditional order or even to go back to some lost golden age. In the last two centuries, the currency of politics is that it promises to destroy the old world and build a better one in its place. Not even the most conservative of political parties vows merely to keep things as they are. Everybody promises social reform, educational reform, economic reform – and they often fulfil those promises.

Just as geologists expect that tectonic movements will result in earthquakes and volcanic eruptions, so might we expect that drastic social movements will result in bloody outbursts of violence. The political history of the nineteenth and twentieth centuries is often told as a series of deadly wars, holocausts and revolutions. Like a child in new boots leaping from puddle to puddle, this view sees history as leapfrogging from one bloodbath to the next, from the First World War to the Second World War to the Cold War, from the Armenian genocide to the Jewish genocide to the Rwandan genocide, from Robespierre to Lenin to Hitler.

There is truth here, but this all too familiar list of calamities is somewhat misleading. We focus too much on the puddles and forget about the dry land separating them. The late modern era has seen unprecedented levels not only of violence and horror, but also of peace and tranquillity. Charles Dickens wrote of the French

Revolution that 'It was the best of times, it was the worst of times.' This may be true not only of the French Revolution, but of the entire era it heralded.

It is especially true of the seven decades that have elapsed since the end of the Second World War. During this period humankind has for the first time faced the possibility of complete self-annihilation and has experienced a fair number of actual wars and genocides. Yet these decades were also the most peaceful era in human history – and by a wide margin. This is surprising because these very same decades experienced more economic, social and political change than any previous era. The tectonic plates of history are moving at a frantic pace, but the volcanoes are mostly silent. The new elastic order seems to be able to contain and even initiate radical structural changes without collapsing into violent conflict.[3]

Peace in Our Time

Most people don't appreciate just how peaceful an era we live in. None of us was alive a thousand years ago, so we easily forget how much more violent the world used to be. And as wars become more rare they attract more attention. Many more people think about the wars raging today in Afghanistan and Iraq than about the peace in which most Brazilians and Indians live.

Even more importantly, it's easier to relate to the suffering of individuals than that of entire populations. However, in order to understand macro-historical processes, we need to examine mass statistics rather than individual stories. In the year 2000, wars caused the deaths of 310,000 individuals, and violent crime killed another 520,000. Each and every victim is a world destroyed, a family ruined, friends and relatives scarred for life. Yet from a macro perspective these 830,000 victims comprised only 1.5 per cent of the 56 million people who died in 2000. That year 1.26 million people died in car accidents (2.25 per cent of total mortality) and 815,000 people committed suicide (1.45 per cent).[4]

The figures for 2002 are even more surprising. Out of 57 million dead, only 172,000 people died in war and 569,000 died of violent crime (a total of 741,000 victims of human violence). In contrast, 873,000 people committed suicide.[5] It turns out that in the year following the 9/11 attacks, despite all the talk of terrorism and war, the average person was more likely to kill himself than to be killed by a terrorist, a soldier or a drug dealer.

In most parts of the world, people go to sleep without fearing that in the middle of the night a neighbouring tribe might surround their village and slaughter everyone. Well-off British subjects travel daily from Nottingham to London through Sherwood Forest without fear that a gang of merry green-clad brigands will ambush them and take their money to give to the poor (or, more likely, murder them and take the money for themselves). Students brook no canings from their teachers, children need not fear that they will be sold into slavery when their parents can't pay their bills, and women know that the law forbids their husbands from beating them and forcing them to stay at home. Increasingly, around the world, these expectations are fulfilled.

The decline of violence is due largely to the rise of the state. Throughout history, most violence resulted from local feuds between families and communities. (Even today, as the above figures indicate, local crime is a far deadlier threat than international wars.) As we have seen, early farmers, who knew no political organisations larger than the local community, suffered rampant violence.[6] As kingdoms and empires became stronger, they reined in communities and the level of violence decreased. In the decentralised kingdoms of medieval Europe, about twenty to forty people were murdered each year for every 100,000 inhabitants. In recent decades, when states and markets have become all-powerful and communities have vanished, violence rates have dropped even further. Today the global average is only nine murders a year per 100,000 people, and most of these murders take place in weak states such as Somalia and Colombia. In the centralised states of Europe, the average is one murder a year per 100,000 people.[7]

There are certainly cases where states use their power to kill their own citizens, and these often loom large in our memories and fears. During the twentieth century, tens of millions if not hundreds of millions of people were killed by the security forces of their own states. Still, from a macro perspective, state-run courts and police forces have probably increased the level of security worldwide. Even in oppressive dictatorships, the average modern person is far less likely to die at the hands of another person than in premodern societies. In 1964 a military dictatorship was established in Brazil. It ruled the country until 1985. During these twenty years, several thousand Brazilians were murdered by the regime. Thousands more were imprisoned and tortured. Yet even in the worst years, the average Brazilian in Rio de Janeiro was far less likely to die at human hands than the average Waorani, Arawete or Yanomamo. The Waorani, Arawete and Yanomamo are indigenous people who live in the depths of the Amazon forest, without army, police or prisons. Anthropological studies have indicated that between a quarter and a half of their menfolk die sooner or later in violent conflicts over property, women or prestige.[8]

Imperial Retirement

It is perhaps debatable whether violence within states has decreased or increased since 1945. What nobody can deny is that international violence has dropped to an all-time low. Perhaps the most obvious example is the collapse of the European empires. Throughout history empires have crushed rebellions with an iron fist, and when its day came, a sinking empire used all its might to save itself, usually collapsing into a bloodbath. Its final demise generally led to anarchy and wars of succession. Since 1945 most empires have opted for peaceful early retirement. Their process of collapse became relatively swift, calm and orderly.

In 1945 Britain ruled a quarter of the globe. Thirty years later it ruled just a few small islands. In the intervening decades it retreated from most of its colonies in a peaceful and orderly manner. Though in some places such as Malaya and Kenya the British tried to hang on by force of arms, in most places they accepted the end of empire with a sigh rather than with a temper tantrum. They focused their efforts not on retaining power, but on transferring it as smoothly as possible. At least some of the praise usually heaped on Mahatma Gandhi for his non-violent creed is actually owed to the British Empire. Despite many years of bitter and often violent struggle, when the end of the Raj came, the Indians did not have to fight the British in the streets of Delhi and Calcutta. The empire's place was taken by a slew of independent states, most of which have since enjoyed stable borders and have for the most part lived peacefully alongside their neighbours. True, tens of thousands of people perished at the hands of the threatened British Empire, and in several hot spots its retreat led to the eruption of ethnic conflicts that claimed hundreds of thousands of lives (particularly in India). Yet when compared to the long-term historical average, the British withdrawal was an exemplar of peace and order. The French Empire was more stubborn. Its collapse involved bloody rearguard actions in Vietnam and Algeria that cost hundreds of thousands of lives. Yet the French, too, retreated from the rest of their dominions quickly and peacefully, leaving behind orderly states rather than a chaotic free-for-all.

The Soviet collapse in 1989 was even more peaceful, despite the eruption of ethnic conflict in the Balkans, the Caucasus and Central Asia. Never before has such a mighty empire disappeared so swiftly and so quietly. The Soviet Empire of 1989 had suffered no military defeat except in Afghanistan, no external invasions, no rebellions, nor even large-scale Martin Luther King-style campaigns of civil disobedience. The Soviets still had millions of soldiers, tens of thousands of tanks and aeroplanes, and enough nuclear weapons to wipe out the whole of humankind several times over. The Red Army and the other Warsaw Pact armies remained loyal. Had the last Soviet ruler, Mikhail Gorbachev, given the order, the Red Army would have opened fire on the subjugated masses.

Yet the Soviet elite, and the Communist regimes through most of eastern Europe (Romania and Serbia were the exceptions), chose not to use even a tiny fraction of this military power. When its members realised that Communism was bankrupt, they renounced force, admitted their failure, packed their suitcases and went home. Gorbachev and his colleagues gave up without a struggle not only the Soviet conquests of the Second World War, but also the much older tsarist conquests in the Baltic, the Ukraine, the Caucasus and Central Asia. It is chilling to contemplate what might have happened if Gorbachev had behaved like the Serbian leadership – or like the French in Algeria.

Pax Atomica

The independent states that came after these empires were remarkably uninterested in war. With very few exceptions, since 1945 states no longer invade other states in order to conquer and swallow them up. Such conquests had been the bread and butter of political history since time immemorial. It was how most great empires were established, and how most rulers and populations expected things to stay. But campaigns of conquest like those of the Romans, Mongols and Ottomans cannot take place today anywhere in the world. Since 1945, no independent country recognised by the UN has been conquered and wiped off the map. Limited international wars still occur from time to time, and millions still die in wars, but wars are no longer the norm.

Many people believe that the disappearance of international war is unique to the rich democracies of western Europe. In fact, peace reached Europe after it prevailed in other parts of the world. Thus the last serious international wars between South American countries were the Peru–Ecuador War of 1941 and the Bolivia–Paraguay War of 1932–5. And before that there hadn't been a serious war between South American countries since 1879–84, with Chile on one side and Bolivia and Peru on the other.

We seldom think of the Arab world as particularly peaceful. Yet only once since the Arab countries won their independence has one of them mounted a full-scale invasion of another (the Iraqi invasion of Kuwait in 1990). There have been quite a few border clashes (e.g. Syria vs Jordan in 1970), many armed interventions of one in the affairs of another (e.g. Syria in Lebanon), numerous civil wars (Algeria, Yemen, Libya) and an abundance of coups and revolts. Yet there have been no full-scale international wars among the Arab states except the Gulf War. Even widening the scope to include the entire Muslim world adds only one more example, the Iran–Iraq War. There was no Turkey–Iran War, Pakistan–Afghanistan War, or Indonesia–Malaysia War.

In Africa things are far less rosy. But even there, most conflicts are civil wars and coups. Since African states won their independence in the 1960s and 1970s, very few countries have invaded one another in the hope of conquest.

There have been periods of relative calm before, as, for example, in Europe between 1871 and 1914, and they always ended badly. But this time it is different. For real peace is not the mere absence of war. Real peace is the implausibility of war. There has never been real peace in the world. Between 1871 and 1914, a European war remained a plausible eventuality, and the expectation of war dominated the thinking of armies, politicians and ordinary citizens alike.

This foreboding was true for all other peaceful periods in history. An iron law of international politics decreed, 'For every two nearby polities, there is a plausible scenario that will cause them to go to war against one another within one year.' This law of the jungle was in force in late nineteenth-century Europe, in medieval Europe, in ancient China and in classical Greece. If Sparta and Athens were at peace in 450 BC, there was a plausible scenario that they would be at war by 449 BC.

Today humankind has broken the law of the jungle. There is at last real peace, and not just absence of war. For most polities, there is no plausible scenario leading to full-scale conflict within one year. What could lead to war between Germany and France next year? Or between China and Japan? Or between Brazil and Argentina? Some minor border clash might occur, but only a truly apocalyptic scenario could result in an old-fashioned full-scale war between Brazil and Argentina in 2014, with Argentinian armoured divisions sweeping to the gates of Rio, and Brazilian carpet-bombers pulverising the neighbourhoods of Buenos Aires. Such wars might still erupt between several pairs of states, e.g. between Israel and Syria, Ethiopia and Eritrea, or the USA and Iran, but these are only the exceptions that prove the rule.

This situation might of course change in the future and, with hindsight, the world of today might seem incredibly naive. Yet from a historical perspective, our very naivety is fascinating. Never before has peace been so prevalent that people could not even imagine war.

Scholars have sought to explain this happy development in more books and articles than you would ever want to read yourself, and they have identified several contributing factors. First and foremost, the price of war has gone up dramatically. The Nobel Peace Prize to end all peace prizes should have been given to Robert Oppenheimer and his fellow architects of the atomic bomb. Nuclear weapons have turned war between superpowers into collective suicide, and made it impossible to seek world domination by force of arms.

Secondly, while the price of war soared, its profits declined. For most of history, polities could enrich themselves by looting or annexing enemy territories. Most wealth consisted of material things like fields, cattle, slaves and gold, so it was easy to loot it or occupy it. Today, wealth consists mainly of human capital and organisational know-how. Consequently it is difficult to carry it off or conquer it by military force.

Consider California. Its wealth was initially built on gold mines. But today it is built on silicon and celluloid – Silicon Valley and the celluloid hills of Hollywood. What would happen if the Chinese were to mount an armed invasion of California, land a million soldiers on the beaches of San Francisco and storm inland? They would gain little. There are no silicon mines in Silicon Valley. The wealth resides in the minds of Google engineers and Hollywood script doctors, directors and special-effects wizards, who would be on the first plane to Bangalore or Mumbai long before the Chinese tanks rolled into Sunset Boulevard. It is not coincidental that the few full-scale international wars that still take place in the world, such as the Iraqi invasion of Kuwait, occur in places where wealth is old-fashioned material wealth. The Kuwaiti sheikhs could flee abroad, but the oil fields stayed put and were occupied.

While war became less profitable, peace became more lucrative than ever. In traditional agricultural economies long-distance trade and foreign investment were sideshows. Consequently, peace brought little profit, aside from avoiding the costs of war. If, say, in 1400 England and France were at peace, the French did not have to pay heavy war taxes and to suffer destructive English invasions, but otherwise it did not benefit their wallets. In modern capitalist economies, foreign trade and investments have become all-important. Peace therefore brings unique dividends. As long as China and the USA are at peace, the Chinese can prosper by selling products to the USA, trading in Wall Street and receiving US investments.

Last but not least, a tectonic shift has taken place in global political culture. Many elites in history – Hun chieftains, Viking noblemen and Aztec priests, for example – viewed war as a positive good. Others viewed it as evil, but an inevitable one, which we had better turn to our own advantage. Ours is the first time in history that the world is dominated by a peace-loving elite – politicians, business people, intellectuals and artists who genuinely see war as both evil and avoidable. (There were pacifists in the past, such as the early Christians, but in the rare cases that they gained power, they tended to forget about their requirement to 'turn the other cheek'.)

There is a positive feedback loop between all these four factors. The threat of nuclear holocaust fosters pacifism; when pacifism spreads, war recedes and trade flourishes; and trade increases both the profits of peace and the costs of war. Over time, this feedback loop creates another obstacle to war, which may ultimately prove the most important of all. The tightening web of international connections erodes the independence of most countries, lessening the chance that any one of them might single-handedly let slip the dogs of war. Most countries no longer engage in full-scale war for the simple reason that they are no longer independent. Though

44. and 45. Gold miners in California during the Gold Rush, and Facebook's headquarters near San Francisco. In 1849 California built its fortunes on gold. Today, California builds its fortunes on silicon. But whereas in 1849 the gold actually lay there in the Californian soil, the real treasures of Silicon Valley are locked inside the heads of high-tech employees.

citizens in Israel, Italy, Mexico or Thailand may harbour illusions of independence, the fact is that their governments cannot conduct independent economic or foreign policies, and they are certainly incapable of initiating and conducting full-scale war on their own. As explained in Chapter 11, we are witnessing the formation of a global empire. Like previous empires, this one, too, enforces peace within its borders. And since its borders cover the entire globe, the World Empire effectively enforces world peace.

So, is the modern era one of mindless slaughter, war and oppression, typified by the trenches of the First World War, the nuclear mushroom cloud over Hiroshima, and the gory manias of Hitler and Stalin? Or is it an era of peace, epitomised by the trenches never dug in South America, the mushroom clouds that never appeared over Moscow and New York, and the serene visages of Mahatma Gandhi and Martin Luther King?

The answer is a matter of timing. It is sobering to realise how often our view of the past is distorted by events of the last few years. If this chapter had been written in 1945 or 1962, it would probably have been much more glum. Since it was written in 2014, it takes a relatively buoyant approach to modern history.

To satisfy both optimists and pessimists, we may conclude by saying that we are on the threshold of both heaven and hell, moving nervously between the gateway of the one and the anteroom of the other. History has still not decided where we will end up, and a string of coincidences might yet send us rolling in either direction.

19

And They Lived Happily Ever After

THE LAST 500 YEARS HAVE WITNESSED A breathtaking series of revolutions. The earth has been united into a single ecological and historical sphere. The economy has grown exponentially, and humankind today enjoys the kind of wealth that used to be the stuff of fairy tales. Science and the Industrial Revolution have given humankind superhuman powers and practically limitless energy. The social order has been completely transformed, as have politics, daily life and human psychology.

But are we happier? Did the wealth humankind accumulated over the last five centuries translate into a new-found contentment? Did the discovery of inexhaustible energy resources open before us inexhaustible stores of bliss? Going further back, have the seventy or so turbulent millennia since the Cognitive Revolution made the world a better place to live? Was the late Neil Armstrong, whose footprint remains intact on the windless moon, happier than the nameless hunter-gatherer who 30,000 years ago left her handprint on a wall in Chauvet Cave? If not, what was the point of developing agriculture, cities, writing, coinage, empires, science and industry?

Historians seldom ask such questions. They do not ask whether the citizens of Uruk and Babylon were happier than their foraging ancestors, whether the rise of Islam made Egyptians more pleased with their lives, or how the collapse of the European empires in Africa have influenced the happiness of countless millions. Yet these are the most important questions one can ask of history. Most current ideologies and political programmes are based on rather flimsy ideas concerning the real source of human happiness. Nationalists believe that political self-determination is essential for our happiness. Communists postulate that everyone would be blissful under the dictatorship of the proletariat. Capitalists maintain that only the free market can ensure the greatest happiness of the greatest number, by creating economic growth and material abundance and by teaching people to be self-reliant and enterprising.

What would happen if serious research were to disprove these hypotheses? If economic growth and self-reliance do not make people happier, what's the benefit of capitalism? What if it turns out that the subjects of large empires are generally happier than the citizens of independent states and that, for example, Ghanaians were happier under British colonial rule than under their own homegrown dictators? What would that say about the process of decolonisation and the value of national self-determination?

These are all hypothetical possibilities, because so far historians have avoided raising these questions – not to mention answering them. They have researched the history of just about everything – politics, society, economics, gender, diseases, sexuality, food, clothing – yet they have seldom stopped to ask how these influence human happiness.

Though few have studied the long-term history of happiness, almost every scholar and layperson has some vague preconception about it. In one common view, human capabilities have increased throughout history. Since humans generally use their capabilities to alleviate miseries and fulfil aspirations, it follows that we must be happier than our medieval ancestors, and they must have been happier than Stone Age hunter-gatherers.

But this progressive account is unconvincing. As we have seen, new aptitudes, behaviours and skills do not necessarily make for a better life. When humans learned to farm in the Agricultural

Revolution, their collective power to shape their environment increased, but the lot of many individual humans grew harsher. Peasants had to work harder than foragers to eke out less varied and nutritious food, and they were far more exposed to disease and exploitation. Similarly, the spread of European empires greatly increased the collective power of humankind, by circulating ideas, technologies and crops, and opening new avenues of commerce. Yet this was hardly good news for millions of Africans, Native Americans and Aboriginal Australians. Given the proven human propensity for misusing power, it seems naive to believe that the more clout people have, the happier they will be.

Some challengers of this view take a diametrically opposed position. They argue for a reverse correlation between human capabilities and happiness. Power corrupts, they say, and as humankind gained more and more power, it created a cold mechanistic world ill-suited to our real needs. Evolution moulded our minds and bodies to the life of hunter-gatherers. The transition first to agriculture and then to industry has condemned us to living unnatural lives that cannot give full expression to our inherent inclinations and instincts, and therefore cannot satisfy our deepest yearnings. Nothing in the comfortable lives of the urban middle class can approach the wild excitement and sheer joy experienced by a forager band on a successful mammoth hunt. Every new invention just puts another mile between us and the Garden of Eden.

Yet this romantic insistence on seeing a dark shadow behind each invention is as dogmatic as the belief in the inevitability of progress. Perhaps we are out of touch with our inner hunter-gatherer, but it's not all bad. For instance, over the last two centuries modern medicine has decreased child mortality from 33 per cent to less than 5 per cent. Can anyone doubt that this made a huge contribution to the happiness not only of those children who would otherwise have died, but also of their families and friends?

A more nuanced position takes the middle road. Until the

Scientific Revolution there was no clear correlation between power and happiness. Medieval peasants may indeed have been more miserable than their hunter-gatherer forebears. But in the last few centuries humans have learned to use their capacities more wisely. The triumphs of modern medicine are just one example. Other unprecedented achievements include the steep drop in violence, the virtual disappearance of international wars, and the near elimination of large-scale famines.

Yet this, too, is an oversimplification. Firstly, it bases its optimistic assessment on a very small sample of years. The majority of humans began to enjoy the fruits of modern medicine no earlier than 1850, and the drastic drop in child mortality is a twentieth-century phenomenon. Mass famines continued to blight much of humanity up to the middle of the twentieth century. During Communist China's Great Leap Forward of 1958–61, somewhere between 10 and 50 million human beings starved to death. International wars became rare only after 1945, largely thanks to the new threat of nuclear annihilation. Hence, though the last few decades have been an unprecedented golden age for humanity, it is too early to know whether this represents a fundamental shift in the currents of history or an ephemeral eddy of good fortune. When judging modernity, it is all too tempting to take the viewpoint of a twenty-first-century middle-class Westerner. We must not forget the viewpoints of a nineteenth-century Welsh coal miner, Chinese opium addict or Tasmanian Aborigine. Truganini is no less important than Homer Simpson.

Secondly, even the brief golden age of the last half-century may turn out to have sown the seeds of future catastrophe. Over the last few decades, we have been disturbing the ecological equilibrium of our planet in myriad new ways, with what seem likely to be dire consequences. A lot of evidence indicates that we are destroying the foundations of human prosperity in an orgy of reckless consumption.

Finally, we can congratulate ourselves on the unprecedented accomplishments of modern Sapiens only if we completely ignore the fate of all other animals. Much of the vaunted material wealth that shields us from disease and famine was accumulated at the expense of laboratory monkeys, dairy cows and conveyor-belt chickens. Over the last two centuries tens of billions of them have been subjected to a regime of industrial exploitation whose cruelty has no precedent in the annals of planet Earth. If we accept a mere tenth of what animal-rights activists are claiming, then modern industrial agriculture might well be the greatest crime in history. When evaluating global happiness, it is wrong to count the happiness only of the upper classes, of Europeans or of men. Perhaps it is also wrong to consider only the happiness of humans.

Counting Happiness

So far we have discussed happiness as if it were largely a product of material factors, such as health, diet and wealth. If people are richer and healthier, then they must also be happier. But is that really so obvious? Philosophers, priests and poets have brooded over the nature of happiness for millennia, and many have concluded that social, ethical and spiritual factors have as great an impact on our happiness as material conditions. Perhaps people in modern affluent societies suffer greatly from alienation and meaninglessness despite their prosperity. And perhaps our less well-to-do ancestors found much contentment in community, religion and a bond with nature.

In recent decades, psychologists and biologists have taken up the challenge of studying scientifically what really makes people happy. Is it money, family, genetics or perhaps virtue? The first step is to define what is to be measured. The generally accepted definition of happiness is 'subjective well-being'. Happiness, according to this view, is something I feel inside myself, a sense of either immediate pleasure or long-term contentment with the way my life is going. If it's something felt inside, how can it be measured from outside?

Presumably, we can do so by asking people to tell us how they feel. So psychologists or biologists who want to assess how happy people feel give them questionnaires to fill out and tally the results.

A typical subjective well-being questionnaire asks interviewees to grade on a scale of zero to ten their agreement with statements such as 'I feel pleased with the way I am', 'I feel that life is very rewarding', 'I am optimistic about the future' and 'Life is good'. The researcher then adds up all the answers and calculates the interviewee's general level of subjective well-being.

Such questionnaires are used in order to correlate happiness with various objective factors. One study might compare a thousand people who earn $100,000 a year with a thousand people who earn $50,000. If the study discovers that the first group has an average subjective well-being level of 8.7, while the latter has an average of only 7.3, the researcher may reasonably conclude that there is a positive correlation between wealth and subjective well-being. To put it in simple English, money brings happiness. The same method can be used to examine whether people living in democracies are happier than people living in dictatorships, and whether married people are happier than singles, divorcees or widowers.

This provides a grounding for historians, who can examine wealth, political freedom and divorce rates in the past. If people are happier in democracies and married people are happier than divorcees, a historian has a basis for arguing that the democratisation process of the last few decades contributed to the happiness of humankind, whereas the growing rates of divorce indicate an opposite trend.

This way of thinking is not flawless, but before pointing out some of the holes, it is worth considering the findings.

One interesting conclusion is that money does indeed bring happiness. But only up to a point, and beyond that point it has little significance. For people stuck at the bottom of the economic ladder, more money means greater happiness. If you are an American single mother earning $12,000 a year cleaning houses and you suddenly win $500,000 on the lottery, you will probably experience a significant and long-term surge in your subjective well-being. You'll be able to feed and clothe your children without sinking further into debt. However, if you're a top executive earning $250,000 a year and you win $1 million on the lottery, or your company board suddenly decides to double your salary, your surge is likely to last only a few weeks. According to the empirical findings, it's almost certainly not going to make a big difference to the way you feel over the long run. You'll buy a snazzier car, move into a palatial home, get used to drinking Chateau Pétrus instead of California Cabernet, but it'll soon all seem routine and unexceptional.

Another interesting finding is that illness decreases happiness in the short term, but is a source of long-term distress only if a person's condition is constantly deteriorating or if the disease involves ongoing and debilitating pain. People who are diagnosed with chronic illness such as diabetes are usually depressed for a while, but if the illness does not get worse they adjust to their new condition and rate their happiness as highly as healthy people do. Imagine that Lucy and Luke are middle-class twins, who agree to take part in a subjective well-being study. On the way back from the psychology laboratory, Lucy's car is hit by a bus, leaving her with a number of broken bones and a permanently lame leg. Just as the rescue crew is cutting her out of the wreckage, the phone rings and Luke shouts that he has won the lottery's $10 million jackpot. Two years later she'll be limping and he'll be a lot richer, but when the psychologist comes around for a follow-up study, they are both likely to give the same answers they did on the morning of that fateful day.

Family and community seem to have more impact on our happiness than money and health. People with strong families who live in tight-knit and supportive communities are significantly happier than people whose families are dysfunctional and who have never found (or never sought) a community to be part of. Marriage is particularly important. Repeated studies have found that there is a very close correlation between good marriages and high subjective well-being, and between bad marriages and misery. This holds true irrespective of economic or even physical conditions. An impecunious invalid surrounded by a loving spouse, a devoted family and a warm community may well feel better than an alienated billionaire, provided that the invalid's poverty is not too severe and that his illness is not degenerative or painful.

This raises the possibility that the immense improvement in material conditions over the last two centuries was offset by the collapse of the family and the community. If so, the average person might well be no happier today than in 1800. Even the freedom we value so highly may be working against us. We can choose our spouses, friends and neighbours, but they can choose to leave us. With the individual wielding unprecedented power to decide her own path in life, we find it ever harder to make commitments. We thus live in an increasingly lonely world of unravelling communities and families.

But the most important finding of all is that happiness does not really depend on objective conditions of either wealth, health or even community. Rather, it depends on the correlation between objective conditions and subjective expectations. If you want a bullock-cart and get a bullock-cart, you are content. If you want a brand-new Ferrari and get only a second-hand Fiat you feel deprived. This is why winning the lottery has, over time, the same impact on people's happiness as a debilitating car accident. When things improve, expectations balloon, and consequently even dramatic improvements in objective conditions can leave us dissatisfied. When things deteriorate, expectations shrink, and consequently even a severe illness might leave you pretty much as happy as you were before.

You might say that we didn't need a bunch of psychologists and their questionnaires to discover this. Prophets, poets and philosophers realised thousands of years ago that being satisfied with what you already have is far more important than getting more of what you want. Still, it's nice when modern research – bolstered by lots of numbers and charts – reaches the same conclusions the ancients did.

The crucial importance of human expectations has far-reaching implications for understanding the history of happiness. If happiness depended only on objective conditions such as wealth, health and social relations, it would have been relatively easy to investigate its history. The finding that it depends on subjective expectations makes the task of historians far harder. We moderns have an arsenal of tranquillisers and painkillers at our disposal, but our expectations of ease and pleasure, and our intolerance of inconvenience and discomfort, have increased to such an extent that we may well suffer from pain more than our ancestors ever did.

It's hard to accept this line of thinking. The problem is a fallacy of reasoning embedded deep in our psyches. When we try to guess or imagine how happy other people are now, or how people in the past were, we inevitably imagine ourselves in their shoes. But that won't work because it pastes our expectations on to the material conditions of others. In modern affluent societies it is customary to take a shower and change your clothes every day. Medieval peasants went without washing for months on end, and hardly ever changed their clothes. The very thought of living like that, filthy and reeking to the bone, is abhorrent to us. Yet medieval peasants seem not to have minded. They were used to the feel and smell of a long-unlaundered shirt. It's not that they wanted a change of clothes but couldn't get it – they had what they wanted. So, at least as far as clothing goes, they were content.

That's not so surprising, when you think of it. After all, our chimpanzee cousins seldom wash and never change their clothes. Nor are we disgusted by the fact that our pet dogs and cats don't shower or change their coats daily. We pat, hug and kiss them all the same. Small children in affluent societies often dislike showering, and it takes them years of education and parental discipline to adopt this supposedly attractive custom. It is all a matter of expectations.

If happiness is determined by expectations, then two pillars of our society – mass media and the advertising industry – may unwittingly be depleting the globe's reservoirs of contentment. If you were an eighteen-year-old youth in a small village 5,000 years ago you'd probably think you were good-looking because there were only fifty other men in your village and most of them were either old, scarred and wrinkled, or still little kids. But if you are a teenager today you are a lot more likely to feel inadequate. Even if the other guys at school are an ugly lot, you don't measure yourself against them but against the movie stars, athletes and supermodels you see all day on television, Facebook and giant billboards.

So maybe Third World discontent is fomented not merely by poverty, disease, corruption and political oppression but also by mere exposure to the First World standards. The average Egyptian was far less likely to die from starvation, plague or violence under Hosni Mubarak than under Ramses II or Cleopatra. Never had the material condition of most Egyptians been so good. You'd think they would have been dancing in the streets in 2011, thanking Allah for their good fortune. Instead they rose up furiously to overthrow Mubarak. They weren't comparing themselves to their ancestors under the pharaohs, but rather to their contemporaries in Obama's America.

If that's the case, even immortality might lead to discontent. Suppose science comes up with cures for all diseases, effective anti-ageing therapies and regenerative treatments that keep people indefinitely young. In all likelihood, the immediate result will be an unprecedented epidemic of anger and anxiety.

Those unable to afford the new miracle treatments – the vast majority of people – will be beside themselves with rage. Throughout history, the poor and oppressed comforted themselves with the thought that at least death is even-handed – that the rich and powerful will also die. The poor will not be comfortable with the thought that they have to die, while the rich will remain young and beautiful for ever.

46. The Egyptian Revolution, 2011. The Egyptian people revolted against the Mubarak regime even though it provided them with safer and longer lives than any previous regime in the history of the Nile Valley.

But the tiny minority able to afford the new treatments will not be euphoric either. They will have much to be anxious about. Although the new therapies could extend life and youth, they cannot revive corpses. How dreadful to think that I and my loved ones can live for ever, but only if we don't get hit by a truck or blown to smithereens by a terrorist! Potentially a-mortal people are likely to grow averse to taking even the slightest risk, and the agony of losing a spouse, child or close friend will be unbearable.

Chemical Happiness

Social scientists distribute subjective well-being questionnaires and correlate the results with socio-economic factors such as wealth and political freedom. Biologists use the same questionnaires, but correlate the answers people give them with biochemical and genetic factors. Their findings are shocking.

Biologists hold that our mental and emotional world is governed by biochemical mechanisms shaped by millions of years of evolution. Like all other mental states, our subjective well-being is not determined by external parameters such as salary, social relations or political rights. Rather, it is determined by a complex system of nerves, neurons, synapses and various biochemical substances such as serotonin, dopamine and oxytocin.

Nobody is ever made happy by winning the lottery, buying a house, getting a promotion or even finding true love. People are made happy by one thing and one thing only – pleasant sensations in their bodies. A person who just won the lottery or found new love and jumps for joy is not really reacting to the money or the lover. She is reacting to various hormones coursing through her bloodstream, and to the storm of electric signals flashing between different parts of her brain.

Unfortunately for all hopes of creating heaven on earth, our internal biochemical system seems to be programmed to keep happiness levels relatively constant. There's no natural selection for happiness as such – a happy hermit's genetic line will go extinct as the genes of a pair of anxious parents get carried on to the next generation. Happiness and misery play a role in evolution only to the extent that they encourage or discourage survival and reproduction. Perhaps it's not surprising, then, that evolution has moulded us to be neither too miserable nor too happy. It enables us to enjoy a momentary rush of pleasant sensations, but these never last for ever. Sooner or later they subside and give place to unpleasant sensations.

For example, evolution provided pleasant feelings as rewards to males who spread their genes by having sex with fertile females. If sex were not accompanied by such pleasure, few males would bother. At the same time, evolution made sure that these pleasant feelings quickly subsided. If orgasms were to last for ever, the very happy males would die of hunger for lack of interest in food, and would not take the trouble to look for additional fertile females.

Some scholars compare human biochemistry to an air-conditioning system that keeps the temperature constant, come heatwave or snowstorm. Events might momentarily change the temperature, but the air-conditioning system always returns the temperature to the same set point.

Some air-conditioning systems are set at twenty-five degrees Celsius. Others are set at twenty degrees. Human happiness conditioning systems also differ from person to person. On a scale from one to ten, some people are born with a cheerful biochemical system that allows their mood to swing between levels six and ten, stabilising with time at eight. Such a person is quite happy even if she lives in an alienating big city, loses all her money in a stock-exchange crash and is diagnosed with diabetes. Other people are cursed with a gloomy biochemistry that swings between three and seven and stabilises at five. Such an unhappy person remains depressed even if she enjoys the support of a tight-knit community, wins millions in the lottery and is as healthy as an Olympic athlete. Indeed, even if our gloomy friend wins $50 million in the morning, discovers the cure for both AIDS and cancer by noon, makes peace between Israelis and Palestinians that afternoon, and then in the evening reunites with her long-lost child who disappeared years ago – she would still be incapable of experiencing anything beyond level seven happiness. Her brain is simply not built for exhilaration, come what may.

Think for a moment of your family and friends. You know some people who remain relatively joyful, no matter what befalls them. And then there are those who are always disgruntled, no matter what gifts the world lays at their feet. We tend to believe that if we could just change our workplace, get married, finish writing that novel, buy a new car or repay the mortgage, we would be on top of the world. Yet when we get what we desire we don't seem to be any happier. Buying cars and writing novels do not change our biochemistry. They can startle it for a fleeting moment, but it is soon back to its set point.

How can this be squared with the above-mentioned psychological and sociological findings that, for example, married people are happier on average than singles? First, these findings are correlations – the direction of causation may be the opposite of what some researchers have assumed. It is true that married people are happier than singles and divorcees, but that does not necessarily mean that marriage produces happiness. It could be that happiness causes marriage. Or more correctly, that serotonin, dopamine and oxytocin bring about and maintain a marriage. People who are born with a cheerful biochemistry are generally happy and content. Such people are more attractive spouses, and consequently they have a greater chance of getting married. They are also less likely to divorce, because it is far easier to live with a happy and content spouse than with a depressed and dissatisfied one. Consequently, it's true that married people are happier on average than singles, but a single woman prone to gloom because of her biochemistry would not necessarily become happier if she were to hook up with a husband.

In addition, most biologists are not fanatics. They maintain that happiness is determined *mainly* by biochemistry, but they agree that psychological and sociological factors also have their place. Our mental air-conditioning system has some freedom of movement within predetermined borders. It is almost impossible to exceed the upper and lower emotional boundaries, but marriage and divorce can have an impact in the area between the two. Somebody born with an average of level five happiness would never dance wildly in the streets. But a good marriage should enable her to enjoy level seven from time to time, and to avoid the despondency of level three.

If we accept the biological approach to happiness, then history turns out to be of minor importance, since most historical events have had no impact on our biochemistry. History can change the external stimuli that cause serotonin to be secreted, yet it does not change the resulting serotonin levels, and hence it cannot make people happier.

Compare a medieval French peasant to a modern Parisian banker. The peasant lived in an unheated mud hut overlooking the local pigsty, while the banker goes home to a splendid penthouse with all the latest technological gadgets and a view to the Champs-Elysées. Intuitively, we would expect the banker to be much happier than the peasant. However, mud huts, penthouses and the Champs-Elysées don't really determine our mood. Serotonin does. When the medieval peasant completed the construction of his mud hut, his brain neurons secreted serotonin, bringing it up to level X. When in 2014 the banker made the last payment on his wonderful penthouse, brain neurons secreted a similar amount of serotonin, bringing it up to a similar level X. It makes no difference to the brain that the penthouse is far more comfortable than the mud hut. The only thing that matters is that at present the level of serotonin is X. Consequently the banker would not be one iota happier than his great-great-great-grandfather, the poor medieval peasant.

This is true not only of private lives, but also of great collective events. Take, for example, the French Revolution. The revolutionaries were busy: they executed the king, gave lands to the peasants, declared the rights of man, abolished noble privileges and waged war against the whole of Europe. Yet none of that changed French biochemistry. Consequently, despite all the political, social, ideological and economic upheavals brought about by the revolution, its impact on French happiness was small. Those who won a cheerful biochemistry in the genetic lottery were just as happy before the revolution as after. Those with a gloomy biochemistry complained about Robespierre and Napoleon with the same bitterness with which they earlier complained about Louis XVI and Marie Antoinette.

If so, what good was the French Revolution? If people did not become any happier, then what was the point of all that chaos, fear, blood and war? Biologists would never have stormed the Bastille. People think that this political revolution or that social reform will make them happy, but their biochemistry tricks them time and again.

There is only one historical development that has real significance. Today, when we finally realise that the keys to happiness are in the hands of our biochemical system, we can stop wasting our time on politics and social reforms, putsches and ideologies, and focus instead on the only thing that can make us truly happy: manipulating our biochemistry. If we invest billions in understanding our brain chemistry and developing appropriate treatments, we can make people far happier than ever before, without any need of revolutions. Prozac, for example, does not change regimes, but by raising serotonin levels it lifts people out of their depression.

Nothing captures the biological argument better than the famous New Age slogan: 'Happiness begins within.' Money, social status, plastic surgery, beautiful houses, powerful positions – none of these will bring you happiness. Lasting happiness comes only from serotonin, dopamine and oxytocin.[1]

In Aldous Huxley's dystopian novel *Brave New World*, published in 1932 at the height of the Great Depression, happiness is the supreme value and psychiatric drugs replace the police and the ballot as the foundation of politics. Every day, each person takes a dose of 'soma', a synthetic drug which makes people happy without harming their productivity and efficiency. The World State that governs the entire globe is never threatened by wars, revolutions, strikes or demonstrations, because all people are supremely content with their current conditions, whatever they may be. Huxley's vision of the future is far more troubling than George Orwell's *Nineteen Eighty-Four*. Huxley's world seems monstrous to most readers, but it is hard to explain why. Everybody is happy all the time – what could be wrong with that?

The Meaning of Life

Huxley's disconcerting world is based on the biological assumption that happiness equals pleasure. To be happy is no more and no less than experiencing pleasant bodily sensations. Since our biochemistry limits the volume and duration of these sensations, the only way to make people experience a high level of happiness over an extended period of time is to manipulate their biochemical system.

But that definition of happiness is contested by some scholars. In a famous study, Daniel Kahneman, winner of the Nobel Prize in economics, asked people to recount a typical work day, going through it episode by episode and evaluating how much they enjoyed or disliked each moment. He discovered what seems to be a paradox in most people's view of their lives. Take the work involved in raising a child. Kahneman found that when counting moments of joy and moments of drudgery, bringing up a child turns out to be a rather unpleasant affair. It consists largely of changing nappies, washing dishes and dealing with temper tantrums, which nobody likes to do. Yet most parents declare that their children are their chief source of happiness. Does it mean that people don't really know what's good for them?

That's one option. Another is that the findings demonstrate that happiness is not the surplus of pleasant over unpleasant moments. Rather, happiness consists in seeing one's life in its entirety as meaningful and worthwhile. There is an important cognitive and ethical component to happiness. Our values make all the difference to whether we see ourselves as 'miserable slaves to a baby dictator' or as 'lovingly nurturing a new life'.[2] As Nietzsche put it, if you have a why to live, you can bear almost any how. A meaningful life can be extremely satisfying even in the midst of hardship, whereas a meaningless life is a terrible ordeal no matter how comfortable it is.

Though people in all cultures and eras have felt the same type of pleasures and pains, the meaning they have ascribed to their experiences has probably varied widely. If so, the history of happiness might have been far more turbulent than biologists imagine. It's a conclusion that does not necessarily favour modernity. Assessing life minute by minute, medieval people certainly had it rough. However, if they believed the promise of everlasting bliss in the afterlife, they may well have viewed their lives as far more meaningful and worthwhile than modern secular people, who in the long term can expect nothing but complete and meaningless oblivion. Asked 'Are you satisfied with your life as a whole?', people in the Middle Ages might have scored quite highly in a subjective well-being questionnaire.

So our medieval ancestors were happy because they found meaning to life in collective delusions about the afterlife? Yes. As long as nobody punctured their fantasies, why shouldn't they? As far as we can tell, from a purely scientific viewpoint, human life has absolutely no meaning. Humans are the outcome of blind evolutionary processes that operate without goal or purpose. Our actions are not part of some divine cosmic plan, and if planet Earth were to blow up tomorrow morning, the universe would probably keep going about its business as usual. As far as we can tell at this point, human subjectivity would not be missed. Hence *any* meaning that people ascribe to their lives is just a delusion. The other-worldly meanings medieval people found in their lives were no more deluded than the humanist, nationalist and capitalist meanings modern people find. The scientist who says her life is meaningful because she increases the store of human knowledge, the soldier who declares that his life is meaningful because he fights to defend his homeland, and the entrepreneur who finds meaning in building a new company are no less delusional than their medieval counterparts who found meaning in reading scriptures, going on a crusade or building a new cathedral.

So perhaps happiness is synchronising one's personal delusions of meaning with the prevailing collective delusions. As long as my personal narrative is in line with the narratives of the people around me, I can convince myself that my life is meaningful, and find happiness in that conviction.

This is quite a depressing conclusion. Does happiness really depend on self-delusion?

Know Thyself

If happiness is based on feeling pleasant sensations, then in order to be happier we need to re-engineer our biochemical system. If happiness is based on feeling that life is meaningful, then in order to be happier we need to delude ourselves more effectively. Is there a third alternative?

Both the above views share the assumption that happiness is some sort of subjective feeling (of either pleasure or meaning), and that in order to judge people's happiness, all we need to do is ask them how they feel. To many of us, that seems logical because the dominant religion of our age is liberalism. Liberalism sanctifies the subjective feelings of individuals. It views these feelings as the supreme source of authority. What is good and what is bad, what is beautiful and what is ugly, what ought to be and what ought not to be, are all determined by what each one of us feels.

Liberal politics is based on the idea that the voters know best, and there is no need for Big Brother to tell us what is good for us. Liberal economics is based on the idea that the customer is always right. Liberal art declares that beauty is in the eye of the beholder. Students in liberal schools and universities are taught to think for themselves. Commercials urge us to 'Just do it.' Action films, stage dramas, soap operas, novels and catchy pop songs indoctrinate us constantly: 'Be true to yourself', 'Listen to yourself', 'Follow your heart'. Jean-Jacques Rousseau stated this view most classically: 'What I feel to be good – is good. What I feel to be bad – is bad.'

People who have been raised from infancy on a diet of such slogans are prone to believe that happiness is a subjective feeling and that each individual best knows whether she is happy or miserable. Yet this view is unique to liberalism. Most religions and ideologies throughout history stated that there are objective yardsticks for goodness and beauty, and for how things ought to be. They were suspicious of the feelings and preferences of the ordinary person. At the entrance of the temple of Apollo at Delphi, pilgrims were greeted by the inscription: 'Know thyself!' The implication was that the average person is ignorant of his true self, and is therefore likely to be ignorant of true happiness. Freud would probably concur.*

And so would Christian theologians. St Paul and St Augustine knew perfectly well that if you asked people about it, most of them would prefer to have sex than pray to God. Does that prove that having sex is the key to happiness? Not according to Paul and Augustine. It proves only that humankind is sinful by nature, and that people are easily seduced by Satan. From a Christian viewpoint, the vast majority of people are in more or less the same situation as heroin addicts. Imagine that a psychologist embarks on a study of happiness among drug users. He polls them and finds that they declare, every single one of them, that they are only happy when they shoot up. Would the psychologist publish a paper declaring that heroin is the key to happiness?

The idea that feelings are not to be trusted is not restricted to Christianity. At least when it comes to the value of feelings, even Darwin and Richard Dawkins might find common ground with St Paul and St Augustine. According to the selfish gene theory, natural selection makes people, like other organisms, choose what is good for the reproduction of their genes, even if it is bad for them as individuals.

* Paradoxically, while psychological studies of subjective well-being rely on people's ability to diagnose their happiness correctly, the basic *raison d'être* of psychotherapy is that people don't really know themselves and that they sometimes need professional help to free themselves of self-destructive behaviours.

Most males spend their lives toiling, worrying, competing and fighting, instead of enjoying peaceful bliss, because their DNA manipulates them for its own selfish aims. Like Satan, DNA uses fleeting pleasures to tempt people and place them in its power.

Most religions and philosophies have consequently taken a very different approach to happiness than liberalism does.[3] The Buddhist position is particularly interesting. Buddhism has assigned the question of happiness more importance than perhaps any other human creed. For 2,500 years, Buddhists have systematically studied the essence and causes of happiness, which is why there is a growing interest among the scientific community both in their philosophy and their meditation practices.

Buddhism shares the basic insight of the biological approach to happiness, namely that happiness results from processes occurring within one's body, and not from events in the outside world. However, starting from the same insight, Buddhism reaches very different conclusions.

According to Buddhism, most people identify happiness with pleasant feelings, while identifying suffering with unpleasant feelings. People consequently ascribe immense importance to what they feel, craving to experience more and more pleasures, while avoiding pain. Whatever we do throughout our lives, whether scratching our leg, fidgeting slightly in the chair, or fighting world wars, we are just trying to get pleasant feelings.

The problem, according to Buddhism, is that our feelings are no more than fleeting vibrations, changing every moment, like the ocean waves. If five minutes ago I felt joyful and purposeful, now these feelings are gone, and I might well feel sad and dejected. So if I want to experience pleasant feelings, I have to constantly chase them, while driving away the unpleasant feelings. Even if I succeed, I immediately have to start all over again, without ever getting any lasting reward for my troubles.

What is so important about obtaining such ephemeral prizes?

Why struggle so hard to achieve something that disappears almost as soon as it arises? According to Buddhism, the root of suffering is neither the feeling of pain nor of sadness nor even of meaninglessness. Rather, the real root of suffering is this never-ending and pointless pursuit of ephemeral feelings, which causes us to be in a constant state of tension, restlessness and dissatisfaction. Due to this pursuit, the mind is never satisfied. Even when experiencing pleasure, it is not content, because it fears this feeling might soon disappear, and craves that this feeling should stay and intensify.

People are liberated from suffering not when they experience this or that fleeting pleasure, but rather when they understand the impermanent nature of all their feelings, and stop craving them. This is the aim of Buddhist meditation practices. In meditation, you are supposed to closely observe your mind and body, witness the ceaseless arising and passing of all your feelings, and realise how pointless it is to pursue them. When the pursuit stops, the mind becomes very relaxed, clear and satisfied. All kinds of feelings go on arising and passing – joy, anger, boredom, lust – but once you stop craving particular feelings, you can just accept them for what they are. You live in the present moment instead of fantasising about what might have been.

The resulting serenity is so profound that those who spend their lives in the frenzied pursuit of pleasant feelings can hardly imagine it. It is like a man standing for decades on the seashore, embracing certain 'good' waves and trying to prevent them from disintegrating, while simultaneously pushing back 'bad' waves to prevent them from getting near him. Day in, day out, the man stands on the beach, driving himself crazy with this fruitless exercise. Eventually, he sits down on the sand and just allows the waves to come and go as they please. How peaceful!

This idea is so alien to modern liberal culture that when Western New Age movements encountered Buddhist insights, they translated them into liberal terms, thereby turning them on their head. New

Age cults frequently argue: 'Happiness does not depend on external conditions. It depends only on what we feel inside. People should stop pursuing external achievements such as wealth and status, and connect instead with their inner feelings.' Or more succinctly, 'Happiness begins within.' This is exactly what biologists argue, but more or less the opposite of what Buddha said.

Buddha agreed with modern biology and New Age movements that happiness is independent of external conditions. Yet his more important and far more profound insight was that true happiness is also independent of our inner feelings. Indeed, the more significance we give our feelings, the more we crave them, and the more we suffer. Buddha's recommendation was to stop not only the pursuit of external achievements, but also the pursuit of inner feelings.

To sum up, subjective well-being questionnaires identify our well-being with our subjective feelings, and identify the pursuit of happiness with the pursuit of particular emotional states. In contrast, for many traditional philosophies and religions, such as Buddhism, the key to happiness is to know the truth about yourself – to understand who, or what, you really are. Most people wrongly identify themselves with their feelings, thoughts, likes and dislikes. When they feel anger, they think, 'I am angry. This is my anger.' They consequently spend their life avoiding some kinds of feelings and pursuing others. They never realise that they are not their feelings, and that the relentless pursuit of particular feelings just traps them in misery.

If this is so, then our entire understanding of the history of happiness might be misguided. Maybe it isn't so important whether people's expectations are fulfilled and whether they enjoy pleasant feelings. The main question is whether people know the truth about themselves. What evidence do we have that people today understand this truth any better than ancient foragers or medieval peasants?

Scholars began to study the history of happiness only a few years ago, and we are still formulating initial hypotheses and searching for appropriate research methods. It's much too early to adopt rigid conclusions and end a debate that's hardly yet begun. What is important is to get to know as many different approaches as possible and to ask the right questions.

Most history books focus on the ideas of great thinkers, the bravery of warriors, the charity of saints and the creativity of artists. They have much to tell about the weaving and unravelling of social structures, about the rise and fall of empires, about the discovery and spread of technologies. Yet they say nothing about how all this influenced the happiness and suffering of individuals. This is the biggest lacuna in our understanding of history. We had better start filling it.

20

The End of *Homo Sapiens*

THIS BOOK BEGAN BY PRESENTING HISTORY as the next stage in the continuum of physics to chemistry to biology. Sapiens are subject to the same physical forces, chemical reactions and natural-selection processes that govern all living beings. Natural selection may have provided *Homo sapiens* with a much larger playing field than it has given to any other organism, but the field has still had its boundaries. The implication has been that, no matter what their efforts and achievements, Sapiens are incapable of breaking free of their biologically determined limits.

But at the dawn of the twenty-first century, this is no longer true: *Homo sapiens* is transcending those limits. It is now beginning to break the laws of natural selection, replacing them with the laws of intelligent design.

For close to 4 billion years, every single organism on the planet evolved subject to natural selection. Not even one was designed by an intelligent creator. The giraffe, for example, got its long neck thanks to competition between archaic giraffes rather than to the whims of a super-intelligent being. Proto-giraffes who had longer necks had access to more food and consequently produced more offspring than did those with shorter necks. Nobody, certainly not the giraffes, said, 'A long neck would enable giraffes to munch leaves off the treetops. Let's extend it.' The beauty of Darwin's theory is that it does not need to assume an intelligent designer to explain how giraffes ended up with long necks.

For billions of years, intelligent design was not even an option, because there was no intelligence which could design things. Microorganisms, which until quite recently were the only living things around, are capable of amazing feats. A microorganism belonging to one species can incorporate genetic codes from a completely different species into its cell and thereby gain new capabilities, such as resistance to antibiotics. Yet, as best we know, microorganisms have no consciousness, no aims in life, and no ability to plan ahead.

At some stage organisms such as giraffes, dolphins, chimpanzees and Neanderthals evolved consciousness and the ability to plan ahead. But even if a Neanderthal fantasised about fowls so fat and slow-moving that he could just scoop them up whenever he was hungry, he had no way of turning that fantasy into reality. He had to hunt the birds that had been naturally selected.

The first crack in the old regime appeared about 10,000 years ago, during the Agricultural Revolution. Sapiens who dreamed of fat, slow-moving chickens discovered that if they mated the fattest hen with the slowest cock, some of their offspring would be both fat and slow. If you mated those offspring with each other, you could produce a line of fat, slow birds. It was a race of chickens unknown to nature, produced by the intelligent design not of a god but of a human.

Still, compared to an all-powerful deity, *Homo sapiens* had limited design skills. Sapiens could use selective breeding to detour around and accelerate the natural-selection processes that normally affected chickens, but they could not introduce completely new characteristics that were absent from the genetic pool of wild chickens. In a way, the relationship between *Homo sapiens* and chickens was similar to many other symbiotic relationships that have so often arisen on their own in nature. Sapiens exerted peculiar selective pressures on chickens that caused the fat and slow ones to proliferate, just as pollinating bees select flowers, causing the bright colourful ones to proliferate.

Today, the 4-billion-year-old regime of natural selection is facing a completely different challenge. In laboratories throughout the world, scientists are engineering living beings. They break the laws of natural selection with impunity, unbridled even by an organism's original characteristics. Eduardo Kac, a Brazilian bio-artist, decided in 2000 to create a new work of art: a fluorescent green rabbit. Kac contacted a French laboratory and offered it a fee to engineer a radiant bunny according to his specifications. The French scientists took a run-of-the-mill white rabbit embryo, implanted in its DNA a gene taken from a green fluorescent jellyfish, and *voilà*! One green fluorescent rabbit for *le monsieur*. Kac named the rabbit Alba.

It is impossible to explain the existence of Alba through the laws of natural selection. She is the product of intelligent design. She is also a harbinger of things to come. If the potential Alba signifies is realised in full – and if humankind doesn't annihilate itself meanwhile – the Scientific Revolution might prove itself far greater than a mere historical revolution. It may turn out to be the most important *biological* revolution since the appearance of life on earth. After 4 billion years of natural selection, Alba stands at the dawn of a new cosmic era, in which life will be ruled by intelligent design. If this happens, the whole of human history up to that point might, with hindsight, be reinterpreted as a process of experimentation and apprenticeship that revolutionised the game of life. Such a process should be understood from a cosmic perspective of billions of years, rather than from a human perspective of millennia.

Biologists the world over are locked in battle with the intelligent-design movement, which opposes the teaching of Darwinian evolution in schools and claims that biological complexity proves there must be a creator who thought out all biological details in advance. The biologists are right about the past, but the proponents of intelligent design might, ironically, be right about the future.

At the time of writing, the replacement of natural selection by intelligent design could happen in any of three ways: through biological engineering, cyborg engineering (cyborgs are beings that combine organic with non-organic parts) or the engineering of inorganic life.

Of Mice and Men

Biological engineering is deliberate human intervention on the biological level (e.g. implanting a gene) aimed at modifying an organism's shape, capabilities, needs or desires, in order to realise some preconceived cultural idea, such as the artistic predilections of Eduardo Kac.

There is nothing new about biological engineering per se. People have been using it for millennia in order to reshape themselves and other organisms. A simple example is castration. Humans have been castrating bulls for perhaps 10,000 years in order to create oxen. Oxen are less aggressive, and are thus easier to train to pull ploughs. Humans also castrated their own young males to create soprano singers with enchanting voices and eunuchs who could safely be entrusted with overseeing the sultan's harem.

But recent advances in our understanding of how organisms work, down to the cellular and nuclear levels, have opened up previously unimaginable possibilities. For instance, we can today not merely castrate a man, but also change his sex through surgical and hormonal treatments. But that's not all. Consider the surprise, disgust and consternation that ensued when, in 1996, the following photograph appeared in newspapers and on television:

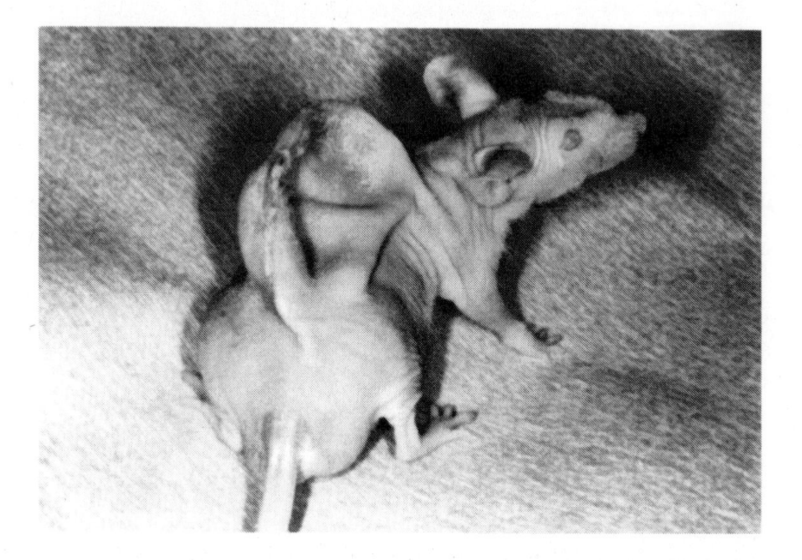

47. A mouse on whose back scientists grew an 'ear' made of cattle cartilage cells. It is an eerie echo of the lion-man statue from the Stadel Cave. Thirty thousand years ago, humans were already fantasising about combining different species. Today, they can actually produce such chimeras.

No, Photoshop was not involved. It's an untouched photo of a real mouse on whose back scientists implanted cattle cartilage cells. The scientists were able to control the growth of the new tissue, shaping it in this case into something that looks like a human ear. The process may soon enable scientists to manufacture artificial ears, which could then be implanted in humans.[1]

Even more remarkable wonders can be performed with genetic engineering, which is why it raises a host of ethical, political and ideological issues. And it's not just pious monotheists who object that man should not usurp God's role. Many confirmed atheists are no less shocked by the idea that scientists are stepping into nature's shoes. Animal-rights activists decry the suffering caused to lab animals in genetic engineering experiments, and to the farmyard animals that are engineered in complete disregard of their needs and desires. Human-rights activists are afraid that genetic engineering might be used to create supermen who will make serfs of the rest of us. Jeremiahs offer apocalyptic visions of bio-dictatorships that will clone fearless soldiers and obedient workers. The prevailing feeling is that too many opportunities are opening too quickly and that our ability to modify genes is outpacing our capacity for making wise and far-sighted use of the skill.

The result is that we're at present using only a fraction of the potential of genetic engineering. Most of the organisms now being engineered are those with the weakest political lobbies – plants, fungi, bacteria and insects. For example, lines of *E. coli*, a bacterium that lives symbiotically in the human gut (and which makes headlines when it gets out of the gut and causes deadly infections), have been genetically engineered to produce biofuel.[2] *E. coli* and several species of fungi have also been engineered to produce insulin, thereby lowering the cost of diabetes treatment.[3] A gene extracted from an Arctic fish has been inserted into potatoes, making the plants more frost-resistant.[4]

A few mammals have also been subject to genetic engineering. Every year the dairy industry suffers billions of dollars in damages due to mastitis, a disease that strikes dairy-cow udders. Scientists are currently experimenting with genetically engineered cows whose milk contains lysostaphin, a biochemical that attacks the bacteria responsible for the disease.[5] The pork industry, which has suffered from falling sales because consumers are wary of the unhealthy fats in ham and bacon, has hopes for a still-experimental line of pigs implanted with genetic material from a worm. The new genes cause the pigs to turn bad omega-6 fatty acid into its healthy cousin, omega-3.[6]

The next generation of genetic engineering will make pigs with good fat look like child's play. Geneticists have managed not merely to extend sixfold the average life expectancy of worms, but also to engineer genius mice that display much-improved memory and learning skills.[7] Voles are small, stout rodents resembling mice, and most varieties of voles are promiscuous. But there is one species in which boy and girl voles form lasting and monogamous relationships. Geneticists claim to have isolated the genes responsible for vole monogamy. If the addition of a gene can turn a vole Don Juan into a loyal and loving husband, are we far off from being able to genetically engineer not only the individual abilities of rodents (and humans), but also their social structures?[8]

The Return of the Neanderthals

But geneticists do not only want to transform living lineages. They aim to revive extinct creatures as well. And not just dinosaurs, as in *Jurassic Park*. A team of Russian, Japanese and Korean scientists has recently mapped the genome of ancient mammoths, found frozen in the Siberian ice. They now plan to take a fertilised egg-cell of a present-day elephant, replace the elephantine DNA with a reconstructed mammoth DNA, and implant the egg in the womb of an elephant. After about twenty-two months, they expect the first mammoth in 5,000 years to be born.[9]

But why stop at mammoths? Professor George Church of Harvard University recently suggested that, with the completion of the Neanderthal Genome Project, we can now implant reconstructed Neanderthal DNA into a Sapiens ovum, thus producing the first Neanderthal child in 30,000 years. Church claimed that he could do the job for a paltry $30 million. Several women have already volunteered to serve as surrogate mothers.[10]

What do we need Neanderthals for? Some argue that if we could study live Neanderthals, we could answer some of the most nagging questions about the origins and uniqueness of *Homo sapiens*. By comparing a Neanderthal to a *Homo sapiens* brain, and mapping out where their structures differ, perhaps we could identify what biological change produced consciousness as we experience it. There's an ethical reason, too – some have argued that if *Homo sapiens* was responsible for the extinction of the Neanderthals, it has a moral duty to resurrect them. And having some Neanderthals around might be useful. Lots of industrialists would be glad to pay one Neanderthal to do the menial work of two Sapiens.

But why stop even at Neanderthals? Why not go back to God's drawing board and design a better Sapiens? The abilities, needs and desires of *Homo sapiens* have a genetic basis, and the Sapiens genome is no more complex than that of voles and mice. (The mouse genome contains about 2.5 billion nucleobases, the Sapiens genome about 2.9 billion bases – meaning the latter is only 14 per cent larger.[11]) In the medium range – perhaps in a few decades – genetic engineering and other forms of biological engineering might enable us to make far-reaching alterations not only to our physiology, immune system and life expectancy, but also to our intellectual and emotional capacities. If genetic engineering can create genius mice, why not genius humans? If it can create monogamous voles, why not humans hard-wired to remain faithful to their partners?

The Cognitive Revolution that turned *Homo sapiens* from an insignificant ape into the master of the world did not require any noticeable change in physiology or even in the size and external shape of the Sapiens brain. It apparently involved no more than a few small changes to internal brain structure. Perhaps another small change would be enough to ignite a Second Cognitive Revolution, create a completely new type of consciousness, and transform *Homo sapiens* into something altogether different.

True, we still don't have the acumen to achieve this, but there seems to be no insurmountable technical barrier preventing us from producing superhumans. The main obstacles are the ethical and political objections that have slowed down research on humans. And no matter how convincing the ethical arguments may be, it is hard to see how they can hold back the next step for long, especially if what is at stake is the possibility of prolonging human life indefinitely, conquering incurable diseases, and upgrading our cognitive and emotional abilities.

What would happen, for example, if we developed a cure for Alzheimer's disease that, as a side benefit, could dramatically improve the memories of healthy people? Would anyone be able to halt the relevant research? And when the cure is developed, could any law enforcement agency limit it to Alzheimer's patients and prevent healthy people from using it to acquire super-memories?

It's unclear whether bioengineering could really resurrect the Neanderthals, but it would very likely bring down the curtain on *Homo sapiens.* Tinkering with our genes won't necessarily kill us. But we might fiddle with *Homo sapiens* to such an extent that we would no longer be *Homo sapiens.*

Bionic Life

There is another new technology which could change the laws of life: cyborg engineering. Cyborgs are beings that combine organic and inorganic parts, such as a human with bionic hands. In a sense, nearly all of us are bionic these days, since our natural senses and functions are supplemented by devices such as eyeglasses, pacemakers, orthotics, and even computers and mobile phones (which relieve our brains of some of their data storage and processing burdens). We stand poised on the brink of becoming true cyborgs, of having inorganic features that are inseparable from our bodies, features that modify our abilities, desires, personalities and identities.

The Defense Advanced Research Projects Agency (DARPA), a US military research agency, is developing cyborgs out of insects. The idea is to implant electronic chips, detectors and processors in the body of a fly or cockroach, which will enable either a human or an automatic operator to control the insect's movements remotely and to absorb and transmit information. Such a fly could be sitting on the wall at enemy headquarters, eavesdrop on the most secret conversations, and if it isn't caught first by a spider, could inform us exactly what the enemy is planning.[12] In 2006 the US Naval Undersea Warfare Center reported its intention to develop cyborg sharks, declaring, 'NUWC is developing a fish tag whose goal is behaviour control of host animals via neural implants.' The developers hope to identify underwater electromagnetic fields made by submarines and mines, by exploiting the natural magnetic detecting capabilities of sharks, which are superior to those of any man-made detectors.[13]

Sapiens, too, are being turned into cyborgs. The newest generation of hearing aids are sometimes referred to as 'bionic ears'. The device consists of an implant that absorbs sound through a microphone located in the outer part of the ear. The implant filters the sounds, identifies human voices, and translates them into electric signals that are sent directly to the central auditory nerve and from there to the brain.[14]

Retina Implant, a government-sponsored German company, is developing a retinal prosthesis that may allow blind people to gain partial vision. It involves implanting a small microchip inside the patient's eye. Photocells absorb light falling on the eye and transform it into electrical energy, which stimulates the intact nerve cells in the retina. The nervous impulses from these cells stimulate the brain, where they are translated into sight. At present the technology allows patients to orientate themselves in space, identify letters, and even recognise faces.[15]

Jesse Sullivan, an American electrician, lost both arms up to the shoulder in a 2001 accident. Today he uses two bionic arms, courtesy of the Rehabilitation Institute of Chicago. The special feature of Jesse's new arms is that they are operated by thought alone. Neural signals arriving from Jesse's brain are translated by micro-computers into electrical commands, and the arms move. When Jesse wants to raise his arm, he does what any normal person unconsciously does – and the arm rises. These arms can perform a much more limited range of movements than organic arms, but they enable

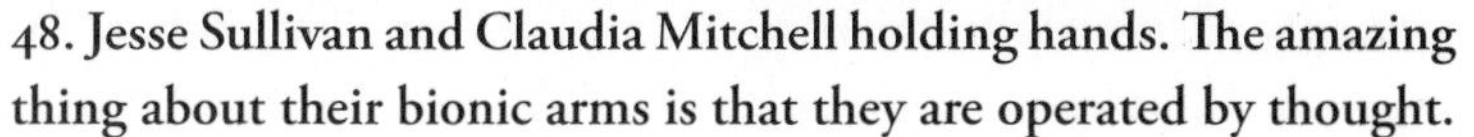

48. Jesse Sullivan and Claudia Mitchell holding hands. The amazing thing about their bionic arms is that they are operated by thought.

Jesse to carry out simple daily functions. A similar bionic arm has recently been outfitted for Claudia Mitchell, an American soldier who lost her arm in a motorcycle accident. Scientists believe that we will soon have bionic arms that will not only move when willed to move, but will also be able to transmit signals back to the brain, thereby enabling amputees to regain even the sensation of touch![16]

At present these bionic arms are a poor replacement for our organic originals, but they have the potential for unlimited development. Bionic arms, for example, can be made far more powerful than their organic kin, making even a boxing champion feel like a weakling. Moreover, bionic arms have the advantage that they can be replaced every few years, or detached from the body and operated at a distance.

Scientists at Duke University in North Carolina have recently demonstrated this with rhesus monkeys whose brains have been implanted with electrodes. The electrodes gather signals from the brain and transmit them to external devices. The monkeys have been trained to control detached bionic arms and legs through thought alone. One monkey, named Aurora, learned to thought-control a detached bionic arm while simultaneously moving her two organic arms. Like some Hindu goddess, Aurora now has three arms, and her arms can be located in different rooms – or even cities. She can sit in her North Carolina lab, scratch her back with one hand, scratch her head with a second hand, and simultaneously steal a banana in New York (although the ability to eat a purloined fruit at a distance remains a dream). Another rhesus monkey, Idoya, won world fame in 2008 when she thought-controlled a pair of bionic legs in Kyoto, Japan, from her North Carolina chair. The legs were twenty times Idoya's weight.[17]

Locked-in syndrome is a condition in which a person loses all or nearly all her ability to move any part of her body, while her cognitive abilities remain intact. Patients suffering from the syndrome have up till now been able to communicate with the outside world only through small eye movements. However, a few patients have had brain-signal-gathering electrodes implanted in their brains. Efforts are being made to translate such signals not merely into movements but also into words. If the experiments succeed, locked-in patients could finally speak directly with the outside world, and we might eventually be able to use the technology to read other people's minds.[18]

Yet of all the projects currently under development, the most revolutionary is the attempt to devise a direct two-way brain–computer interface that will allow computers to read the electrical signals of a human brain, simultaneously transmitting signals that the brain can read in turn. What if such interfaces are used to directly link a brain to the Internet, or to directly link several brains to each other, thereby creating a sort of Inter-brain-net? What might happen to human memory, human consciousness and human identity if the brain has direct access to a collective memory bank? In such a situation, one cyborg could, for example, retrieve the memories of another – not hear about them, not read about them in an autobiography, not imagine them, but directly remember them as if they were his own. Or her own. What happens to concepts such as the self and gender identity when minds become collective? How could you know thyself or follow your dream if the dream is not in your mind but in some collective reservoir of aspirations?

Such a cyborg would no longer be human, or even organic. It would be something completely different. It would be so fundamentally another kind of being that we cannot even grasp the philosophical, psychological or political implications.

Another Life

The third way to change the laws of life is to engineer completely inorganic beings. The most obvious examples are computer programs that can undergo independent evolution.

Recent advances in machine learning already enable present-day computer programs to evolve by themselves. Though the program is initially coded by human engineers, it can subsequently acquire new information on its own, teach itself new skills, and gain insights that go beyond those of its human creators. The computer program is therefore free to evolve in directions its makers could never have envisaged.

Such computer programs can learn to play chess, to drive cars, to diagnose diseases and to invest money in the stock market. In all these fields they might increasingly outperform old-fashioned humans, but will have to compete against each other. They will thus confront new forms of evolutionary pressures. As a thousand computer programs invest money in the stock market, each adopting different strategies, many will go bust but some will become billionaires. In the process, they will evolve remarkable skills that humans can neither rival nor understand. Such a program could not explain its investment strategy to a Sapiens, for the same reason that a Sapiens could not explain Wall Street to a chimpanzee. Many of us might eventually work for such programs, which will decide not only where to invest money, but also whom to hire for a particular job, whom to give a mortgage to, and whom to send to prison.

Are these living creatures? It depends on what you mean by 'living creatures'. They have certainly been produced by a new evolutionary process, completely independent of the laws and limitations of organic evolution.

Imagine another possibility – suppose you could back up your brain to a portable hard drive and then run it on your laptop. Would your laptop be able to think and feel just like a Sapiens? If so, would it be you or someone else? What if computer programmers could create an entirely new but digital mind, composed of computer code, complete with a sense of self, consciousness and memory? If you ran the program on your computer, would it be a person? If you deleted it could you be charged with murder?

We might soon have the answer to such questions. The Human Brain Project, founded in 2005, hopes to recreate a complete human brain inside a computer, with electronic circuits in the computer emulating neural networks in the brain. The project's director has claimed that, if funded properly, within a decade or two we could have an artificial human brain inside a computer that could talk and behave very much as a human does. If successful, that would mean that after 4 billion years of milling around inside the small world of organic compounds, life will suddenly break out into the vastness of the inorganic realm, ready to take up shapes beyond our wildest dreams. Not all scholars agree that the mind works in a manner analogous to today's digital computers – and if it doesn't, present-day computers would not be able to simulate it. Yet it would be foolish to categorically dismiss the possibility before giving it a try. In 2013 the project received a grant of €1 billion from the European Union.[19]

The Singularity

Presently, only a tiny fraction of these new opportunities have been realised. Yet the world of 2014 is already a world in which culture is releasing itself from the shackles of biology. Our ability to engineer not merely the world around us, but above all the world inside our bodies and minds, is developing at breakneck speed. More and more spheres of activity are being shaken out of their complacent ways. Lawyers need to rethink issues of privacy and identity; governments are faced with rethinking matters of health care and equality; sports associations and educational institutions need to redefine fair play and achievement; pension funds and labour markets should readjust to a world in which sixty might be the new thirty. They must all deal with the conundrums of bioengineering, cyborgs and inorganic life.

Mapping the first human genome required fifteen years and $3 billion. Today you can map a person's DNA within a few weeks and at the cost of a few hundred dollars.[20] The era of personalised medicine – medicine that matches treatment to DNA – has begun. The family doctor could soon tell you with greater certainty that you face high risks of liver cancer, whereas you needn't worry too much about heart attacks. She could determine that a popular medication that helps 92 per cent of people is useless to you, and you should instead take another pill, fatal to many people but just right for you. The road to near-perfect medicine stands before us.

However, with improvements in medical knowledge will come new ethical conundrums. Ethicists and legal experts are already wrestling with the thorny issue of privacy as it relates to DNA. Would insurance companies be entitled to ask for our DNA scans and to raise premiums if they discover a genetic tendency to reckless behaviour? Would we be required to fax our DNA, rather than our CV, to potential employers? Could an employer favour a candidate because his DNA looks better? Or could we sue in such cases for 'genetic discrimination'? Could a company that develops a new creature or a new organ register a patent on its DNA sequences? It is obvious that one can own a particular chicken, but can one own an entire species?

Such dilemmas are dwarfed by the ethical, social and political implications of the quest for immortality and of our potential new abilities to create superhumans. The Universal Declaration of Human Rights, government medical programmes throughout the world, national health insurance programmes and national constitutions worldwide recognise that a humane society ought to give all its members fair medical treatment and keep them in relatively good health. That was all well and good as long as medicine was chiefly concerned with preventing illness and healing the sick. What might happen once medicine becomes preoccupied with enhancing human abilities? Would all humans be entitled to such enhanced abilities, or would there be a new superhuman elite?

Our late modern world prides itself on recognising, for the first time in history, the basic equality of all humans, yet it might be poised to create the most unequal of all societies. Throughout history, the upper classes always claimed to be smarter, stronger and generally better than the underclass. They were usually deluding themselves. A baby born to a poor peasant family was likely to be as intelligent as the Crown prince. With the help of new medical capabilities, the pretensions of the upper classes might soon become an objective reality.

This is not science fiction. Most science-fiction plots describe a world in which Sapiens – identical to us – enjoy superior technology such as light-speed spaceships and laser guns. The ethical and political dilemmas central to these plots are taken from our own world, and they merely recreate our emotional and social tensions against a futuristic backdrop. Yet the real potential of future technologies is to change *Homo sapiens* itself, including our emotions and desires, and not merely our vehicles and weapons. What is a spaceship compared to an eternally young cyborg who does not breed and has no sexuality, who can share thoughts directly with other beings, whose abilities to focus and remember are a thousand times greater than our own, and who is never angry or sad, but has emotions and desires that we cannot begin to imagine?

Science fiction rarely describes such a future, because an accurate description is by definition incomprehensible. Producing a film about the life of some super-cyborg is akin to producing *Hamlet* for an audience of Neanderthals. Indeed, the future masters of the world will probably be more different from us than we are from Neanderthals. Whereas we and the Neanderthals are at least human, our inheritors will be godlike.

Physicists define the Big Bang as a singularity. It is a point at which all the known laws of nature did not exist. Time too did not exist. It is thus meaningless to say that anything existed 'before' the Big Bang. We may be fast approaching a new singularity, when all the concepts that give meaning to our world – me, you, men, women, love and hate – will become irrelevant. Anything happening beyond that point is meaningless to us.

The Frankenstein Prophecy

In 1818 Mary Shelley published *Frankenstein*, the story of a scientist who tries to create a superior being and instead creates a monster. In the last two centuries, this story has been told over and over again in countless variations. It has become a central pillar of our new scientific mythology. At first sight, the Frankenstein story appears to warn us that if we try to play God and engineer life we will be punished severely. Yet the story has a deeper meaning.

The Frankenstein myth confronts *Homo sapiens* with the fact that the last days are fast approaching. Unless some nuclear or ecological catastrophe destroys us first, the pace of technological development will soon lead to the replacement of *Homo sapiens* by completely different beings who possess not only different physiques, but also very different cognitive and emotional worlds. This is something most Sapiens find extremely disconcerting. We like to believe that in the future people just like us will travel from planet to planet in fast spaceships. We don't like to contemplate the possibility that in the future, beings with emotions and identities like ours will no longer exist, and our place will be taken by alien life forms whose abilities dwarf our own.

We seek comfort in the fantasy that Dr Frankenstein can create only terrible monsters, whom we would have to destroy in order to save the world. We like to tell the story that way because it implies that we are the best of all beings, that there never was and never will be something better than us. Any attempt to improve us will inevitably fail, because even if our bodies might be improved, you cannot touch the human spirit.

We would have a hard time swallowing the fact that scientists could engineer spirits as well as bodies, and that future Dr Frankensteins could therefore create something truly superior to us, something that will look at us as condescendingly as we look at the Neanderthals.

We cannot be certain whether today's Frankensteins will indeed fulfil this prophecy. The future is unknown, and it would be surprising if the forecasts of the last few pages were realised in full. History teaches us that what seems to be just around the corner may never materialise due to unforeseen barriers, and that other unimagined scenarios will in fact come to pass. When the nuclear age erupted in the 1940s, many forecasts were made about the future nuclear world of the year 2000. When sputnik and Apollo 11 fired the imagination of the world, everyone began predicting that by the end of the century, people would be living in space colonies on Mars and Pluto. Few of these forecasts came true. On the other hand, nobody foresaw the Internet.

So don't go out just yet to buy liability insurance to indemnify you against lawsuits filed by digital beings. The above fantasies – or nightmares – are just stimulants for your imagination. What we should take seriously is the idea that the next stage of history will include not only technological and organisational transformations, but also fundamental transformations in human consciousness and identity. And these could be transformations so fundamental that they will call the very term 'human' into question. How long do we have? No one really knows. As already mentioned, some say that by 2050 a few humans will already be a-mortal. Less radical forecasts speak of the next century, or the next millennium. Yet from the perspective of 70,000 years of Sapiens history, what are a few centuries?

If the curtain is indeed about to drop on Sapiens history, we members of one of its final generations should devote some time to answering one last question: what do we want to become? This question, sometimes known as the Human Enhancement question, dwarfs the debates that currently preoccupy politicians, philosophers, scholars and ordinary people. After all, today's debate between today's religions, ideologies, nations and classes will in all likelihood disappear along with *Homo sapiens*. If our successors indeed function on a different level of consciousness (or perhaps possess something beyond consciousness that we cannot even conceive), it seems doubtful that Christianity or Islam will be of interest to them, that their social organisation could be Communist or capitalist, or that their genders could be male or female.

And yet the great debates of history are important because at least the first generation of these gods would be shaped by the cultural ideas of their human designers. Would they be created in the image of capitalism, of Islam, or of feminism? The answer to this question might send them careening in entirely different directions.

Most people prefer not to think about it. Even the field of bioethics prefers to address another question: 'What is it forbidden to do?' Is it acceptable to carry out genetic experiments on living human beings? On aborted fetuses? On stem cells? Is it ethical to clone sheep? And chimpanzees? And what about humans? All of these are important questions, but it is naive to imagine that we might simply hit the brakes and stop the scientific projects that are upgrading *Homo sapiens* into a different kind of being. For these projects are inextricably meshed together with the Gilgamesh Project. Ask scientists why they study the genome, or try to connect a brain to a computer, or try to create a mind inside a computer. Nine out of ten times you'll get the same standard answer: we are doing it to cure diseases and save human lives. Even though the implications of creating a mind inside a computer are far more dramatic than curing psychiatric illnesses, this is the standard justification given, because nobody can argue with it. This is why the Gilgamesh Project is the flagship of science. It serves to justify everything science does. Dr Frankenstein piggybacks on the shoulders of Gilgamesh. Since it is impossible to stop Gilgamesh, it is also impossible to stop Dr Frankenstein.

The only thing we can try to do is to influence the direction scientists are taking. But since we might soon be able to engineer our desires too, the real question facing us is not 'What do we want to become?', but 'What do we want to want?' Those who are not spooked by this question probably haven't given it enough thought.

Afterword: The Animal that Became a God

SEVENTY THOUSAND YEARS AGO, *HOMO sapiens* was still an insignificant animal minding its own business in a corner of Africa. In the following millennia it transformed itself into the master of the entire planet and the terror of the ecosystem. Today it stands on the verge of becoming a god, poised to acquire not only eternal youth, but also the divine abilities of creation and destruction.

Unfortunately, the Sapiens regime on earth has so far produced little that we can be proud of. We have mastered our surroundings, increased food production, built cities, established empires and created far-flung trade networks. But did we decrease the amount of suffering in the world? Time and again, massive increases in human power did not necessarily improve the well-being of individual Sapiens, and usually caused immense misery to other animals.

In the last few decades we have at last made some real progress as far as the human condition is concerned, with the reduction of famine, plague and war. Yet the situation of other animals is deteriorating more rapidly than ever before, and the improvement in the lot of humanity is too recent and fragile to be certain of.

Moreover, despite the astonishing things that humans are capable of doing, we remain unsure of our goals and we seem to be as discontented as ever. We have advanced from canoes to galleys to steamships to space shuttles – but nobody knows where we're going.

We are more powerful than ever before, but have very little idea what to do with all that power. Worse still, humans seem to be more irresponsible than ever. Self-made gods with only the laws of physics to keep us company, we are accountable to no one. We are consequently wreaking havoc on our fellow animals and on the surrounding ecosystem, seeking little more than our own comfort and amusement, yet never finding satisfaction.

Is there anything more dangerous than dissatisfied and irresponsible gods who don't know what they want?

Notes

Due to limitations of space, this section includes only a small number of the sources on which the book is based. For a full list of sources and references, please visit https://www.ynharari.com/sapiens-references/.

1 An Animal of No Significance

1 Ann Gibbons, 'Food for Thought: Did the First Cooked Meals Help Fuel the Dramatic Evolutionary Expansion of the Human Brain?', *Science* 316:5831 (2007), 1,558–60.

2 The Tree of Knowledge

1 Robin Dunbar, *Grooming, Gossip and the Evolution of Language* (Cambridge, Mass.: Harvard University Press, 1998).
2 Frans de Waal, *Chimpanzee Politics: Power and Sex Among Apes* (Baltimore: Johns Hopkins University Press, 2000); Frans de Waal, *Our Inner Ape: A Leading Primatologist Explains Why We Are Who We Are* (New York: Riverhead Books, 2005); Michael L. Wilson and Richard W. Wrangham, 'Intergroup Relations in Chimpanzees', *Annual Review of Anthropology* 32 (2003), 363–92; M. McFarland Symington, 'Fission-Fusion Social Organization in *Ateles* and *Pan*', *International Journal of Primatology* 11:1 (1990), 49; Colin A. Chapman and Lauren J. Chapman, 'Determinants of Group Size in Primates: The Importance of Travel Costs', in *On the Move: How and Why Animals Travel in Groups*, ed. Sue Boinsky and Paul A. Garber (Chicago: University of Chicago Press, 2000), 26.
3 Dunbar, *Grooming, Gossip and the Evolution of Language*, 69–79; Leslie C. Aiello and R. I. M. Dunbar, 'Neocortex Size, Group Size, and the Evolution of Language', *Current Anthropology* 34:2 (1993), 189. For criticism of this approach see: Christopher McCarthy et al., 'Comparing Two Methods for Estimating Network Size', *Human Organization* 60:1 (2001), 32; R. A. Hill and R. I. M. Dunbar, 'Social Network Size in Humans', *Human Nature* 14:1 (2003), 65.
4 Yvette Taborin, 'Shells of the French Aurignacian and Perigordian', in *Before Lascaux: The Complete Record of the Early Upper Paleolithic*, ed. Heidi Knecht, Anne Pike-Tay and Randall White (Boca Raton: CRC Press, 1993), 211–28.
5 G. R. Summerhayes, 'Application of PIXE-PIGME to Archaeological Analysis

of Changing Patterns of Obsidian Use in West New Britain, Papua New Guinea', in *Archaeological Obsidian Studies: Method and Theory*, ed. Steven M. Shackley (New York: Plenum Press, 1998), 129–58.

3 A Day in the Life of Adam and Eve

1 Christopher Ryan and Cacilda Jethá, *Sex at Dawn: The Prehistoric Origins of Modern Sexuality* (New York: Harper, 2010); S. Beckerman and P. Valentine (eds), *Cultures of Multiple Fathers: The Theory and Practice of Partible Paternity in Lowland South America* (Gainesville: University Press of Florida, 2002).
2 Noel G. Butlin, *Economics and the Dreamtime: A Hypothetical History* (Cambridge: Cambridge University Press, 1993), 98–101; Richard Broome, *Aboriginal Australians* (Sydney: Allen & Unwin, 2002), 15; William Howell Edwards, *An Introduction to Aboriginal Societies* (Wentworth Falls, NSW: Social Science Press, 1988), 52.
3 Fekri A. Hassan, *Demographic Archaeology* (New York: Academic Press, 1981), 196–9; Lewis Robert Binford, *Constructing Frames of Reference: An Analytical Method for Archaeological Theory Building Using Hunter-Gatherer and Environmental Data Sets* (Berkeley: University of California Press, 2001), 143.
4 Brian Hare and Vanessa Woods, *The Genius of Dogs: How Dogs Are Smarter Than You Think* (New York: Penguin, 2013).
5 Christopher B. Ruff, Erik Trinkaus and Trenton W. Holliday, 'Body Mass and Encephalization in Pleistocene *Homo*', *Nature* 387 (1997), 173–6; M. Henneberg and M. Steyn, 'Trends in Cranial Capacity and Cranial Index in Subsaharan Africa During the Holocene', *American Journal of Human Biology* 5:4 (1993), 473–9; Drew H. Bailey and David C. Geary, 'Hominid Brain Evolution: Testing Climatic, Ecological and Social Competition Models', *Human Nature* 20 (2009), 67–79; Daniel J. Wescott and Richard L. Jantz, 'Assessing Craniofacial Secular Change in American Blacks and Whites Using Geometric Morphometry', in *Modern Morphometrics in Physical Anthropology: Developments in Primatology: Progress and Prospects*, ed. Dennis E. Slice (New York: Plenum Publishers, 2005), 231–45.
6 Nicholas G. Blurton Jones et al., 'Antiquity of Postreproductive Life: Are There Modern Impacts on Hunter-Gatherer Postreproductive Life Spans?', *American Journal of Human Biology* 14 (2002), 184–205.
7 Kim Hill and A. Magdalena Hurtado, *Aché Life History: The Ecology and Demography of a Foraging People* (New York: Aldine de Gruyter, 1996), 164, 236.
8 Ibid., 78.
9 Vincenzo Formicola and Alexandra P. Buzhilova, 'Double Child Burial from Sunghir (Russia): Pathology and Inferences for Upper Paleolithic Funerary Practices', *American Journal of Physical Anthropology* 124:3 (2004), 189–98; Giacomo Giacobini, 'Richness and Diversity of Burial Rituals in the Upper Paleolithic', *Diogenes* 54:2 (2007), 19–39.

10 I. J. N. Thorpe, 'Anthropology, Archaeology and the Origin of Warfare', *World Archaeology* 35:1 (2003), 145–65; Raymond C. Kelly, *Warless Societies and the Origin of War* (Ann Arbor: University of Michigan Press, 2000); Azar Gat, *War in Human Civilization* (Oxford: Oxford University Press, 2006); Lawrence H. Keeley, *War Before Civilization: The Myth of the Peaceful Savage* (Oxford: Oxford University Press, 1996); Slavomil Vencl, 'Stone Age Warfare', in *Ancient Warfare: Archaeological Perspectives*, ed. John Carman and Anthony Harding (Stroud: Sutton Publishing, 1999), 57–73.

4 The Flood

1 James F. O'Connell and Jim Allen, 'Pre-LGM Sahul (Pleistocene Australia–New Guinea) and the Archaeology of Early Modern Humans', in *Rethinking the Human Revolution: New Behavioural and Biological Perspectives on the Origin and Dispersal of Modern Humans*, ed. Paul Mellars, Ofer Bar-Yosef and Katie Boyle (Cambridge: McDonald Institute for Archaeological Research, 2007), 395–410; James F. O'Connell and Jim Allen, 'When Did Humans First Arrive in Greater Australia and Why Is It Important to Know?', *Evolutionary Anthropology* 6:4 (1998), 132–46; James F. O'Connell and Jim Allen, 'Dating the Colonization of Sahul (Pleistocene Australia–New Guinea): A Review of Recent Research', *Journal of Radiological Science* 31:6 (2004), 835–53; Jon M. Erlandson, 'Anatomically Modern Humans, Maritime Voyaging and the Pleistocene Colonization of the Americas', in *The First Americans: The Pleistocene Colonization of the New World*, ed. Nina G. Jablonski (San Francisco: University of California Press, 2002), 59–60, 63–4; Jon M. Erlandson and Torben C. Rick, 'Archaeology Meets Marine Ecology: The Antiquity of Maritime Cultures and Human Impacts on Marine Fisheries and Ecosystems', *Annual Review of Marine Science* 2 (2010), 231–51; Atholl Anderson, 'Slow Boats from China: Issues in the Prehistory of Indo-China Seafaring', *Modern Quaternary Research in Southeast Asia* 16 (2000), 13–50; Robert G. Bednarik, 'Maritime Navigation in the Lower and Middle Paleolithic', *Earth and Planetary Sciences* 328 (1999), 559–60; Robert G. Bednarik, 'Seafaring in the Pleistocene', *Cambridge Archaeological Journal* 13:1 (2003), 41–66.

2 Timothy F. Flannery, *The Future Eaters: An Ecological History of the Australasian Lands and Peoples* (Port Melbourne: Reed Books Australia, 1994); Anthony D. Barnosky et al., 'Assessing the Causes of Late Pleistocene Extinctions on the Continents', *Science* 306:5693 (2004), 70–5; Barry W. Brook and David M. J. S. Bowman, 'The Uncertain Blitzkrieg of Pleistocene Megafauna', *Journal of Biogeography* 31:4 (2004), 517–23; Gifford H. Miller et al., 'Ecosystem Collapse in Pleistocene Australia and a Human Role in Megafaunal Extinction', *Science* 309:5732 (2005), 287–90; Richard G. Roberts et al., 'New Ages for the Last Australian Megafauna: Continent-Wide Extinction about 46,000 Years Ago', *Science* 292:5523 (2001), 1,888–92.

3 Stephen Wroe and Judith Field, 'A Review of Evidence for a Human Role in the

Extinction of Australian Megafauna and an Alternative Explanation', *Quaternary Science Reviews* 25:21–2 (2006), 2,692–703; Barry W. Brook et al., 'Would the Australian Megafauna Have Become Extinct if Humans Had Never Colonised the Continent? Comments on "A Review of the Evidence for a Human Role in the Extinction of Australian Megafauna and an Alternative Explanation" by S. Wroe and J. Field', *Quaternary Science Reviews* 26:3–4 (2007), 560–4; Chris S. M. Turney et al., 'Late-Surviving Megafauna in Tasmania, Australia, Implicate Human Involvement in their Extinction', *Proceedings of the National Academy of Sciences* 105:34 (2008), 12,150–3.

4 John Alroy, 'A Multispecies Overkill Simulation of the End-Pleistocene Megafaunal Mass Extinction', *Science*, 292:5523 (2001), 1,893–6; O'Connell and Allen, 'Pre-LGM Sahul', 400–1.

5 L. H. Keeley, 'Proto-Agricultural Practices Among Hunter-Gatherers: A Cross-Cultural Survey', in *Last Hunters, First Farmers: New Perspectives on the Prehistoric Transition to Agriculture*, ed. T. Douglas Price and Anne Birgitte Gebauer (Santa Fe: School of American Research Press, 1995), 243–72; R. Jones, 'Firestick Farming', *Australian Natural History* 16 (1969), 224–8.

6 David J. Meltzer, *First Peoples in a New World: Colonizing Ice Age America* (Berkeley: University of California Press, 2009).

7 Paul L. Koch and Anthony D. Barnosky, 'Late Quaternary Extinctions: State of the Debate', *Annual Review of Ecology, Evolution, and Systematics* 37 (2006), 215–50; Barnosky et al., 'Assessing the Causes of Late Pleistocene Extinctions on the Continents', 70–5.

5 History's Biggest Fraud

1 The map is based mainly on: Peter Bellwood, *First Farmers: The Origins of Agricultural Societies* (Malden: Blackwell Publishing, 2005).

2 Jared Diamond, *Guns, Germs, and Steel: The Fates of Human Societies* (New York: W. W. Norton, 1997).

3 Gat, *War in Human Civilization*, 130–1; Robert S. Walker and Drew H. Bailey, 'Body Counts in Lowland South American Violence', *Evolution and Human Behavior* 34 (2013), 29–34.

4 Katherine A. Spielmann, 'A Review: Dietary Restriction on Hunter-Gatherer Women and the Implications for Fertility and Infant Mortality', *Human Ecology* 17:3 (1989), 321–45. See also: Bruce Winterhalder and Eric Alder Smith, 'Analyzing Adaptive Strategies: Human Behavioral Ecology at Twenty-Five', *Evolutionary Anthropology* 9:2 (2000), 51–72.

5 Alain Bideau, Bertrand Desjardins and Hector Perez-Brignoli (eds), *Infant and Child Mortality in the Past* (Oxford: Clarendon Press, 1997); Edward Anthony Wrigley et al., *English Population History from Family Reconstitution, 1580–1837* (Cambridge: Cambridge University Press, 1997), 295–6, 303.

6 Manfred Heun et al., 'Site of Einkorn Wheat Domestication Identified by DNA Fingerprints', *Science* 278:5341 (1997), 1,312–14.
7 Charles Patterson, *Eternal Treblinka: Our Treatment of Animals and the Holocaust* (New York: Lantern Books, 2002), 9–10; Peter J. Ucko and G. W. Dimbleby (eds), *The Domestication and Exploitation of Plants and Animals* (London: Duckworth, 1969), 259.
8 Avi Pinkas (ed.), *Farmyard Animals in Israel – Research, Humanism and Activity* (Rishon Le-Ziyyon: The Association for Farmyard Animals, 2009 [Hebrew]), 169–99; 'Milk Production – the Cow' [Hebrew], previously accessed here: The Dairy Council, http://www.milk.org.il/cgi-webaxy/sal/sal.pl?lang=he&ID=645657_milk&act=show&dbid=katavot&dataid=cow.htm.
9 Edward Evan Evans-Pritchard, *The Nuer: A Description of the Modes of Livelihood and Political Institutions of a Nilotic People* (Oxford: Oxford University Press, 1969); E. C. Amoroso and P. A. Jewell, 'The Exploitation of the Milk-Ejection Reflex by Primitive People', in *Man and Cattle: Proceedings of the Symposium on Domestication at the Royal Anthropological Institute, 24–26 May 1960*, ed. A. E. Mourant and F. E. Zeuner (London: The Royal Anthropological Institute, 1963), 129–34.
10 Johannes Nicolaisen, *Ecology and Culture of the Pastoral Tuareg* (Copenhagen: National Museum, 1963), 63.

6 Building Pyramids

1 Angus Maddison, *The World Economy*, vol. 2 (Paris: Development Centre of the Organization of Economic Co-operation and Development, 2006), 636; 'Historical Estimates of World Population', U.S. Census Bureau, previously accessed here: http://www.census.gov/ipc/www/worldhis.html.
2 Robert B. Mark, *The Origins of the Modern World: A Global and Ecological Narrative* (Lanham, MD: Rowman & Littlefield Publishers, 2002), 24.
3 Raymond Westbrook, 'Old Babylonian Period', in *A History of Ancient Near Eastern Law*, vol. 1, ed. Raymond Westbrook (Leiden: Brill, 2003), 361–430; Martha T. Roth, *Law Collections from Mesopotamia and Asia Minor*, 2nd edn (Atlanta: Scholars Press, 1997), 71–142; M. E. J. Richardson, *Hammurabi's Laws: Text, Translation and Glossary* (London: T & T Clark International, 2000).
4 Roth, *Law Collections from Mesopotamia*, 76.
5 Ibid., 121.
6 Ibid., 122–3.
7 Ibid., 133–4.
8 Constance Brittaine Bouchard, *Strong of Body, Brave and Noble: Chivalry and Society in Medieval France* (New York: Cornell University Press, 1998), 99; Mary Martin McLaughlin, 'Survivors and Surrogates: Children and Parents from the Ninth to Thirteenth Centuries', in *Medieval Families: Perspectives on Marriage,*

Household and Children, ed. Carol Neel (Toronto: University of Toronto Press, 2004), 81*n*.; Lise E. Hull, *Britain's Medieval Castles* (Westport: Praeger, 2006), 144.

7 Memory Overload

1 Andrew Robinson, *The Story of Writing* (New York: Thames & Hudson, 1995), 63; Hans J. Nissen, Peter Damerow and Robert K. Englund, *Archaic Bookkeeping: Writing and Techniques of Economic Administration in the Ancient Near East* (Chicago: The University of Chicago Press, 1993), 36.
2 Marcia and Robert Ascher, *Mathematics of the Incas – Code of the Quipu* (New York: Dover Publications, 1981).
3 Gary Urton, *Signs of the Inka Khipu* (Austin: University of Texas Press, 2003); Galen Brokaw, *A History of the Khipu* (Cambridge: Cambridge University Press, 2010).
4 Stephen D. Houston (ed.), *The First Writing: Script Invention as History and Process* (Cambridge: Cambridge University Press, 2004), 222.

8 There Is No Justice in History

1 Sheldon Pollock, 'Axialism and Empire', in *Axial Civilizations and World History*, ed. Johann P. Arnason, S. N. Eisenstadt and Björn Wittrock (Leiden: Brill, 2005), 397–451.
2 Harold M. Tanner, *China: A History* (Indianapolis: Hackett Pub. Co., 2009), 34.
3 Ramesh Chandra, *Identity and Genesis of Caste System in India* (Delhi: Kalpaz Publications, 2005); Michael Bamshad et al., 'Genetic Evidence on the Origins of Indian Caste Population', *Genome Research* 11 (2001), 904–1,004; Susan Bayly, *Caste, Society and Politics in India from the Eighteenth Century to the Modern Age* (Cambridge: Cambridge University Press, 1999).
4 Houston, *First Writing*, 196.
5 The Secretary General, United Nations, *Report of the Secretary General on the In-depth Study on All Forms of Violence Against Women,* delivered to the General Assembly, UN Doc. A/16/122/Add.1 (6 July 2006), 89.
6 Sue Blundell, *Women in Ancient Greece* (Cambridge, Mass.: Harvard University Press, 1995), 113–29, 132–3.

10 The Scent of Money

1 Francisco López de Gómara, *Historia de la Conquista de Mexico*, vol. 1, ed. D. Joaquin Ramirez Cabañes (Mexico City: Editorial Pedro Robredo, 1943), 106.

2 Andrew M. Watson, 'Back to Gold – and Silver', *Economic History Review* 20:1 (1967), 11–12; Jasim Alubudi, *Repertorio Bibliográfico del Islam* (Madrid: Vision Libros, 2003), 194.
3 Watson, 'Back to Gold – and Silver', 17–18.
4 David Graeber, *Debt: The First 5,000 Years* (Brooklyn, NY: Melville House, 2011).
5 Glyn Davies, *A History of Money: From Ancient Times to the Present Day* (Cardiff: University of Wales Press, 1994), 15.
6 Szymon Laks, *Music of Another World*, trans. Chester A. Kisiel (Evanston, Ill.: Northwestern University Press, 1989), 88–9. The Auschwitz 'market' was restricted to certain classes of prisoners and conditions changed dramatically across time.
7 See also Niall Ferguson, *The Ascent of Money* (New York: The Penguin Press, 2008), 4.
8 For information on barley money I have relied on an unpublished PhD thesis: Refael Benvenisti, 'Economic Institutions of Ancient Assyrian Trade in the Twentieth to Eighteenth Centuries BC' (Hebrew University of Jerusalem, unpublished PhD thesis, 2011). See also Norman Yoffee, 'The Economy of Ancient Western Asia', in *Civilizations of the Ancient Near East*, vol. 1, ed. J. M. Sasson (New York: C. Scribner's Sons, 1995), 1,387–99; R. K. Englund, 'Proto-Cuneiform Account-Books and Journals', in *Creating Economic Order: Record-Keeping, Standardization and the Development of Accounting in the Ancient Near East*, ed. Michael Hudson and Cornelia Wunsch (Bethesda, Md.: CDL Press, 2004), 21–46; Marvin A. Powell, 'A Contribution to the History of Money in Mesopotamia Prior to the Invention of Coinage', in *Festschrift Lubor Matouš*, ed. B. Hruška and G. Komoróczy (Budapest: Eötvös Loránd Tudományegyetem, 1978), 211–43; Marvin A. Powell, 'Money in Mesopotamia', *Journal of the Economic and Social History of the Orient* 39:3 (1996), 224–42; John F. Robertson, 'The Social and Economic Organization of Ancient Mesopotamian Temples', in *Civilizations of the Ancient Near East*, vol. 1, ed. Sasson, 443–500; M. Silver, 'Modern Ancients', in *Commerce and Monetary Systems in the Ancient World: Means of Transmission and Cultural Interaction*, ed. R. Rollinger and U. Christoph (Stuttgart: Steiner, 2004), 65–87; Daniel C. Snell, 'Methods of Exchange and Coinage in Ancient Western Asia', in *Civilizations of the Ancient Near East*, vol. 1, ed. Sasson, 1,487–97.

11 Imperial Visions

1 Nahum Megged, *The Aztecs* (Tel Aviv: Dvir, 1999 [Hebrew]), 103.
2 Tacitus, *Agricola*, ch. 30 (Cambridge, Mass.: Harvard University Press, 1958), 220–1.
3 A. Fienup-Riordan, *The Nelson Island Eskimo: Social Structure and Ritual Distribution* (Anchorage: Alaska Pacific University Press, 1983), 10.
4 Yuri Pines, 'Nation-State, Globalization and a United Empire: The Chinese Experience (fifth to third centuries BCE)', *Historia* 15 (1995)[Hebrew], 54.
5 Alexander Yakobson, 'Us and Them: Empire, Memory and Identity in Claudius'

Speech on Bringing Gauls into the Roman Senate', in *On Memory: An Interdisciplinary Approach*, ed. Doron Mendels (Oxford: Peter Land, 2007), 23–4.

12 The Law of Religion

1 W. H. C. Frend, *Martyrdom and Persecution in the Early Church* (Cambridge: James Clarke & Co., 2008), 536–7.
2 Robert Jean Knecht, *The Rise and Fall of Renaissance France, 1483–1610* (London: Fontana Press, 1996), 424.
3 Marie Harm and Hermann Wiehle, *Lebenskunde für Mittelschulen – Fünfter Teil. Klasse 5 für Jungen* (Halle: Hermann Schroedel Verlag, 1942), 152–7.

13 The Secret of Success

1 Susan Blackmore, *The Meme Machine* (Oxford: Oxford University Press, 1999).

14 The Discovery of Ignorance

1 David Christian, *Maps of Time: An Introduction to Big History* (Berkeley: University of California Press, 2004), 344–5; Maddison, *The World Economy*, vol. 2, 636; 'Historical Estimates of World Population', U.S. Census Bureau, previously accessed here: http://www.census.gov/ipc/www/worldhis.html.
2 Maddison, *The World Economy*, vol. 1, 261.
3 'Gross Domestic Product 2009', the World Bank, Data and Statistics, http://siteresources.worldbank.org/DATASTATISTICS/Resources/GDP.pdf.
4 Christian, *Maps of Time*, 141.
5 The largest contemporary cargo ship can carry about 100,000 tons. In 1470 all the world's fleets could together carry no more than 320,000 tons. By 1570 total global tonnage was up to 730,000 tons (Maddison, *The World Economy*, vol. 1, 97).
6 The world's largest bank – the Royal Bank of Scotland – has reported in 2007 deposits worth $1.3 trillion. That's five times the annual global production in 1500. See 'Annual Report and Accounts 2008', the Royal Bank of Scotland, 35, http://files.shareholder.com/downloads/RBS/626570033x0x278481/eb7a003a-5c9b-41ef-bad3-81fb98a6c823/RBS_GRA_2008_09_03_09.pdf.
7 Ferguson, *Ascent of Money*, 185–98.
8 Maddison, *The World Economy*, vol. 1, 31; Wrigley et al., *English Population History*, 295; Christian, *Maps of Time*, 450, 452; 'World Health Statistic Report 2009', 35–45, World Health Organization, http://www.who.int/whosis/whostat/EN_WHS09_Full.pdf.

9 Wrigley et al., *English Population History*, 296.
10 'England, Interim Life Tables, 1980–82 to 2007–09', Office for National Statistics, http://www.ons.gov.uk/ons/publications/re-reference-tables.html?edition=tcm%3A77-61850.
11 Michael Prestwich, *Edward I* (Berkeley: University of California Press, 1988), 125–6.
12 Jennie B. Dorman et al., 'The *age-1* and *daf-2* Genes Function in a Common Pathway to Control the Lifespan of *Caenorhabditis elegans*', *Genetics* 141:4 (1995), 1,399–406; Koen Houthoofd et al., 'Life Extension via Dietary Restriction is Independent of the Ins/IGF-1 Signalling Pathway in *Caenorhabditis elegans*', *Experimental Gerontology* 38:9 (2003), 947–54.
13 Shawn M. Douglas, Ido Bachelet and George M. Church, 'A Logic-Gated Nanorobot for Targeted Transport of Molecular Payloads', *Science* 335:6070 (2012), 831–4; Dan Peer et al., 'Nanocarriers as an Emerging Platform for Cancer Therapy', *Nature Nanotechnology* 2 (2007), 751–60; Dan Peer et al., 'Systemic Leukocyte-Directed siRNA Delivery Revealing Cyclin D1 as an Anti-Inflammatory Target', *Science* 319:5863 (2008), 627–30.

15 The Marriage of Science and Empire

1 Stephen R. Bown, *Scurvy: How a Surgeon, a Mariner and a Gentleman Solved the Greatest Medical Mystery of the Age of Sail* (New York: Thomas Dunne Books, St. Martin's Press, 2004); Kenneth John Carpenter, *The History of Scurvy and Vitamin C* (Cambridge: Cambridge University Press, 1986).
2 James Cook, *The Explorations of Captain James Cook in the Pacific, as Told by Selections of his Own Journals 1768–1779*, ed. Archibald Grenfell Price (New York: Dover Publications, 1971), 16–17; Gananath Obeyesekere, *The Apotheosis of Captain Cook: European Mythmaking in the Pacific* (Princeton: Princeton University Press, 1992), 5; J. C. Beaglehole (ed.), *The Journals of Captain James Cook on His Voyages of Discovery*, vol. 1 (Cambridge: Cambridge University Press, 1968), 588.
3 Mark, *Origins of the Modern World*, 81.
4 Christian, *Maps of Time*, 436.
5 John Darwin, *After Tamerlane: The Global History of Empire Since 1405* (London: Allen Lane, 2007), 239.
6 Soli Shahvar, 'Railroads i. First Railroad Built and Operated in Persia', in the Online Edition of *Encyclopaedia Iranica*, last modified 7 April 2008, http://www.iranicaonline.org/articles/railroads-i; Charles Issawi, 'The Iranian Economy 1925–1975: Fifty Years of Economic Development', in *Iran Under the Pahlavis*, ed. George Lenczowski (Stanford: Hoover Institution Press, 1978), 156.
7 Mark, *Origins of the Modern World*, 46.
8 Kirkpatrick Sale, *Christopher Columbus and the Conquest of Paradise* (London: Tauris Parke Paperbacks, 2006), 7–13.

9 Edward M. Spiers, *The Army and Society: 1815–1914* (London: Longman, 1980), 121; Robin Moore, 'Imperial India, 1858–1914', in *The Oxford History of the British Empire: The Nineteenth Century*, vol. 3, ed. Andrew Porter (New York: Oxford University Press, 1999), 442.
10 Vinita Damodaran, 'Famine in Bengal: A Comparison of the 1770 Famine in Bengal and the 1897 Famine in Chotanagpur', *The Medieval History Journal* 10:1–2 (2007), 151.

16 The Capitalist Creed

1 Maddison, *The World Economy*, vol. 1, 261, 264; 'Gross National Income per Capita 2009, Atlas Method and PPP', the World Bank, previously accessed here: http://siteresources.worldbank.org/DATASTATISTICS/Resources/GNIPC.pdf.
2 The mathematics of my bakery example are not as accurate as they could be. Since banks are allowed to loan $10 for every dollar they keep in their possession, of every million dollars deposited in the bank, the bank can loan out to entrepreneurs only about $909,000 while keeping $91,000 in its vaults. But to make life easier for the readers I preferred to work with round numbers. Besides, banks do not always follow the rules.
3 Carl Trocki, *Opium, Empire and the Global Political Economy* (New York: Routledge, 1999), 91.
4 Georges Nzongola-Ntalaja, *The Congo from Leopold to Kabila: A People's History* (London: Zed Books, 2002), 22.

17 The Wheels of Industry

1 Mark, *Origins of the Modern World*, 109.
2 Nathan S. Lewis and Daniel G. Nocera, 'Powering the Planet: Chemical Challenges in Solar Energy Utilization', *Proceedings of the National Academy of Sciences* 103:43 (2006), 15,731.
3 Kazuhisa Miyamoto (ed.), 'Renewable Biological Systems for Alternative Sustainable Energy Production', *FAO Agricultural Services Bulletin* 128 (Osaka: Osaka University, 1997), Chapter 2.1.1, http://www.fao.org/docrep/W7241E/w7241e06.htm#2.1.1percent20solarpercent20energy; James Barber, 'Biological Solar Energy', *Philosophical Transactions of the Royal Society A* 365:1853 (2007), 1,007.
4 'International Energy Outlook 2010', U.S. Energy Information Administration, 9, http://www.eia.doe.gov/oiaf/ieo/pdf/0484(2010).pdf.
5 S. Venetsky, '"Silver" from Clay', *Metallurgist* 13:7 (1969), 451; Fred Aftalion, *A History of the International Chemical Industry* (Philadelphia: University of Pennsylvania

Press, 1991), 64; A. J. Downs, *Chemistry of Aluminium, Gallium, Indium and Thallium* (Glasgow: Blackie Academic & Professional, 1993), 15.

6 Jan Willem Erisman et al., 'How a Century of Ammonia Synthesis Changed the World', *Nature Geoscience* 1 (2008), 637.

7 G. J. Benson and B. E. Rollin (eds), *The Well-Being of Farm Animals: Challenges and Solutions* (Ames, IA: Blackwell, 2004); M. C. Appleby, J. A. Mench and B. O. Hughes, *Poultry Behaviour and Welfare* (Wallingford: CABI Publishing, 2004); J. Webster, *Animal Welfare: Limping Towards Eden* (Oxford: Blackwell Publishing, 2005); C. Druce and P. Lymbery, *Outlawed in Europe: How America Is Falling Behind Europe in Farm Animal Welfare* (New York: Archimedean Press, 2002).

8 Harry Harlow and Robert Zimmermann, 'Affectional Responses in the Infant Monkey', *Science* 130:3373 (1959), 421–32; Harry Harlow, 'The Nature of Love', *American Psychologist* 13 (1958), 673–85; Laurens D. Young et al., 'Early Stress and Later Response to Separation in Rhesus Monkeys', *American Journal of Psychiatry* 130:4 (1973), 400–5; K. D. Broad, J. P. Curley and E. B. Keverne, 'Mother–Infant Bonding and the Evolution of Mammalian Social Relationships', *Philosophical Transactions of the Royal Society B* 361:1476 (2006), 2,199–214; Florent Pittet et al., 'Effects of Maternal Experience on Fearfulness and Maternal Behaviour in a Precocial Bird', *Animal Behaviour* 85:4 (April 2013), 797–805.

9 'National Institute of Food and Agriculture', United States Department of Agriculture, previously accessed here: http://www.csrees.usda.gov/qlinks/extension.html.

18 A Permanent Revolution

1 Vaclav Smil, *The Earth's Biosphere: Evolution, Dynamics and Change* (Cambridge, Mass.: MIT Press, 2002); Sarah Catherine Walpole et al., 'The Weight of Nations: An Estimation of Adult Human Biomass', *BMC Public Health* 12:439 (2012), http://www.biomedcentral.com/1471-2458/12/439.

2 William T. Jackman, *The Development of Transportation in Modern England* (London: Frank Cass & Co., 1966), 324–7; H. J. Dyos and D. H. Aldcroft, *British Transport – An Economic Survey from the Seventeenth Century to the Twentieth* (Leicester: Leicester University Press, 1969), 124–31; Wolfgang Schivelbusch, *The Railway Journey: The Industrialization of Time and Space in the 19th Century* (Berkeley: University of California Press, 1986).

3 For a detailed discussion of the unprecedented peacefulness of the last few decades, see in particular Steven Pinker, *The Better Angels of Our Nature: Why Violence Has Declined* (New York: Viking, 2011); Joshua S. Goldstein, *Winning the War on War: The Decline of Armed Conflict Worldwide* (New York: Dutton, 2011); Gat, *War in Human Civilization.*

4 'World Report on Violence and Health: Summary, Geneva 2002', World Health Organization, http://www.who.int/whr/2001/en/whr01_annex_en.pdf. For mortality rates in previous eras see: Lawrence H. Keeley, *War Before Civilization: The Myth of the Peaceful Savage* (New York: Oxford University Press, 1996).

5 'World Health Report, 2004', World Health Organization, 124, http://www.who.int/whr/2004/en/report04_en.pdf.

6 Raymond C. Kelly, *Warless Societies and the Origin of War* (Ann Arbor: University of Michigan Press, 2000), 21. See also Gat, *War in Human Civilization*, 129–31; Keeley, *War Before Civilization*.

7 Manuel Eisner, 'Modernization, Self-Control and Lethal Violence', *British Journal of Criminology* 41:4 (2001), 618–38; Manuel Eisner, 'Long-Term Historical Trends in Violent Crime', *Crime and Justice: A Review of Research* 30 (2003), 83–142; 'World Report on Violence and Health: Summary, Geneva 2002', World Health Organization, http://www.who.int/whr/2001/en/whr01_annex_en.pdf; 'World Health Report, 2004', World Health Organization, 124, http://www.who.int/whr/2004/en/report04_en.pdf.

8 Walker and Bailey, 'Body Counts in Lowland South American Violence', 30.

19 And They Lived Happily Ever After

1 For both the psychology and biochemistry of happiness, the following are good starting points: Jonathan Haidt, *The Happiness Hypothesis: Finding Modern Truth in Ancient Wisdom* (New York: Basic Books, 2006); R. Wright, *The Moral Animal: Evolutionary Psychology and Everyday Life* (New York: Vintage Books, 1994); M. Csikszentmihalyi, 'If We Are So Rich, Why Aren't We Happy?', *American Psychologist* 54:10 (1999), 821–7; F. A. Huppert, N. Baylis and B. Keverne (eds), *The Science of Well-Being* (Oxford: Oxford University Press, 2005); Michael Argyle, *The Psychology of Happiness*, 2nd edn (New York: Routledge, 2001); Ed Diener (ed.), *Assessing Well-Being: The Collected Works of Ed Diener* (New York: Springer, 2009); Michael Eid and Randy J. Larsen (eds), *The Science of Subjective Well-Being* (New York: Guilford Press, 2008); Richard A. Easterlin (ed.), *Happiness in Economics* (Cheltenham: Edward Elgar Publishing, 2002); Richard Layard, *Happiness: Lessons from a New Science* (New York: Penguin, 2005).

2 Daniel Kahneman, *Thinking, Fast and Slow* (New York: Farrar, Straus and Giroux, 2011); Ronald F. Inglehart et al., 'Development, Freedom and Rising Happiness: A Global Perspective (1981–2007)', *Perspectives on Psychological Science* 3:4, 278–81.

3 D. M. McMahon, *The Pursuit of Happiness: A History from the Greeks to the Present* (London: Allen Lane, 2006).

20 The End of *Homo Sapiens*

1 Keith T. Paige et al., 'De Novo Cartilage Generation Using Calcium Alginate–Chondrocyte Constructs', *Plastic and Reconstructive Surgery* 97:1 (1996), 168–78.
2 David Biello, 'Bacteria Transformed into Biofuel Refineries', *Scientific American*, 27 January 2010, http://www.scientificamerican.com/article.cfm?id=bacteria-transformed-into-biofuel-refineries.
3 Gary Walsh, 'Therapeutic Insulins and Their Large-Scale Manufacture', *Applied Microbiology and Biotechnology* 67:2 (2005), 151–9.
4 James G. Wallis et al., 'Expression of a Synthetic Antifreeze Protein in Potato Reduces Electrolyte Release at Freezing Temperatures', *Plant Molecular Biology* 35:3 (1997), 323–30.
5 Robert J. Wall et al., 'Genetically Enhanced Cows Resist Intramammary *Staphylococcus aureus* Infection', *Nature Biotechnology* 23:4 (2005), 445–51.
6 Liangxue Lai et al., 'Generation of Cloned Transgenic Pigs Rich in Omega-3 Fatty Acids', *Nature Biotechnology* 24:4 (2006), 435–6.
7 Ya-Ping Tang et al., 'Genetic Enhancement of Learning and Memory in Mice', *Nature* 401 (1999), 63–9.
8 Zoe R. Donaldson and Larry J. Young, 'Oxytocin, Vasopressin and the Neurogenetics of Sociality', *Science* 322:5903 (2008), 900–4; Zoe R. Donaldson, 'Production of Germline Transgenic Prairie Voles (*Microtus ochrogaster*) Using Lentiviral Vectors', *Biology of Reproduction* 81:6 (2009), 1,189–95.
9 Terri Pous, 'Siberian Discovery Could Bring Scientists Closer to Cloning Woolly Mammoth', *Time*, 17 September 2012; Pasqualino Loi et al., 'Biological Time Machines: A Realistic Approach for Cloning an Extinct Mammal', *Endangered Species Research* 14 (2011), 227–33; Leon Huynen, Craig D. Millar and David M. Lambert, 'Resurrecting Ancient Animal Genomes: The Extinct Moa and More', *Bioessays* 34 (2012), 661–9.
10 Nicholas Wade, 'Scientists in Germany Draft Neanderthal Genome', *New York Times*, 12 February 2009, http://www.nytimes.com/2009/02/13/science/13neanderthal.html?_r=2&ref=science; Zack Zorich, 'Should We Clone Neanderthals?', *Archaeology* 63:2 (2009), http://www.archaeology.org/1003/etc/neanderthals.html.
11 Robert H. Waterston et al., 'Initial Sequencing and Comparative Analysis of the Mouse Genome', *Nature* 420:6915 (2002), 520.
12 'Hybrid Insect Micro Electromechanical Systems (HI-MEMS)', Microsystems Technology Office, DARPA, previously accessed here: http://www.darpa.mil/Our_Work/MTO/Programs/Hybrid_Insect_Micro_Electromechanical_Systems_percent28HI-MEMSpercent29.aspx. See also: Sally Adee, 'Nuclear-Powered Transponder for Cyborg Insect', *IEEE Spectrum*, December 2009, http://spectrum.ieee.org/semiconductors/devices/nuclearpowered-transponder-for-cyborg-insect; Jessica Marshall, 'The Fly Who

Bugged Me', *New Scientist* 197:2646 (2008), 40–3; Emily Singer, 'Send in the Rescue Rats', *New Scientist* 183:2466 (2004), 21–2.

13 Bill Christensen, 'Military Plans Cyborg Sharks', *Live Science*, 7 March 2006, http://www.livescience.com/technology/060307_shark_implant.html; Susan Brown, 'Stealth Sharks to Patrol the High Seas', *New Scientist* 189:2541 (2006), 30–1.

14 'Cochlear Implants', National Institute on Deafness and Other Communication Disorders, http://www.nidcd.nih.gov/health/hearing/pages/coch.aspx.

15 Retina Implant, http://www.retina-implant.de/en/doctors/technology/default.aspx.

16 David Brown, 'For 1st Woman with Bionic Arm, a New Life Is Within Reach', *Washington Post*, 14 September 2006, http://www.washingtonpost.com/wp-dyn/content/article/2006/09/13/AR2006091302271.html.

17 Miguel Nicolelis, *Beyond Boundaries: The New Neuroscience of Connecting Brains and Machines – and How It Will Change Our Lives* (New York: Times Books, 2011).

18 Chris Berdik, 'Turning Thought into Words', *BU Today*, 15 October 2008, http://www.bu.edu/today/2008/turning-thoughts-into-words/.

19 Jonathan Fildes, 'Artificial Brain "10 Years Away"', *BBC News*, 22 July 2009, http://news.bbc.co.uk/2/hi/8164060.stm.

20 Radoje Drmanac et al., 'Human Genome Sequencing Using Unchained Base Reads on Self-Assembling DNA Nanoarrays', *Science* 327:5961 (2010), 78–81; 'Complete Genomics' website: http://www.completegenomics.com/; Rob Waters, 'Complete Genomics Gets Gene Sequencing under $5,000 (Update 1)', *Bloomberg*, 5 November 2009, http://www.bloomberg.com/apps/news?pid=newsarchive&sid=aWutnyE4SoWw; Fergus Walsh, 'Era of Personalised Medicine Awaits', *BBC News*, last updated 8 April 2009, http://news.bbc.co.uk/2/hi/health/7954968.stm; Leena Rao, 'PayPal Co-Founder and Founders Fund Partner Joins DNA Sequencing Firm Halcyon Molecular', *TechCrunch*, 24 September 2009, http://techcrunch.com/2009/09/24/paypal-co-founder-and-founders-fund-partner-joins-dna-sequencing-firm-halcyon-molecular/.

Acknowledgements

For their advice and assistance, thanks to: Sarai Aharoni, Dorit Aharonov, Amos Avisar, Tzafrir Barzilai, Noah Beninga, Suzanne Dean, Caspian Dennis, Tirza Eisenberg, Amir Fink, Sara Holloway, Benjamin Z. Kedar, Yossi Maurey, Eyal Miller, David Milner, John Purcell, Simon Rhodes, Shmuel Rosner, Rami Rotholz, Michal Shavit, Michael Shenkar, Idan Sherer, Ellie Steel, Ofer Steinitz, Haim Watzman, Guy Zaslavsky and all the teachers and students in the World History programme of the Hebrew University of Jerusalem.

Special thanks to Jared Diamond, who taught me to see the big picture; to Diego Olstein, who inspired me to write a story; and to Itzik Yahav and Deborah Harris, who helped spread the story around.

Image Credits

1. © ImageBank/Getty Images Israel.
2. © Visual/Corbis.
3. © Anthropologisches Institut und Museum, Universität Zürich.
4. Photo: Thomas Stephan © Ulmer Museum.
5. © magiccarpics.co.uk.
6. © Andreas Solaro/AFP/Getty Images.
7. Photo: The Upper Galilee Museum of Prehistory.
8. © Visual/Corbis.
9. © Visual/Corbis.
10. Poster: Waterhouse Hawkins, *c.*1862 © The Trustees of the Natural History Museum.
11. © Visual/Corbis.
12. Photo: Karl G. Heider © President and Fellows of Harvard College, Peabody Museum of Archaeology and Ethnology, PM# 2006.17.1.89.2 (digital file# 98770053).
13. Photos and © Deutsches Archäologisches Institut.
14. © Visual/Corbis.
15. Photo and © Anonymous for Animal Rights (Israel).
16. © De Agostini Picture Library/G. Dagli Orti/The Bridgeman Art Library.
17. Engraving: William J. Stone, 1823 © The Art Archive/National Archives Washington DC (ref: AA399024).
18. © Adam Jones/Corbis.
19. © The Schøyen Collection, Oslo and London, MS 1717. http://www.schoyen collection.com/.
20. Manuscript: History of the Inca Kingdom, Nueva Coronica y buen Gobierno, *c.*1587, illustrations by Guaman Poma de Ayala, Peru © The Art Archive/ Archaeological Museum Lima/Gianni Dagli Orti (ref: AA365957).
21. Photo: Guy Tillim/Africa Media Online, 1989 © africanpictures/akg.
22. © Réunion des musées nationaux/Gérard Blot.
23. © Visual/Corbis.
24. © Visual/Corbis.
25. © Universal History Archive/UIG/The Bridgeman Art Library.
26. Illustration based on: Joe Cribb (ed.), *Money: From Cowrie Shells to Credit Cards* (London: Published for the Trustees of the British Museum by British Museum Publications, 1986), 27.
27. © akg/Bible Land Pictures.

28. © Stuart Black/Robert Harding World Imagery/Getty Images.
29. © The Art Archive/Gianni Dagli Orti (ref: AA423796).
30. Library of Congress, Bildarchiv Preussischer Kulturbesitz, United States Holocaust Memorial Museum © courtesy of Roland Klemig.
31. Photo: Boaz Neumann. From *Kladderadatsch* 49 (1933), 7.
32. © Visual/Corbis.
33. © Ria Novosti/Science Photo Library.
34. Painting: *Franklin's Experiment, June 1752*, published by Currier & Ives © Museum of the City of New York/Corbis.
35. Portrait: C. A. Woolley, 1866, National Library of Australia (ref: an23378504).
36. © British Library Board (shelfmark add. 11267).
37. © Firenze, Biblioteca Medicea Laurenziana, Ms. Laur. Med. Palat. 249 (mappa Salviati).
38. Illustration © Neil Gower.
39. Redraft of the Castello Plan, John Wolcott Adams, 1916 © Collection of the New-York Historical Society/The Bridgeman Art Library.
40. © National Maritime Museum, Greenwich, London.
41. Photo and © Anonymous for Animal Rights (Israel).
42. © Photo Researchers/Visualphotos.com.
43. © Chaplin/United Artists/The Kobal Collection/Max Munn Autrey.
44. Lithograph from a photo by Fishbourne & Gow, San Francisco, 1850s © Corbis.
45. © Proehl Studios/Corbis.
46. © Khaled El Fiqi/epa/Corbis.
47. Photo and © Charles Vacanti.
48. © ImageBank/Getty Images Israel.

Index

Page numbers in *italics* indicate images.

Abbasid caliphate 222, 407
Aboriginal Australians 17, 27, 50, 66, 260, 310, 313, 336, 423, 424
Achaemenid Persian Empire 247
Aché people 59–60
Aemilianus, Scipio 210–11, 293
Afghanistan 187, 292, 351, 410, 414, 415
Africa ix, 4, 6, 8, 15–21, *16*, 22, 23, 49, 54, 71, 72, 75, 77, 79, 80, 87, 88, 110, 125, *151*, 156–7, 175, 186, 193, 194, 198, 217, 223, 225, 226, 227, 234, 238, 243, 247, 259, 269, 292, 307, 311, 313, 314, 317, 320, 321, 322, 324, 325, 326, 331, 355, 369, 370, 371, 372, 384, 415, 421, 423, 465
Afro-Asian World 70, 71–2, 75, 80, 104, 172, 186, 187, 188, *188*, 189, 193, 205–6, 243, 249, 272, 293, 319
Agricultural Revolution x, 3, 44, 47, 50, 51, 53, 54, 58, 65–6, 67, 80, 82, 85–178, 181, 195, 236, 237, 372, 382, 398, 422–3, 446
Ahura Mazda 247
Akhenaten, Pharaoh 242
Akkadian Empire of Sargon the Great x, 116, 145, 145*n*, 217, *218*
Alabama 158–9, 173
Alamogordo, first atomic bomb detonated at, 1945 *273*, 277, 305
Alaska 77, 78, 88, 217, 219, 331
al-Assad, Hafez 407
Alba (green fluorescent rabbit) 447
Aldrin, Buzz 318
Alexander the Great 126, 164, 175, 219, 323
Algeria 175, 331, 413, 414, 415
'alpha male' 28, 37–8, *39*, 129, 173, 190
Altamira, cave art of 112
Alyattes of Lydia, King 203
Amazon, river 69, 78, 412
America ix, x, 20, 33, 66, 70, 71, 74, 76–9, 80, 87, 88, 90, 110, 117–18, 121–3, 125, 126, 132, 149–50, 152, 154, 156–61, 163, 169, 172, 183, 186, 187, 189, 192, 205, 206, 213, 217, 219, 221, 222, 243, 256, 260, 277, 278, 290, 291–2, 304, 307, 311, 315, 316, 317, 318–22, 323, 325–32, *327*, 340, 352, 354, 363, 369, 370, 426–7, 430 *see also* United States
Anatolia 116, 203
Andean World 186, 187, 219
Angra Mainyu 247
animals:
  biological engineering of 448–51, *449*
  cruelty to 102–9, *106*, *108*, 382–8, *384*, *386*, 392, 425, 448–51, *449*
  domestication of x, 51–2, 58, 87–8, 102–9, *106*, *108*
  extinction of ix, x, 70–83, *81*, 94, 109, 341, 392, 393
  industrial agriculture and 382–8, *384*, 425
animism 60–2, 235–8, 244, 248
Apollo 11 71, 318, 320, 462–3

Arab Empire 146, 171, 194, 205, 217, 222, 224–5, 226, 227, 243, 267, 269, 281, 292, 316, 317, 415
Arab Spring, 2011 268
Arabian peninsula 15, 243
Arabic numerals 146
Arctic 40, 66, 74, 77, 78, 82, 354, 450
Argentina 65, 78, 141, 187, 189, 416
Aristotle 150, 152
Armenians 215, 409
arms race 271
Armstrong, Neil 318, 340, 421
Arthur, King 128, 182–3
artificial intelligence 231–2, 352, 457–8
Aryan race 155, 157, 258, 259, 261, *261*, 337–8
Asia 6–7, 8, 16, 17, 23, 70, 71, 72, 75, 79, 80, 87, 104, 157, 172, 184, 186, 187, 188, 189, 193, 198, 205–6, 217, 233–4, 240, 243, 247, 249, 253, 272, 293, 311, 312, 313, 314, 315, 319, 320, 330, 331, 334, 337, 353, 354, 355, 359, 413, 414
  see also Afro-Asian World
Assyrian Empire 116, 171, 215, 217, 218, 334, 397
Atahualpa 330
Athens, ancient 163, 167, 170, 213, 324, 416
Atman 238, 239
atomic bomb 273, *273*, 277, 291, 292, 378, 416 *see also* nuclear physics; weapons
Augustine, St 216, 440
Augustus, Emperor 175
Aurelius, Emperor Marcus 224
Australia xi, 17, 23, 27, 50, 54, 66, 69, 70, 71–6, 77, 80, 87, 88, 110, 132, 185, 186, 187, 189, 259, 260, 308, 309, 310, 311, 313, 336, 340, 380, 423
Australian World 186
Australopithecus 6
Aztec Empire 62, 172, 187, 193, 213, 239–40, 245, 317, 325–30, *327*, 418

Babylon 118, 121, 129, 130, 421
Babylonian Empire 116, 117, 118–22, 125, 129, 130, 135, 215, 217, 218, 333, 334, 407, 421
Bacon, Francis 288–9
Banks, Joseph 308, 311, 336
barbarians 191, 211, 221, 222, 223, 225, 237, 399
Barí Indians 46
Battuta, Ibn 188
Beagle, HMS 318
bees 24, 27, 135, 190, 447
Behistun Inscription 333
Berbers 224, 225, 226, 227
Bernoulli, Jacob 286
Bible 27, 142, 162, 280, 284, 285, 318, 320, 322
Big Bang 3, 281, 461
bin Laden, Osama 191–2, 292
binary script 148
bio-dictatorships 450
bioengineering 231–2, 453–7, *455*, 459
biofuel 277, 450
biology:
  birth of xi, 3
  biological determinism 164–5
  biological engineering 448–51, *449*, 452
  equality and 122–3
  gender and 163–5, 166, 167, *168*, *169*, 170, 172, 176, 178
  happiness and 425–6, 431–6, 438, 441, 443
  history of 41–4
  race and 150–1, *151*, 152, 155, 157, 161, 163, 259, 261–2, 263, 337–9
bionic arms 454–6, *455*
biotechnology 352
bonobos 37, 46, 63, 177
Brahmins 152, 153, 160, 161

brains 9–10, 11, 14, 15, 21, 22, 23, 33, 45, 55, 89, 134–7, 138, 143, 145, 147, 152, 161, 199, 281, 292, 432, 433, 435, 436, 451, 452, 453, 454, 455–7, 458, 464
British East India Company 229, 320, 363, 364, 370
British Empire 50, 121, 154, 187, 198, 212, 213, 215, 221, 223, 228–31, *229*, 309–10, 314, 315, 332–3, 335–7, 360, 362, 363, 364–6, 370, 413
Buddhism xii, 11, 37, 142, 189, 191, 221, 235, 249–53, *251*, 254, 255, 256, 265, 279, 288, 294, 391, 441, 442, 443
Buka 71
Byron, Lord 365
Byzantines 267, 292

Caesar, Julius 175, 189
Caledonian tribes 216
Calgacus 216
California 307, 342, 417, *419*
Caligula, Emperor 107
capitalism x, 11, 126, 128, 151, 187, 221, 225, 254, 256, 267, 279, 283, 294, 302, 305, 306, 315, 340, 341–73, 374, 388, 390–1, 417, 422, 438, 463, 464 *see also* money
Caribbean Islands 79–80, 325, 326, 329, 349 *see also under individual island name*
Carthage 210, 213, 292, 293, 324
Çatalhöyük, Anatolia 116
Catholic Church 30, 34, 37, 38, 39, *39*, 174, 175, 194, 199, 211, 240, 241, 245, 355
Celts 210, 211, 222, 223, 245, 335, 337
Central America 78, 88, 99, 142, 186, 219, 326
Cervantes, Miguel de: *The Siege of Numantia* 211
Chak Tok Ich'aak of Tikal, King 186
chaotic systems 267–8
Chauvet-Pont-d'Arc Cave, France *1*, 112, 139, 421
chemistry, beginning of ix, 3
Chhatrapati Shivaji train station, Mumbai 229, *229*
child mortality 10–11, 57, 58, 97, 299, 372, 423, 424
childbirth 10–11, 166
childrearing 10–11, 107, 120
chimpanzees ix, 4, 5, 10, 13, 27–8, 29, 35, 37–8, 42, 46, 63, 124, 129, 166, 173, 177, 190, 263, 392, 429, 446, 464
China x, 16, 20, 24*n*, 37, 54, 56, 57, 62, 88, 93, 116, 142, 144, 152, 154, 162, 171, 175, 185, *188*, *198*, 205, 206, 219–20, 221, 222, 225, 226, 227, 231, 233, 247, 249, 259, 266, 272, 292, 293, 312, 313–14, 315, 316, 324, 325, 330, 331, 353, 359, 364, 376, 378, 400, 406, 416, 417, 424
chivalry 182–4
Christianity x, 11, 22, 42, 122, 126, 165, 182, 183, 184–5, 191, 194, 207, 208, 224, 240–1, 242–3, 244, *244*, 245, 247, 248, 255, 257, 263, 264, 265, 266, 269, 270, 272, 279, 280, 296, 297, 310, 322, 368, 370, 391, 418, 440, 463
Church, Professor George 451
Cicero 216
Claudius, Emperor 223–4
Cleopatra of Egypt 171, 430
Code of Hammurabi, 1776 BC 117, *117*, 118–19, 120–2, 124, 126, 135, 143, 149, 150, 203, 407
cognitive dissonance 181–4
Cognitive Revolution ix, 1–69, 70, 80, 82, 190, 279, 398, 421, 452

coinage x, 33, 194, 197–9, *198*, 201, 202–5, *204*, 206, 208, 233, 272, 343, 349, 356, 357, 358, 421
Columbus, Christopher 71, 275, 303, 317, 319–20, 324, *325*, 326, 340, 352–4
Communism 38, 162, 183, 225, 254–5, *255*, 261, 270, 282, 302, 305, 372, 414, 422, 424, 463
communities:
 collapse of 398–405, 428
 imagined 405–8
Confucianism 249, 279, 284, 288, 294, 391
Congo Free State 371, 372
conquest, the mentality of 316–19
Constantine, Emperor 240, 265, 266, 293
consumerism 129–30, 388–91, 406, 408
Cook Islands 82
Cook, Captain James 308, 309–10, 311, 313, 317, 336, 340
cooperation, social 26–8, 29–31, 36–40, 41, 43, 46, 52, 102, 115–18, 121, 124, 131, 132, 134–5, 149, 175, 177, 178, 181, 190, 195, 207–8, 282
Copernicus, Nicolaus 307
corporations 31, 33, 34, 36, 40, 305, 347, 360, 368, 383
Cortés, Hernan 193, 206, 326–7, 328–30
cowry shells 198, *198*, 200, 201, 205, 206, 207, 208
credit 344–6, 347–8, 352–3, 354, 355, 356, 358, 362, 364, 366, 367, 368
Crusades 182
Cuba 79, 80, 329, 330
cultures, human:
 'authentic' 188
 biological laws and 42, 163–70
 birth of 3, 19, 41, 181
 clash of 183–4, 187, 339
 constant flux of 181–2
 contradictions in 182–4
 empires spread a common culture 210–32, 264 *see also* empire
 global culture, emergence of a single 189–92, 231, 312–13
 history and 41, 181–92, *188*, 264, 269–72
 ideal of progress and 294–7
 memetics (cultures as mental infections) 269–72
 universal orders and 189–92 *see also under individual order*
culturism 338–9
cuneiform 142, 333
cyborg engineering 448, 453–7, *455*, 459, 461
Cynics 126, 249
Cyrus the Great of Persia 217, 218, 219, 221, 232

da Gama, Vasco 317
Dani, the 93
Danube Valley 67–8, *67n*
Daoism 249, 256, 293
Darius I, King 333
Darwin, Charles 20, 261, 281, 287, 303, 316, 318, 338, 440, 445–6, 447
David, King 215
Declaration of Independence, US, 1776 20, 117–18, *118*, 121–3, 150
Defense Advanced Research Projects Agency (DARPA) 453–4
demography 53, 77, 99, 100, 287, 312, 341
denarius coin 204, 205
Denisova Cave, Siberia 8
Denisovans 8, 17, 18–19, 20
Department of Defense, US 292
determinism 265–7
Dickens, Charles 183, 409–10
dinar 205
Dinka people 219
Diogenes 126
diprotodon 72, 73, 75, 76, 82, 83

DNA 4, 17, 20, 23–4, 36, 37, 38, 42, 46, 65, 94, 104, 135, 138, 393, 441, 447, 451, 459–60
Dutch West Indies Company (WIC) 360
dwarfing 7

East Africa ix, 4, 6, 8, 15, 16, 22, 54, 87, 324, 331
East Asia 6, 16, 23, 157, 198, 206, 217, 243, 253, 320, 353
Easter Island 82
ecological disasters 76–83, 266, 393
Ecuador 93, 141, 415
Edward I, King 300, 301
Edward II, King 301
Egypt 85, *85*, 106, *106*, 116, 129, 130, *130*, 131, 140, 142, 144, 171, 173, 191, 217, 224, 225, 226, 227, 239–40, 242, 268, 297, 314, 317, 364–5, 421, 430, 431, *431*
Einstein, Albert 24, 43, 283, 378
Eisenhower, Dwight 290
Eleanor, Queen 300, 301
electricity 278, *295*, 378, 379
elephant bird 81
elephants 4, 5, 8, 24, 177, 238, 249, 392, 451
elites 37, 90, 114, 119, 125, 126, 129, 131, 175, 214, 216, 221, 222, 223, 224, 227, 231, 232, 240, 259, 316, 329, 338, 350, 353, 390, 414, 418–19, 460
Elizabeth I, Queen 171
energy 373, 374–82, 389, 392, 393
empire:
  capitalism and 363–73
  common culture, spreads 210–32
  cultural assimilation under 210–32
  cycle, imperial 226–7
  definition 212–13
  'evil' nature of 214–17
  first x, 217–20, *218*
  global 231–2
  language, emergence of and 138–48
  majority of cultures as offspring of 227–8
  positive legacies of 220–31
  religion and 234, 235, 239–41, 243, 245, 247, 265–6, 269–70, 272
  modern, collapse of 412–14
  science and 307–40
  as universal order 189–92
Enga 93
Epicureanism 249
equality 121–6, 127, 149, 150, 183, 225, 258, 260, 315, 339, 459, 460
Euphrates valley 113, 148
Eurasia ix, 7, 15, 74, 313
Europe ix, x, 6, 8, 15, 17, 23, 31, 38, 50, 66, 87, 127, 146, 152, 156, 157, 161, 168, 182, 183, 184, 185, 186, 187, 188–9, 205, 206, 215, 225, 226, 227, 233, 240, 243, 257, 269, 272, 279, 280, 287, 293, 298, 300, 303, 307, 308, 310, 311–15, 316–17, 319, 320, 321, 322, 324, 325, 328, 330, 331, 332, 333, 334, 335, 336, 337, 338, 339, 340, 352, 353, 354, 355, 356, 359, 360, 364, 368, 369, 370, 371, 372, 382, 390, 397, 406, 411, 412, 414, 415, 416, 421, 423, 425, 435, 459
European Union 406, 459
evolution ix, 4, 8, 9, 10, 11, 17, 18, 22–3, 36, 43, 45, 46, 79, 89, 90, 94, 104–5, 108, 109, 115, 122, 123, 134, 136, 165, 173–4, 176, 191, 218, 258, 259, 260, 261, 263, 270, 271, 276, 281, 287, 303, 318, 384–5, 403–4, 423, 432–3, 438, 447, 457, 458 *see also* genetics
evolutionary humanism 258–63
evolutionary psychology 45, 46, 384–5
extinction ix, x, 19, 20, 22, 23, 72, 73–83, *81*, 94, 109, 259, 261, 332, 341, 392–4, 432

family and the local community, collapse of 398–405, *405*, 428
famine 58, 294, 295, 337, 370–1, 424, 425, 465
fictions, evolution of 27–40, 150, 181
see also mythology
Fiji 81
fire, domestication of ix, 13–14, 15, 75
First World War, 1914–18 26, 290, 381–2, 409, 419
fishing villages 54, 71
Flores Island, Indonesia 7–8, 20, 70
food chain, man jumps to top of 12–13, 72–3, 76, 173
France *1*, 33–5, 36, 42, 114, 132, 133, *168*, 174, 175, 183, 213, 215, 217, 225, 226, 241, 245, 301, 314, 315, 316, 331, 332, 335, 338, 345, 354, 356, 357, 360–3, 364, 381, 407, 408, 409–10, 413, 414, 416, 418, 422, 435–6, 447
Frankenstein 461–4
Franklin, Benjamin 295, *295*
free market 126, 256, 366–8, 370, 399, 422
free trade 221, 364
French Empire 175, 215, 362, 413
French Revolution, 1789 42, 114, 183, 363, 408, 410, 435–6
Front national 338–9

Galapagos Islands 82, 280, 318
game theory 271
Gandhi, Mohandas Karamchand 223, 413, 420
Ganges Valley 235
Gauls 221, 226, 227, 323
Gautama, Siddhartha 249–52, 254
gender 120, 161–78, *168*, *169*, 339, 422, 457, 463
genetics 4, 8, 16, 17, 18, *18*, 19, 20, 23–4, 36–40, *39*, 42, 46, 48, 50, 55, 65, 78, 94, 95, 102, 104, 122, 134, 135, 138, 151, 157, 176–8, 259, 263, 266, 270, 271, 287, 298, 301, 305, 316, 352, 393, 425, 432, 435, 440, 441, 446, 447, 448, 449–51, *449*, 452, 453, 459–60, 464
genus ix, 5–6, 8, 9, 12, 20
Germany 25, 38, 163, 214, 215, 221, 261, 286, 290, 291, 314, 335, 338, 356, 381–2, 397, 406, 407, 408, 416, 454
Gilgamesh Project 297–302, 464
global warming 78, 95, 231, 393
Gnosticism 247, 248
Göbekli Tepe 100–1, *102*, 139
gold 33, 193, 194, 201, 203, 204, 205–7, 209, 233, 234, 272, 332, 353, 354, 356, 357, 358, 369, 379, 380, 381, 417, 419, *419*
Gorbachev, Mikhail 414
Great Leap Forward, 1958–61 424
Great Pyramid of Giza 130, *130*
Great Survey of India 332
Greece 62, 126, 142, 163–4, 171, 210, 213, 222, 237, 238, 239, 316, 335, 337, 338, 365, 366, 416
Greek Rebellion, 1821 365–6, *365*
green monkeys 24, 35
Green, Charles 308
Gulf War, 1990–1 415
Gupta Empire 230, 333

Haber, Fritz 381–2
Habsburg Empire 213, 216
Halley, Edmond 286
Ham, son of Noah 157
Han Empire x, 217, 225
happiness 94, 104, 121, 122, 123, 148, 271, 351, 421–44
Harlow, Harry 385–7
Hawaii 70, 82, 186
Henry the Navigator, Prince 317
Hephaestion 164

hierarchy, principle of 28, 29, 37, 59, 61, 62, 63, 120–4, 128, 149–78, 234, 269, 339, 383
hindsight fallacy 265–9
Hindu religion 48, 142, 146, 151–2, 154–5, 156, 160, 169, 229, 230, 238, 239, 253, 256, 305, 335, 456
Hispaniola 79
history:
  biology and 41–4
  birth of ix, 3, 41–4
  direction of 181–92, 265–9
  hindsight fallacy 265–9
  human well-being and 269–72 *see also* happiness
  justice in 149–78
  next stage of 445–6
  prediction of 269–72
  timeline of ix–x
Hitler, Adolf 259, 262, *262*, 263, 409, 419
Hittite Empire 217
Holy Grail 182
home 110–11, 145
Homo denisova 8, 17–19, 20
Homo erectus 6, 7, *7*, 13, 37, 87
Homo ergaster 8, 87
Homo floresiensis x, 8
Homo neanderthalensis see Neanderthals
Homo rudolfensis 6, *6*, 8
Homo sapiens:
  Agricultural Revolution and *see* Agricultural Revolution
  appearance of in Africa ix, 6, 8, 13–21, 23, 26
  becomes a god 465–6
  Cognitive Revolution and *see* Cognitive Revolution
  end of 445–64
  global migrations ix, 5–6, 15–21, *16*, 23, 26, 54, 87
  other human species and 15–21, 22–44
  Scientific Revolution and *see* Scientific Revolution
  unification of humankind and 179–272
  as xenophobic creature 218–19
Homo soloensis 7, 20
Homo: evolution of genus 5, 6 *see also* human
Hong Kong 364
Huitzilopochtli 240, 245
Human Brain Project 458–9
Human Enhancement question 463
human rights 31, 35–6, 41, 123–4, 126, 132, 187, 222, 225, 226, 228, 231, 257, 261, 267, 406, 450, 460
humanism 256–8, 260, 261, 263, 282
humans:
  appearance of 3–4, 5–9
  brains of *see* brain
  common defining characteristics of 9–10
  distinct species of ix, 5–8, *6*, *7* *see also under individual species name*
  fire and cooking, discovery of 13–14
  food chain, jumps to top of 12–13, 72–3, 76, 173
  other apes and ix, 5
  relations between different species of 14–21, 22–44
  spread from Africa to Eurasia ix, 6–7, 15–21, *16*, 23, 26, 54, 87
  superhumans *see* superhumans
  use of tools *see* tools, first stone
  walk upright 10–11
  see also under individual species name
hunter-gatherers/foragers 43, 45–69, 72, 88, 89, 90, 91, 96, 101, 110, 111, 136, 185, 194, 195, 236, 302–3, 421, 422, 423, 424
Hussein, Saddam 407
Huxley, Aldous: *Brave New World* 436, 437

Iberian peninsula 14, 193, 210, 211, 222, 224
Ice Age 7, 73, 95, 185
ignorance, discovery of x, 275–306
Iliad 142, 164
imagined communities 405–8
imagined orders 114–24, 125, 126, 127–33, 149, 191
imagined realities 35, 36, 41, 50, 126–33
Inca Empire 141, *141*, 144, 172, 187, 197, 325–30, *327*, 397
Incitatus 107
India x, 130, 146, 152, 154–6, 157, 161, 163, *188*, 205, 206, 207, 216, 219, 223, 225, 226, 227, 228–30, *229*, *230*, 234, 243, 246, 247, 249, 250, 253, 267, 271, 272, 312, 313, 314, 316, 317, 324, 331, 332–3, 334–8, 353, 359, 363, 370, 381, 413
individualism 127, 128, 258, 260, 263, 280, 402–4, 410, 428, 439, 440
Indonesia 6–7, 14, 54, 71, 188, 231, 320, 324, 325, 359–60, 363, 371, 372, 415
Indus River/Valley 113, 236, 332
Industrial Revolution x, 82, 158, 294, 371, 375, 377, 379, 380, 382, 388, 392, 393, 394, 395–6, 398, 401, 407, 421 *see also* Scientific Revolution
intelligent design x, 445–6, 447–8
inter-subjectivity 131–3, 170, 197, 406
Interbreeding Theory 15, 16, 17
internal combustion engine 352, 378–9
Internet 409, 456, 457, 463
Iran 62, 87, 118, 187, 217, 224, 225, 226, 227, 415, 416
Iraq 118, 205, 217, 407, 410, 415, 417
Isabella of France, Queen 301, 354
Islam x, 191, 224, 226, 227, 233–4, 235, 243, *244*, 254, 255, 267, 270, 279, 297, 316, 331, 406, 421, 463, 464
Israel 53, 67, 68, 187, 242, 389, 416, 419, 433

Jabl Sahaba, Sudan 67–8
Jainism 249
Japan 233, 253, 291, 320, 325, 331, 359, 382, 416, 451, 456
jati (Indian caste groupings) 156
Java, Indonesia 7, 371
Jefferson, Thomas 124, 126
Jericho 94, 96, 97, 116
Jerusalem 143, 215
Jesus of Nazareth 20, 242, 265, 288, 294, 296, 345
Jews 62, 155, 215, 218, 242, 243, 247, 248, 258, 259, 381, 409
'Jim Crow' laws 159
Jones, William 334–5, 337
Jordan 205, 213 407, 415
Judaea 215, 218
Judaism 242, 265

Ka'aba, Mecca *179*, 217, 233
Kac, Eduardo 447, 448
Kahneman, Daniel 437
Kalahari Desert, Africa 49, 56, 66
Karaçadag Hills 101
King, Clennon 160
Kipling, Rudyard 336
Kshatriyas 152
Ku Klux Klan 160
Kublai Khan 233
Kushan Empire 230
'Kushim' 138, *139*
Kuwait 415, 417

language, evolution of x, 21, 23–8, 24*n*, 50, 52, 135–48, 149, 185, 187, 194, 198, 206, 207, 211, 214–15, 217, 219, 221, 223, 224, 233, 280, 284, 285, 288, 313, 318–19, 322, 329, 332, 333–5, 337 *see also under individual language name*

Lascaux Cave 64, *64*, 112
Latin language 5, 6, 33, 34, 60, 91, 140, 142, 185, 211, 222–3, 226, 279, 335, 337
Law of Large Numbers 286–7
Law, John 360, 362
Le Pen, Marine 338
Lebanon 407, 415
legal fiction 32–3
Lenin, Vladimir Ilyich 254, 255, 282, 409
Leopold II of Belgium, King 371
Levant 22, 87, 88, 96, 224
liberal humanism 256–9, 260, 261, 263, 282
liberalism 221, 225, 254, 256, 261, 302, 339, 439, 440, 441
liberty, concept of 121, 122, 123, 125, 149, 150, 183, 257
Libya 36, 224, 415
life expectancy 57, 299, 301, 372, 450, 452
life sciences 263
life insurance 285–7
limited liability companies 32–5, 41, 123, 355, 358–63, 406
Lind, James 309
linguistics 24n, 26, 288, 317, 334–9 *see also* language, evolution of
locked-in syndrome 456
Louis XIV of France, King *168*
Louis XV of France, King 360–1, 362
Louis XVI of France, King 125, 294, 362–3, 435
luxury trap 94–100
Lydia 203, 204, *204*, 205

Macedonian Empire 164, 171, 205, 210, 292
Maclaurin, Colin 285–6
Madagascar 70, 80–1, 307, 325
Magellan, Ferdinand 187, 217, 276, 317
Majapahit, empire of 324
Mali 234
Malthus, Robert 287
mammoths 5, 54, 63, 74, 77–8, 79, 83, 88, 101–2, 423, 451
Mandate of Heaven 219–20, 221, 231
Manhattan Island 356, 361, *361*
Manhattan Project 291, 293
Manichaean creed 247, 248, 264, 265, 269, 284
Manus 71
Maori 74, 225, 310
maps 319–25, *321*, *323*
Mari 143–4
Marquis Islands 82
marriage 47, 156, 158, 160, 178, 213, 224, 286, 427–8, 434
marsupials 72, 75, 76
Marx, Karl 20, 254, 255, 270, 282, 302, 316, 364
masculinity 163, 166, *168*, *169*, 170–8
mathematics 43, 100, 137, 140, 141, 142, 146, 147, 148, 279, 283–8, *284*, 307, 335
Mauryan Empire 221, 230, 333
Mayan Empire 186, 326
meditation 252, 441, 442
Mediterranean x, 38, 82, 115, 116, 205, 206–7, 210, 213, 217, 249, 266, 311, 381
Melanesians 17
memetics 269–71
Menes 190–1
Mesoamerican World 186, 187
Mesopotamia 117, 118–19, 137, 142, 144, 185, 181–2, 202, 217, 224, 233
Mexico 62, 78, 88, 189, 193, 206, 213, 325–30, *327*, 419
microorganisms 277, 278, 446
Middle East ix, 15, 17, 22, 23, 87, 88, 90, 94–5, 96, 99, 116, 146, 157, 162, 215, 247, 268, 311, 333, 334, 339, 369, 407
military-industrial-scientific complex 290, 313, 314

Ming Empire 312, 324, 400–1
Mississippi Bubble 360–3, 368
Mississippi Delta 78
Mitchell, Claudia 455, *455*
Modern Times (movie) *395*
Mohenjo-daro 332
money x, 191–2, 193–209 *see also* capitalism
Mongol Empire 184, 185, 222, 233, 292, 293, 317, 414
monogamy 44, 46, 47, 63, 451, 452
Montezuma II, Emperor 328–9, 332
moon, humans land on the, 20 July 1969 4, 71, 276, 277, 278, 305, 318–19, 320, 340, 421
Mubarak, Hosni 268–9, 430, 431
Mughal Empire 216, 230, 312, 330, 333
Muslims 48, 49, 161, 171, 184, *188*, 193, 194, 205, 206, 207, 219, 221, 224, 225, 226, 227, 230, 234, 247, 248, 264, 269, 270, 313, 314, 315, 316, 332, 335, 338, 339, 353, 406, 415
Mussolini, Benito 290
mythology 27, 30–1, 36, 40, 42, 52, 115, 117, 122, 124, 127, 128, 129, 130, 131, 132, 151–2, 155, 157–8, 159, 161, 164, 166, 167, 176, 178, 181, 215, 236, 282, 284, 294, 297, 314–15, 336, 406, 461, 462

Nader Shah 353, 354
nanotechnology 292, 301, 352
Napoleon Bonaparte 69, 125, 174, 293, 294, 317, 363, 435
Napoleon III of France 381
nationalism 27, 225, 226, 228, 229, 230, 231, 254, 256, 267, 270–1, 302, 363, 364–5, 402, 406, 422, 438
Native Americans 149, 169, 189, 278, 316, 318, 360, 423
Natufian culture 96, 100, 139
natural selection x, 11, 37, 261, 303, 338, 432, 440, 445–8
Navarino, Battle of, 1827 365, *365*, 366
Nazism 38, 254, 258–63, *261*, *262*, 282, 305, 370, 397
Neanderthal Genome Project 451
Neanderthals (Homo neanderthalensis) ix, 6, *7*, 8, 9, 13, 14, 15, 16, 17, 18, *18*, 19, 20, 21, 22, 23, 24, 26, 38–9, 40, 42, 68, 77, 87, 115, 139, 259, 446, 451–3, 461, 462
Netherlands 316, 338, 355–60, 363, 371
New Amsterdam 360, *361*
New Britain 39, 71
New Caledonia 81
New Guinea 39, 88, 92, *92*, 93, 105–6
New Ireland 39, 71
New Testament 38, 296
New Zealand 51, 70, 74, 82, 186, 308, 310
Newton, Isaac 284–5, 287, 316; *The Mathematical Principles of Natural Philosophy* 284–5
Nietzsche, Friedrich 437
Nile Valley 116, 431
non-organic life forms x, 457–9
Nordic gods 238
North Africa 6, 193, 194, 243, 247, 292, 311
North America 66, 74, 79, 88, 90, 121, 186, 189, 307, 363
Nü Wa 152
nuclear family x, 26, 43, 46, 47, 50, 62–3, 398–9, 404, 405
nuclear physics/weapons xii, 26, 43, 187, 271, 273, *273*, 278, 288, 304, 375, 378, 380, 393, 414, 416, 418, 419–20, 424, 448, 462
Nuer tribe 107, 219
Numantia 210–11, 213, 214, 222, 293
Nurhaci 353, 354

Obama, Barack 169, *169*, 430
obsidian 39, 40, 194, 235
Oceanic World 186, 187

Ofnet Cave, Bavaria 68
Old Testament 203, 247, 248
Olympias of Macedon, Queen 164
Opium War, First, 1840–2 364–5
Oppenheimer, Robert 273, 416
organisms, emergence of ix, 3
Orwell, George: *Nineteen Eighty-Four* 436
Ottoman Empire 154, 171, 311–12, 313, 314, 330, 331, 332, 353, 365, 366, 400, 414
'Outer World' 70, 74, 311

Pakistan 246, 271, 415
patriarchy 5, 171–8, 398
Patroclus 164
Paul of Tarsus 242–3, 440
peaceful era, modern times as 410–20
permanent settlements, emergence of x, 54, 58, 98, 110–11
Persian Empire x, 116, 171, 217, 218, *218*, 219, 222, 247, 269, 314, 315, 316, 323, 324–5, 331, 333, 334, 335, 337, 338
personalised medicine 459
Peugeot 28–36, *32*, 117, 132, 133
Peugeot, Armand 33–4
Philip, Emperor 224
Philip of Macedon, King 164
Philippines 317
physics, beginning of ix, 3, 23
Pizarro, Francisco 326, 330
plants:
  domestication of x, 87–8, 90, 91, 92, 94, 96, 109, 111, 113, 236
  genetic engineering of 450
  mechanisation of 382–3
Polynesians 81, 316
population numbers 57, 66, 90, 95, 96, 98, 110, 126, 212, 213, 275, 310, 330, 364, 372, 387, 390, 393–4
Portugal 67, 68, 217, 221, 316, 332, 353, 354
postmodernism 270–1
poverty 148, 161, 239, 249–50, 294, 295, 296–7, 371, 428, 430
progress, the ideal of 294–7
Protestant Church 240–1, 355
Purusa 152

Qin dynasty 116, 117, 400
Qing dynasty 312, 353
Qín Shǐ Huángdì 220, 221
quantum mechanics 23, 147, 281, 285, 287, 288
quipu 140–2, *141*, 143

race 17, 150–1, *151*, 152, 156–61, 163, 219, 259, 261, *261*, 263, 269, 282, 310, 338
railways 229, *229*, 314, 378
raw materials 54, 56, 104, 201, 373, 374–5, 380–2, 383, 407
Rawlinson, Henry 333–4, 335
religion 19–20
  Agricultural Revolution and 100–2, *102*, 236–8
  animism and *see* animism
  birth of 19–20, 23, 25, 27, 30, 48, 61–2, 68, 69
  definition of 235–6
  dualism 245–8
  free will and 246, 263
  happiness and 440–3
  hierarchies and 154–6
  hindsight fallacy and 265–6
  humanist 254–63
  hunter-gatherers and 61–2, 68, 69
  language and emergence of 27, 30, 31, 34
  local and exclusive 235
  monotheistic 238, 240, 241–6, 247–8, 249, 253, 256, 257, 258, 264, 266, 449
  natural-law religions 249–53, 254
  patron saints 245

religion – *cont.*
polytheistic x, 237–41,243, 244, 245, 248, 249, 265
Problem of Evil 245–6
science and 282–3, 284, 297, 302, 303, 304, 305, 317, 351–2, 370, 389–90, 391
syncretism 248, 256
see also mythology *and under individual religion name*
Replacement Theory 16–18
research, funding of scientific 302–6
Roman Empire x, 20, 62, 107, 114–15, 116–17, 171, 172, 186, 204–5, 210, 211, 213, 214, 215, 216, 219, 221, 222, 223–4, 226, 227, 239–40, 241, 243, 244, 247, 265, 266, 272, 292–3, 311, 317, 323, 324, 338, 414
romantic consumerism 129–30
Rousseau, Jacques 439
Royal Navy 171, 309, 318
Royal Society 307–8, 309

sacrifice, human 58, 64
Safavid Empire 312, 330
Salviati World Map, 1525 *323*
Samarkand 233
Samoa 82, 181
Sargon the Great x, 116, 217
Sassanid Persian Empire 247, 269, 292, 397
Scientific Revolution x, 3, 272, 273–464
Scotland 216, 245, 285–7, 330–1
Scottish Widows 285–7
script, partial and full 139–42, *140*, 143
scurvy 308–9
sea levels 7, 71, 77, 185
seafaring societies, first 71
Second World War, 1939–45 259–60, 290, 365, 397, 409, 410, 414
Seleucid Empire 210, 292
Seneca 216
Severus, Emperor Septimius 224
sexual relations 4, 15, 16, 18, 19, 38, 39, 45, 46, 47, 48, 50, 95, 105, 158, 160, 162, 163, 164, 166–70, 178, 199, 252, 258, 404, 432–3, 440, 461
shells, trading in 38–9, 40, 52, 115, 194, 195, 198, *198*, 200–1, 205, 206, 207, 208
Shudras 152, 153, 161
Siberia 8, 69, 74, 77, 78, 219, 307, 380, 451
Silicon Valley 148
silver shekel 120, 135, 202–3
singularity 459–61
skeleton, effect of Agricultural Revolution on human 10, 91
slavery 83, 108, 117, 119, 120, 124, 125, 135, 149, 150, 152, 153, 155, 156–9, 173, 174, 202, 203, 208, 211, 216, 233, 311, 326, 368, 369–70, 371, 372, 379, 384, 404, 411, 417
Smith, Adam 126, 302, 316, 348–9, 368; *The Wealth of Nations* 348–9
social structure 11, 28, 38, 46–7, 68, 87, 106, 135, 161, 315, 408, 444, 451
see also hierarchy, principle of
socialist humanism 258, 260
Solander, Daniel 308
Solomon Islands 81
Solzhenitsyn, Alexander 183
Song Empire 293
South Africa 88, 151, *151*, 217, 223, 243, 307
South America 79, 88, 141–2, 186, 318, 325–30, 355, 415, 419
Soviet Union 197, 219, 221–2, 254–5, 268, 290, 304, 413–14
Spain 141–2, *141*, 187, 189, 193, 211, 217, 221, 225, 275, 276, 316, 317, 319–20, 325–30, 332, 333, 349, 354, 355, 356, 357, 358, 359
species, classification of 4–5

St Bartholomew's Day Massacre, 1572 241
Stadel Cave: lion-man in 23, 25, *25*, 31, 35, 44, 449
statistics 285–8, 299, 410–11
Stoicism 249
Stone Age 13, 44, 46, 48, 62, 75, 422
Sudan 67, 107, 219
Suez Canal 364
Sullivan, Jesse 454, 455, *455*
Sumer/Sumerians 137, 138–42, *139*, 144–5, 145*n*, 148, 202, 297, 334, 344
Sungir, Russia 63–5, 77–8
superhuman order 234, 235, 249, 254, 255
superhumans x, 255, 260, 263, 421, 452, 460
Syria 47, 116, 118, 204, 217, 224–5, 233, 407, 415, 416

Tacitus 216
Tahiti 308
Taj Mahal 216, 230, *230*
Talmud 215
Tasmania 181, 185–6, 187, 310–11, *312*, 424
taxes 32, 116, 135, 137, 138, 140, 141, 142–3, 145, 183, 194, 198, 201, 204, 213, 221, 245, 268, 278, 304, 345, 351, 353, 358, 367, 376, 400–1, 403, 417
Tenochtitlan 328, 329, 330
Teotihuacan 186
Theism 62 *see also* monotheism; polytheism and dualism
Theory of Relativity 147, 255, 285
thought-control 456
Tierra del Fuego 78
time, modern 394–8, *395*
Toltecs 326
Tonga 82
tools, first stone ix, 8, 10, 11, 12, 15, 20, 22, 37, 39, 42–3, 48, 68, 96
tourism 130
trade 39–40, 42, 52, 71, 115, 134, 157, 158, 189, 195, 196, 197, 199, 201, 206, 208, 221, 237, 268, 303, 322, 347, 350, 353, 356, 359, 360, 361, 364, 368, 369–70, 371, 372, 384, 417, 418, 419, 465
Trajan, Emperor 224
Truganini 310–11, *312*, 424
Turkey 87, 100, 217, 233–4, 378, 415

unification of humankind 179–272
United Nations 42, 414
United States 4, 20, 33, 78, 117–18, *118*, 121–2, 123, 125, 126, 132, 149–50, 152, 154, 156–61, 163, 169, 172, 183, 187, 192, 205, 219, 222, 231, 256, 260, 266, 277, 278, 290, 291–2, 307, 312, 314, 315, 318, 331, 369, 370, 387, 389–90, 426–7, 430, 454–5 *see also* America
US Civil War, 1861–5 125, 158
US Constitution 158
universal orders 62, 189–91 *see also* empire, money and religion
University of Mississippi 160

V-2 rocket 290, *291*
Vaishyas 152
Valence, Emperor 186
van Leeuwenhoek, Anton 277
Vereenigde Osstindische Compagnie (VOC) 359–60, 363, 370–1
Verne, Jules 276
Vespucci, Amerigo 320–1
Victoria, Queen 365
Vietnam War, 1956–75 331–2, 413
Voltaire 124

Waldseemüller, Martin 320–1
Wall Street 42, 360, *361*, 417, 458
Wallace, Robert 285–7
war, disappearance of international 410–20

Waterloo, Battle of, 1815 299
weaponry 290–4, *291*, 309
Webster, Alexander 285–7
wheat, Agricultural Revolution and 13, 57, 87, 88, 89, 90–2, 93–4, 95–8, 101–2, 109, 143–4, 196, 200, 204, 350, 376
women:
  Agricultural Revolution and 96–7
  hierarchies and 149, 150, 155, 160, 161–78, 403, 411
  hunter-gatherer 10–11, 46, 57, 59, 80, 91, 95, 96–7
  liberation of the individual and 403
  sex and gender 161–78
Wrangel Island, Arctic Ocean 74
writing, evolution of 137, 138–48
Wu Zetian of China, Empress 171

Yoruba religion 238
Yupik 219

Zheng He, Admiral 324–5, *325*, 331
Zimrilim of Mari, King 143
Zoroastrianism 247, 264, 265
Zulu Empire 217